Lectures on Mathematical Logic
Volume II : Calculi for Derivations and Deductions

Lectures on Mathematical Logic Volume II : Calculi for Derivations and Deductions

Walter Felscher

University of Tuebingen
Germany

CRC Press is an imprint of the
Taylor & Francis Group, an **informa** business

CRC Press
Taylor & Francis Group
6000 Broken Sound Parkway NW, Suite 300
Boca Raton, FL 33487-2742

First issued in paperback 2019

ISBN-13: 978-90-5699-267-5 (hbk)
ISBN-13: 978-0-367-39858-3 (pbk)

Library of Congress Cataloging-in-Publication Data

Catalog record is available from the Library of Congress

Visit the Taylor & Francis Web site at
http://www.taylorandfrancis.com

and the CRC Press Web site at
http://www.crcpress.com

Contents

Dependences

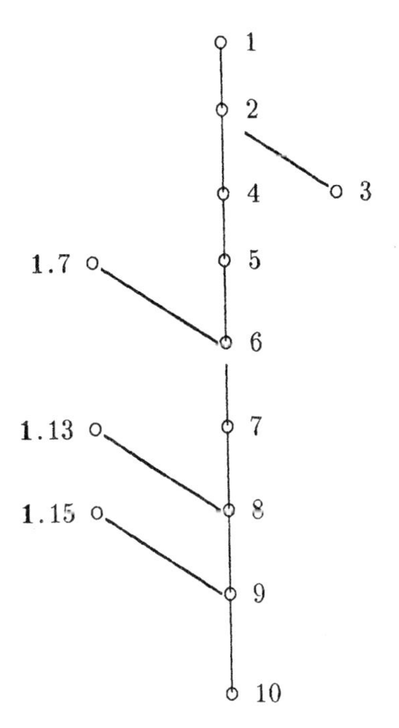

Preface

This is the second of three books of Lectures on Mathematical Logic, destined for students of mathematics or computer science, in their third or fourth year at the university, as well as for their instructors. It is written as the traditional combination of textbook and monograph: while the titles of chapters and sections will sound familiar to those moderately acquainted with logic, their content and its presentation is likely to be new also to the more experienced reader. In particular, concepts have been set up such that the proofs do not act as mousetraps, from which the reader cannot escape without acknowledging the desired result, but make it obvious *why* they entail their result with an inner necessity. In so far, the book expects from the reader a certain maturity; while accessible to students, it also will be helpful to lecturers who look for motivations of procedures which they may only have come to know in the form of dry prescriptions.

The book is a combinatorial study of derivations and deductions. Gentzen's rules of sequential logic are motivated for the positive connectives in Chapter **1** and for quantifiers in Chapter **8** ; the rules for negation are analysed in Chapter **4** , and classical logic is discussed in Chapter **5** with particular attention to the Peirce rule. Chapter **2** is devoted to the familiar cut elimination going down from the top; Chapter **3** presents Mints' algorithm of cut elimination going upwards. In the case of classical logic, the usual superexponential bounds are improved to twofold exponential ones, but reasons of space have prevented me from including a discussion of Hudelmayer's bounds for intuitionistic logic. Chapter **6** is an excursion into algebra, associating to every sequential calculus the class of algebras defined by the equations given by interdeducible formulas; the classes of algebras arising in this manner are then characterized by generators of the unit filter of their free algebras. These are just the axioms for modus ponens calculi generating such filters, and in Chapter **7** the modus ponens calculi corresponding to the sequential calculi are set up — with sharp upper bounds for the lengths of proofs obtained under the transformations between sequential derivations and modus ponens deductions.

The premisses of the non-critical quantifier rules can be formulated in two ways: either referring to replacement instances of the formula to be quantified, in which the quantifying variable is replaced by a term t which is free for that formula, or instead referring to substitution instances which involve a built-in renaming of the bound variables which occur in t. For the first case, Chapter **8** contains the known result that cut elimination works for derivations with an endsequent in which no variable occurs both free and bound. In the second case, a semantical completeness argument would show that cut elimination must always work, and the second part of Chapter **8** contains a syntactical proof for this fact. Chapter **9** develops modus ponens calculi for quantifier logic based on the algorithmic descriptions of consequence in Chapter **1.15** ; Chapter **10** contains various applications such as the conservativeness of equality logic, conservativeness of extensions with predicate and function symbols, the midsequent theorem and Herbrand's theorem.

More about the book's content will be said in its introduction and in the introductory sections of its Chapters.

The book is divided into Chapters **1** to **10**, and each chapter is divided into numbered sections. Theorems, Corollaries and Lemmas are numbered successively within each chapter, and a later reference such as Theorem **j**.k will be one to Theorem k in Chapter **j**.

The Chapters **1** to **7** do not depend on the first book of these Lectures; sections **8**.4–7 depend on Chapter **13** of Book **1**, and sections **9**.1–3 depend on Chapter **15** of Book **1**. A reference to Chapter **1.j** and to Theorem **1.j**.k will be one to Chapter **j** of Book **1** and to Theorem **j**.k of Book **1**.

I owe special thanks to my colleague Lev Gordeev for many enlightening conversations about the subjects treated in this book, in particular about operators acting on derivations. I am indebted to Professor Dov Gabbay and to the publishers of Gordon & Breach whose efforts made the appearance of these Lectures possible.

Introduction

The semantical approach to logic starts from truth and from mathematical structures, and it proceeds to an analysis of the relation of semantical consequence, as well as to various, and sometimes surprising, properties of models of axiom systems. It is a mathematician's approach, talking about a world of – preferably infinite – sets, as it was developed in Book 1 of this work.

The approach to be followed in the present Book 2 may be described as *argumentative* or *linguistic* in that it starts from the rhetorical and literary practise of presenting collections of (usually successive) arguments, connected by logical laws referring to their linguistic form and independent from their contents. In view of its formal linguistic character, this approach may also be called *syntactical,* and the treatment of linguistic objects will use techniques which, in mathematics, are counted as *combinatorial*. The combinatorial rules then describe particular collections of arguments as deductions and derivations in a uniform manner, and the systems of such rules make up what are called logical *calculi*. In so far, that approach can also be called *algorithmic* or *computational* – although the actual design of programs for devices, performing such computations, will not be discussed here.

While there are many ways to define when a collection of arguments may be called a deduction, singling out a particular one among them can hardly avoid being arbitrary and accidental. It was the basic idea of Gentzen to begin, rather, with an abstract notion of *general* deductive situations, and to develop the *foundation* of logic from an analysis of the transformations *between* such deductive situations, as effected by the logical connectives and quantifiers occurring in them. The setting then is that of *deductive situations*, expressed as $M \Longrightarrow v$ and meant to say that a statement v can, in an unspecified manner, be deduced from a collection M of assumptions. Consider now a deductive situation which involves, either as its conclusion v or among its assumptions M, a logically composite statement. An analysis of the linguistic meaning, of the propositional connectives or of the quantifiers governing this composition, will lead to conditions, acceptable to every consensus (fair and reasonable), about the form of one or several other deductive situations $M' \Longrightarrow v'$ which (a) involve the components of that composite statement and (b) are such that the deducibility $M \Longrightarrow v$ may be linguistically *concluded* from the deducibilities $M' \Longrightarrow v'$ as *premisses*. Condensing such relationships between premisses and conclusions into *logical rules*, I arrive at a *calculus* of deductive situations, built from the weakest requirements which those logical transformations between deductions are to obey; for the case of positive connectives this will be carried out immediately at the start Chapter **1**. In this manner, the calculus of *sequents*, as deductive situations are called in more technical terminology, will not care to make explicit the content of deductions, but will describe derivations as producing deductive situations from other deductive situations.

A second basic idea, due to Frege, concerns the deductive meaning of quantifiers in the above analysis, and it will be explained in Chapter **8**. It rests on the observation that a *free variable deduction* of $M \Longrightarrow v(x)$, i.e. a deductive situation in which the variable is free of additional constraints from M, can be subjected to arbitrary substitutions of x by terms and, therefore, gives rise to the new deductive situation $M \longrightarrow \forall x v(x)$. This approach completely avoids the reference to infinitely many instantiations v(t) of v(x) as they were required for the set theoretical semantics of Book 1 .

Presented in this manner, the calculus of sequents receives its semantical meaning from the logical properties of deductive situations. A more careful investigation, however, will be required to discuss the propositional connective *negation* (which, at first sight, seems to call for an exterior semantics of *true* and *false*) in Chapters **4** and **5**; to this end, the linguistic environment is extended such as to consider, besides deductive situations, also statements Δ behaving deductively as *refutable* or even as *absurd* (in the sense of *ex absurdo quodlibet*). Negation then is reduced to *relative* refutability, respectively absurdity, leading to the calculi of minimal and of intuitionistic logic. Classical negation, finally, is discussed in Chapter **5**, and it appears that the rules for that *cannot* be given a deductive meaning, making this logic classical only in the sense that it can be founded upon the exterior truth value semantics. – While the deductive analysis of these different calculi is carried out in the environment of situations $M \Longrightarrow v$, now also called K–sequents, which make use of statements Δ (and possibly of the connective $\neg$), it is a technical observation that the statements Δ may be eliminated and alone the connective $\neg$ be employed, if only I use instead of K–sequents now L–*sequents*, differing from K–sequents in that the succedent formula v may not be present. The resulting calculi for L–sequents then are formally simpler to handle; still, the rule making negation classical violates the useful combinatorial principles otherwise available. It then is another technical observation that this classical rule can be omitted altogether, if only I use instead of L–sequents now M–*sequents* $M \Longrightarrow N$ in which N may be a finite number of formulas. I thus arrive, depending on the kind of sequents used, at three types of calculi, distinguished again by the first letters K, L and M used for their names, and translation operators can be set up which transform derivations belonging to, say, a classical L–calculus into derivations belonging to a classical M–calculus.

Derivations in a calculus of sequents are trees, the nodes of which carry sequents; they proceed to their endsequent at the root in accordance with the calculus' rules. In order to perform tasks requiring the construction of new derivations from given ones, they must be treated as mathematical objects in their own right: e.g. when translating from one calculus into another or when proving the admissibility of inversion rules and cut rules.

The methods employed here are combinatorial, pruning irrelevant subtrees and implanting various new ones, the *length* of a derivation's tree being an important measure of complexity. In particular, admissibility of cuts is treated in Chapters **2** and **3**, establishing *cut elimination algorithms* which, for the calculi enriched by a cut rule, transform derivations employing that rule into derivations avoiding it. The solution in Chapter **2** is a variant of Gentzen's original proof, attacking the given derivation *from above*; in Chapter **3** an algorithm invented by Mints is presented which works *from below* and has the advantage that it can also be applied to the case of infinitary derivations (which are not studied further in the present work). Both these algorithms continue to work if a connective for negation, or if quantifiers are added, in Chapters **4**, **5** and **8**. In the case of quantifiers, the precise formulation of results depends on whether the instantiations v(t) in the premisses of the rules are formed with help of the replacement function *rep* or the substitution function *sub* (which involves a renaming of bound variables), treated already in Book 1, Chapter **13**.

Gentzen introduced sequent calculi only in 1934, but it had been Frege who in 1879 was the first to invent a calculus of *formulas*; Frege's calculus, and the ones developed from it, had the characteristic to use the rule of *modus ponens*. The bridge to this type of calculi is established in the algebraic Chapter **6** where to every propositional sequent calculus is associated the equational class **A** of algebras, obtained by identifying formulas v, w in an equation if they are interdeducible in that calculus. Actually, each of these classes can be defined by only a very few, particular equations, and in Chapter **7** these equations are used to define, for each of the various sequent calculi, an associated modus ponens calculus such that derivations of sequents M $\Longrightarrow$ v translate into modus ponens deductions of v *from* M and vice versa; this correspondence will easily extend to the case of quantifiers once suitable modus ponens calculi for quantifier logic will have been set up.

Such calculi are the subject of Chapter **9** where I start from the aim that, for classical logic, they be complete with respect to the semantical consequence operation ct with respect to truth: it shall v be deducible *from* C if, and only if, $v \in ct(C)$. It had been shown in Book 1, Chapter **15**, that $ct(C) = ct^{\mathbf{B}}(C)$ and that $ct^{\mathbf{B}}(C)$ is a congruence class of a relation R, generated by an explicitly described finitary algorithm. This algorithm can be transformed immediately into a modus ponens calculus complete for ct, giving rise to several variants, as well as to modus ponens calculi complete for the consequence operation cs with respect to satisfaction; it is the latter which will correspond to the appropriate sequent calculi. Moreover, it had been observed in Book 1 that the algorithmic description of $ct^{\mathbf{B}}(C)$ remains in effect for a consequence $ct^{\mathbf{X}}(C)$ which refers, not to Boolean valued structures, but to **X**-valued structures defined with a suitable class **X** of algebras. The classes **A** associated to positive, minimal and intuitionistic logic are suitable, and so I obtain positive, minimal and intuitionistic modus ponens calculi employing the same quantifier rules, and they are complete with respect to the semantical consequence $ct^{\mathbf{A}}$ defined now from **A**-valued structures.

Various applications are discussed in Chapter **10**, beginning with the correspondence between sequent and modus ponens calculi in the presence of quantifiers, and ending with a detailed discussion of Herbrand's theorem.

Chapter 1. Sequent Calculi for Positive Logic

1. Positive Rules for Deductive Situations

It is an experience of life that man can be persuaded to believe certain claims. One way of effecting such persuasion is to offer *arguments* for those claims, and a particular kind of argument is that which is called *logical.*

A chain of arguments, leading from a set M of assumptions to a claim v, is called a *deduction of* v *from* M. I shall not specify here which of the many conceivable kinds of arguments may have been used in such a deduction; I only assume the reader to know that there *are* linguistic objects which in daily conversation are considered deductions. But among those many kinds of arguments a few can be isolated as being connected with the usage of the logical phrases themselves – e.g. with *and* and *or*. In order to talk about them without ambiguity, I shall consider claims and hypotheses to be propositional formulas; I shall abbreviate the presence of a *deductive situation,* leading from M to v, by expressions such as $M \Longrightarrow v$.

For each of the logical phrases, as they are symbolized by the propositional connectives $\wedge$, $\vee$, $\rightarrow$, I may ask the question

(I) What are fair and reasonable hypotheses, about one or two given deductions, in order to conclude from them that there also is a deduction leading to a formula composed with one of these phrases?

An obvious answer for the phrase *and* is

> If there are two deductions $M \Longrightarrow v$ *and* $M \Longrightarrow w$ then it should be possible to conclude that there is a deduction $M \Longrightarrow v\wedge w$

which I abbreviate as a rule

$$(I\wedge) \quad \frac{M \Longrightarrow v \qquad M \Longrightarrow w}{M \Longrightarrow v\wedge w}$$

leading from the given deductive situations as *premisses* to a new one as its *conclusion*. An obvious answer for the phrase *or* is

> If there is a deduction $M \Longrightarrow v$, *or* there is a deduction $M \Longrightarrow w$, then it should be possible to conclude that there is a deduction $M \Longrightarrow v\vee w$

which I abbreviate as the two rules

$$(\mathrm{I}\vee\mathrm{l})\ \frac{M \Longrightarrow v}{M \Longrightarrow v\vee w} \qquad (\mathrm{I}\vee\mathrm{r})\ \frac{M \Longrightarrow w}{M \Longrightarrow v\vee w} \quad .$$

But it is not only the question *when*, say, there is a deduction leading to $v\wedge w$ which can be asked; I also may ask *which* formulas can an be deduced *from* $v\wedge w$:

(E) What are fair and reasonable hypotheses, about one or two given deductions, in order to conclude from them that there also is a deduction leading to a formula u from the composite assumption, say $v\wedge w$, together with further assumptions M ?

Writing the list of assumptions, consisting of v together with M, as v,M , an obvious answer for *and* is

> If there is a deduction $v, M \Longrightarrow u$, *or* there is a deduction $w, M \Longrightarrow u$, then it should be possible to conclude that there is a deduction $v\wedge w, M \Longrightarrow u$

which I abbreviate as the two rules

$$(\mathrm{E}\wedge\mathrm{l})\ \frac{v,M \Longrightarrow u}{v\wedge w,M \Longrightarrow u} \qquad (\mathrm{E}\wedge\mathrm{r})\ \frac{w,M \Longrightarrow u}{v\wedge w,M \Longrightarrow u} \quad .$$

An obvious answer for *or* is

> If there are two deductions $v, M \Longrightarrow u$ *and* $w, M \Longrightarrow u$ then it should be possible to conclude that there is a deduction $v\vee w, M \Longrightarrow u$

which I abbreviate as the rule

$$(\mathrm{E}\vee)\ \frac{v,M \Longrightarrow u \qquad w,M \Longrightarrow u}{v\vee w,M \Longrightarrow u} \quad .$$

Re-reading the preceding discussion, I will notice that I have employed a *meaning* for the phrases "and" and "or" which refers to manipulations with deductions (and to nothing else). In the process of forming propositional formulas (as a special case of term algebras) the connectives $\wedge$ and $\vee$ and were meaningless combinatorial signs; in the *2*-valued interpretation of propositional logic in Book 1 of this work, they received a semantical *meaning* through the Boolean operations on truth values. Forgetting about this (highly special) semantics, I now can observe that here the connectives $\wedge$ and $\vee$ have obtained a (much more specific) meaning through the very rules which describe the manipulations leading to deductions $M \Longrightarrow v\wedge w$ and $v\vee w, M \Longrightarrow u$. In so far, the connectives $\wedge$ and $\vee$ now serve as *internalizations* which mirror, *within* the language of propositional logic, the external usage of *and* and *or* as performed with respect to my (still rather vaguely described) deductions.

While this discussion of the connectives $\wedge$ and $\vee$ as internalizations may appear as a superfluous aside, the aspect of internalization becomes decisive when I look for answers to the questions (I) and (E) for the propositional connective $\rightarrow$. Of course, it is read as *if – then*, but if I wish not to rely on a mechanical use of the classical truth values 0 and 1, then I need to interpret this phrase by describing manipulations on derivations as well. Which *if this – then that* shall $v \rightarrow w$ internalize? As for (I), a modest proposal is

> If there is a deduction $v, M \Longrightarrow w$ of w from v and other assumptions M then it should be possible to conclude that there is a deduction $M \Longrightarrow v \rightarrow w$ from M alone, saying *if* v *then* w under the assumptions M

which I abbreviate as the rule

$$(\mathrm{I}\rightarrow) \quad \frac{v, M \Longrightarrow w}{M \Longrightarrow v \rightarrow w} \quad .$$

Thus the deducibility of $v \rightarrow w$ uses the formula $v \rightarrow w$ as an internalization of the premiss that there is a deduction of w from v – always with a fixed set of further assumptions.

In the question (E) I ask for premisses from which to deduce $v \rightarrow w, M \Longrightarrow u$. Let me first consider the situation of an actual (not internalized) deduction of w from v:

> given a deduction $v, M \Longrightarrow w$, which additional premisses will give rise to a deduction $M \Longrightarrow u$?

Such premisses, obviously, will have to express *some* deducibility of u. Deducibility from M alone would too much since it would give the desired result already without the deducibility of w from v. Deducibility of u from v, M would be of no use either (it would, as will become clear in a moment, be the *wrong* direction). There remains deducibility of u from w, M , and with that indeed the premisses

> $M \Longrightarrow v$ and $w, M \Longrightarrow u$

will suffice. Because then the assumed deduction $v, M \Longrightarrow w$ leads to a deduction $M \Longrightarrow u$: the deduction $M \Longrightarrow v$, prolonged by $v, M \Longrightarrow w$, will first produce an auxiliary deduction $M \Longrightarrow w$, and then in $w, M \Longrightarrow u$ the assumption w, wherever it is used, can be replaced by this auxiliary deduction. Internalizing the conclusion of

> if $M \Longrightarrow v$ and $w, M \Longrightarrow u$
> then $v, M \Longrightarrow w$ gives rise to $M \Longrightarrow u$,

I arrive at the rule

$$(\mathrm{E}\rightarrow) \quad \frac{M \Longrightarrow v \qquad\qquad w, M \Longrightarrow u}{v \rightarrow w, M \Longrightarrow u} \quad .$$

I so have provided some answers to my questions (I) and (E), stating the rules (I∧), (E∧), (I∨), (E∨), (I→), (E→) as sufficient conditions in order to obtain new deductive situations as conclusions from given situations as premisses. There may, of course, be many more conditions by which new deductive situations can be produced, and they may refer to additional logical connectives as well as to particular *contents* expressed by the formulas used here to express linguistic statements. Thus a complete theoretical description of the construction of deductions will depend on additional linguistic as well as contentual properties of the domain of reference under discussion, and these here have been left completely open.

2. The Calculus K_sP

What can be done already at this stage, is the setting up of a calculus which models only the behaviour of deductions as far as they depend on formulas with the connectives ∧, ∨ and → . So I now shall turn the tables: instead of analyzing the essentially undetermined deductive situations, I shall replace them by well defined objects, called *sequents* and consisting of formulas from a given language, and I shall use the rules isolated above as *defining rules* for a calculus whose objects are *derivations* of sequents from given sequents (and *not* deductions of formulas from assumptions).

I consider formulas from a propositional language with the connectives ∧, ∨ and → . A *sequent* shall be an ordered pair, consisting of a (possibly empty) set M of formulas and a formula v ; sequents will be written as $M \Longrightarrow v$, and M is called the *antecedent*, v the *succedent* of the sequent. A sequent $\{m_0, \ldots m_{k-1}\} \cup M \Longrightarrow v$ shall always be written as $m_0, \ldots, m_{k-1}, M \Longrightarrow v$. Observe that it is not requested that the sets M and $\{m_0, \ldots m_{k-1}\}$ be disjoint, nor that M be non-empty if $k>0$.

Rules are the following schemata, listing one or two *premisses* and one *conclusion*:

(W) $\dfrac{M \Longrightarrow u}{v, M \Longrightarrow u}$

(I∧) $\dfrac{M \Longrightarrow v \qquad M \Longrightarrow w}{M \Longrightarrow v \wedge w}$ (E∧l) $\dfrac{v, M \Longrightarrow u}{v \wedge w, M \Longrightarrow u}$ (E∧r) $\dfrac{w, M \Longrightarrow u}{v \wedge w, M \Longrightarrow u}$

(I∨l) $\dfrac{M \Longrightarrow v}{M \Longrightarrow v \vee w}$ (I∨r) $\dfrac{M \Longrightarrow w}{M \Longrightarrow v \vee w}$ (E∨) $\dfrac{v, M \Longrightarrow u \qquad w, M \Longrightarrow u}{v \vee w, M \Longrightarrow u}$

(I→) $\dfrac{v, M \Longrightarrow w}{M \Longrightarrow v \to w}$ (E→) $\dfrac{M \Longrightarrow v \qquad w, M \Longrightarrow u}{v \to w, M \Longrightarrow u}$

Here (W) is called the *weakening* rule, and the other rules are the *logical* ones.

A *derivation* is a finite 2-ary tree T together with functions d, r. The function d assigns sequents to the nodes of T and the function r assigns (names of) rules to the non-maximal nodes such that

> if e is maximal then d(e) is a sequent $x \Longrightarrow x$ where x is some variable;
>
> if e is not maximal then r(e) is a one- or two premiss rule, depending on whether e has one or two upper neighbours e' or e', e'', and the sequents assigned to e, e', e'' are *instances* of r(e), i.e. they can be written in the form indicated in the rule by choosing M, v, w, u appropriately.

The sequents assigned to the maximal nodes are called *axioms.* In instances of (W) the formula v is called the *weakening formula.* In instances of the logical rules, the composite formulas $v \wedge w$, $v \vee w$, $v \rightarrow w$ are called the *principal formulas,* and the formulas v and (if present) w are called the *side formulas* of that instantiation of the rule. The formulas in M and the formula u, as well as v in (W), are called the *parametric* formulas of the instantiation.

Since in sequents of the form $v, M \Longrightarrow u$ the formula v may or may not be in M, an instantiation of a rule distinct from ($I\wedge$) and ($I\vee$) will require the explicit specification not only of its principal formula (except for ($I\rightarrow$)) and its side formula(s), but also that of its parameters. It may well happen that side formulas or the principal formula are also parametric, and instances of ($E\wedge$) and ($E\vee$) are conceivable in which premisses and conclusions are the same sequents.

A precise description, therefore, will require, in addition to the functions d and r, further functions: a function r_1 pointing out the weakening formula or the principal formula $r_1(e)$ in d(e); a function r_2 pointing out the side formulas $r_2(e,e')$, $r_2(e,e'')$ of a logical rule; a function r_3 with $r_3(e)$ as the set M, subjected to appropriate conditions.

It also should be noticed that principal and side formulas, as well as parametric ones, refer to an instantiation of a rule, not to a sequent by itself. Consider, for instance, a sequent which (a) is premiss $v, M \Longrightarrow u$ at the upper neighbour e' of e for an instantiation at e of a rule with v as side formula, and which (b) is conclusion $v, N \Longrightarrow u$ at e' for another instantiation of a rule with v as principal formula. Then M may well be distinct from N, and all that can be said is that $M \cup \{v\}$ equals $N \cup \{v\}$.

The rules and derivations defined are called rules and derivations *of the calculus* K_sP. What the calculus itself may be is left to the reader's imagination. The sequent at the root (of the tree of) a derivation is called its *endsequent.*

Sequents of the form $x, M \Longrightarrow x$ are called *generalized axioms.* If derivations are allowed to start from generalized instead of usual axioms then the weakening rule does not need to be required: every derivation D of the premiss of (W) becomes a derivation D' of the conclusion of (W) from generalized axioms by simply adjoining the formula v to the left side of each of the sequents of D.

Sequents of the form $v, M \Longrightarrow v$ are called *reflexivity axioms* for the formula v. Induction on the complexity of v shows that, for every reflexivity axiom, a K_sP-derivation can be found. Because if v is a variable then I have a generalized axiom, and if v is composite and if derivations of reflexivity axioms for the components p, q of v have already been found, then they may be extended to a derivation of a reflexivity axiom for v :

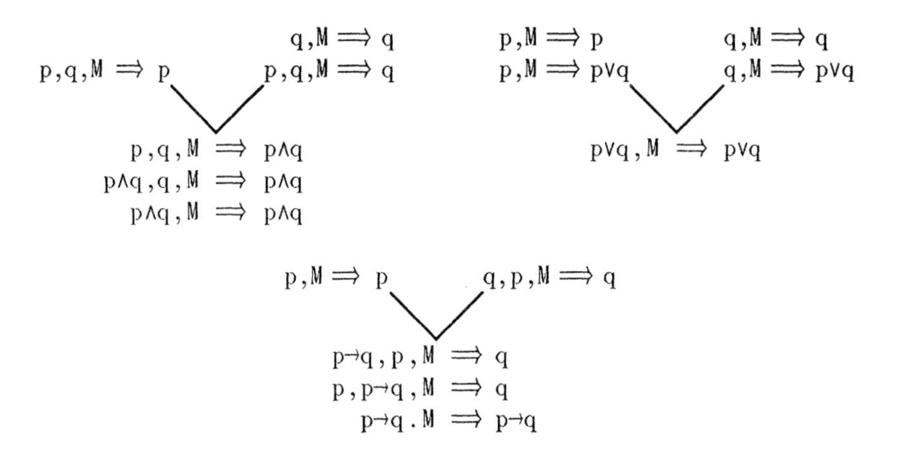

Consequently, derivations may also be assumed to start from reflexivity axioms instead of from simpler ones.

Occasionally the 2-ary rules (I∧), (E∨), (E→) are requested in a different form, namely

(I∧) $\dfrac{M_0 \Longrightarrow v \qquad M_1 \Longrightarrow w}{M_0, M_1 \Longrightarrow v \wedge w}$ (E∨) $\dfrac{v, M_0 \Longrightarrow u \qquad w, M_1 \Longrightarrow u}{v \vee w, M_0, M_1 \Longrightarrow u}$

(E→) $\dfrac{M_0 \Longrightarrow v \qquad w, M_1 \Longrightarrow u}{v \rightarrow w, M_0, M_1 \Longrightarrow u}$

which then are called their *mixed parameter* forms and contrasted to the earlier *pure parameter* forms. Every derivation employing the pure parameter forms is also a derivation with rules of mixed parameter form; conversely, every derivation employing rules in mixed parameter form can be transformed into one using only rules in pure parameter form – either by purifying mixed premisses through appropriate weakenings, or by working with generalized axioms from the outset. In what follows, I shall work exclusively with pure parameter forms of 2-ary rules, but there are occasions (e.g. in Chapter **10**.7) where the mixed parameter forms offer an advantage.

3. The Calculi K_tP and K_uP

The sequents considered so far, having *sets* as their left sides, I shall call *s-sequents*. I now shall present calculi which employ sequents, the left sides of which are *sequences*. So a *t-sequent* $M \Longrightarrow u$ shall be a pair consisting of a (possibly empty) *sequence* M of formulas together with a formula u; the prefix t will be omitted if no ambiguity can arise. In order to improve readability, I sometimes shall write $\blacktriangle \Longrightarrow v$ for $M \Longrightarrow v$ with an empty M.

For sequences $M = \langle u_0,\dots, u_{m-1}\rangle$ and $N = \langle v_0,\dots, v_{n-1}\rangle$ their *concatenation* is the sequence $\langle u_0,\dots, u_{m-1}, v_0,\dots, v_{n-1}\rangle$; I shall write it as M,N as well as $u_0,\dots, u_{m-1},N$ and as $M, v_0,\dots, v_{n-1}$.

Rules of the calculus K_tP are, first, three *structural* rules, namely again the weakening rule (W), the *contraction rule*

$$\text{(RC)}\quad \frac{v,v,M \Longrightarrow u}{v,M \Longrightarrow u}$$

and, for every permutation π of def(M), the *permutation rule*

$$\text{(RP)}\quad \frac{M \Longrightarrow u}{\pi(M) \Longrightarrow u},$$

and since every permutation can be generated from transpositions, it suffices to employ (RP) for transpositions π only. Secondly, the *logical* rules of K_tP are the same schemata as those used for K_tP:

$$(\text{I}\wedge)\quad \frac{M \Longrightarrow v \qquad M \Longrightarrow w}{M \Longrightarrow v\wedge w} \qquad (\text{E}\wedge\text{l})\quad \frac{v,M \Longrightarrow u}{v\wedge w, M \Longrightarrow u} \qquad (\text{E}\wedge\text{r})\quad \frac{w,M \Longrightarrow u}{v\wedge w, M \Longrightarrow u}$$

$$(\text{I}\vee\text{l})\quad \frac{M \Longrightarrow v}{M \Longrightarrow v\vee w} \qquad (\text{I}\vee\text{r})\quad \frac{M \Longrightarrow w}{M \Longrightarrow v\vee w} \qquad (\text{E}\vee)\quad \frac{v,M \Longrightarrow u \qquad w,M \Longrightarrow u}{v\vee w, M \Longrightarrow u}$$

$$(\text{I}\rightarrow)\quad \frac{v,M \Longrightarrow w}{M \Longrightarrow v\rightarrow w} \qquad (\text{E}\rightarrow)\quad \frac{M \Longrightarrow v \qquad w,M \Longrightarrow u}{v\rightarrow w, M \Longrightarrow u}$$

A *derivation* is a finite 2-ary tree T together with functions d, r. The function d assigns t-sequents to the nodes of T and the function r assigns rules to the non-maximal nodes such that

> if e is maximal then d(e) is a sequent $x \Longrightarrow x$ where x is some variable;
>
> if e is not maximal then r(e) is a one- or two premiss rule, depending on whether e has one or two upper neighbours e' or e', e", and the sequents assigned to e, e', e" are *instances* of r(e), i.e. are of the form indicated in the rule.

Weakening, principal, side and parametric formulas for instances of the logical rules are defined as in the case of K_sP. But as I now use sequences and not sets, the ambiguities disappear which arose for K_sP concerning parametric side and principal formulas in instances of logical rules. To express matters easily, I now shall speak of *occurrences* of formulas in a t-sequent: the one occurrence on the right side of a sequent, and occurrence i of a formula in the antecedent where $i<n$ for the length n of the antecedent. The *corresponding* occurrences in sequents forming the instance of rule then are defined as follows:

> For (I∧) and (I∨) the occurrence i in the antecedent of a premiss corresponds to occurrence i in the conclusion,
>
> for (I→) and (RC) the occurrence i+1 in the antecedent of the premiss corresponds to the occurrence i in the conclusion, for (RC) also the occurrence 0 in the in the antecedent of the premiss corresponds to the occurrence 0 in the conclusion,
>
> for (E∧), (E∨) the occurrence i in the antecedent of a premiss corresponds, for $i>0$, to the occurrence i in the conclusion,
>
> for (E→) the occurrence i in the antecedent of the right premiss corresponds, for $i>0$, to the occurrence i in the conclusion, and the occurrence i in the left premiss corresponds to the occurrence i+1 in the conclusion,
>
> for (W) the occurrence i in the antecedent of the premiss corresponds to the occurrence i+1 in the conclusion,
>
> for (RP) the occurrence i in the antecedent of the premiss corresponds to the occurrence $\pi(i)$ in the conclusion,
>
> for all structural rules and E-rules the right occurrence of u in any, respectively for (E→) the right, premiss corresponds to the right occurrence in the conclusion.

For I-rules the right occurrences are *side* or *principal* occurrences of the side- and the principal formulas respectively, and for (I→) the occurrence 0 in the premiss is also a *side* occurrence. For E-rules the occurrences 0 in the antecedents, with exception of the left premiss of (E→), are *side* or *principal* occurrences of the side and the principal formulas respectively, and for the left premiss of (E→) the occurrence to the right is also a side occurrence. All occurrences which are not side or principal ones are *parametric*. It follows from the above that corresponding occurrences are always parametric. Hence if in an instance of rule a formula is replaced by another formula at a pair of corresponding occurrences then the result remains an instance of that rule.

While principal and side formulas always have principal and side occurrences, it is well possible that they also have parametric occurrences. The logical rules lower, from premisses to conclusions, the multiplicity with which the side formulas occur, and they rise the multiplicities with which principal formulas occur. The weakening rule rises the multiplicity of the weakening formula, and the contraction rule lowers the multiplicity of the contracted formula.

In view of the special position of principal, side, weakening and contraction formulas, the necessity of the permutation rule (RP) is obvious. It is usual practice to omit indications of the use of (RP) in pictorial diagrams of derivations.

The same proofs as given for K_sP show that also for K_tP the reflexivity axioms have derivations and that the weakening rule can be omitted if generalized axioms are used.

The length of a derivation is defined as follows. The *depth* of a node e in a derivation is defined analogously to its depth in the underlying tree, but *not* counting the nodes with sequents obtained under the *structural* rules (W), (RC), (RP): it shall be 0 if e is maximal, it shall be 1 plus the maximum of the depths of its upper neighbours if the rule leading to the sequent at e is logical, and it shall be the maximum of the depths of its upper neighbours if that rule is structural. The *length* of a derivation shall be the depth of its minimal node. In particular, derivations of length 0 consist of one sequent only which then is an axiom.

I now turn to the comparison of K_sP and K_tP-derivations. This will show, in particular, the role of the contraction rule (RC).

For every t-sequent $M \Longrightarrow u$, I define its *contraction* to be the s-sequent $M_s \Longrightarrow v$ where M_s is the set of elements of M; in that case, $M \Longrightarrow u$ is said to be a t-*expansion* of $M_s \Longrightarrow u$, and a *canonical* t-expansion of $M_s \Longrightarrow u$ shall be a t-sequent $M \Longrightarrow u$ in which M consists of the members of M_s *without repetitions*.

Given a K_tP-derivation, I clearly obtain a K_sP-derivation if I replace each of the t-sequents by its contraction. Such a construction, transforming derivations (from some calculus) into other derivations (of the same or of another calculus), I shall call an *operator*. In what follows, I shall describe various such operators, some of which will be given special names (giving names to all of them might cause an inflation). I know already the first part of the

LEMMA 1 There is an operator, transforming K_tP-derivations into K_sP-derivations, which acts by replacing every sequent by its contraction.

There is an operator, transforming K_sP-derivations into K_tP-derivations, which replaces sequents by expansions and endsequents by their canonical t-expansions; it preserves the lengths of the derivations.

I define the second operator by recursion on the height, ascending a given K_sP-derivation D from below. Thus I know what to do at the endsequent. Assume that a sequent $A \Longrightarrow d$ has been assigned an expansion $A^* \Longrightarrow d$. If $A \Longrightarrow d$ was an axiom then so is $A^* \Longrightarrow d$; consider next the case that $A \Longrightarrow d$ was conclusion of a logical rule. For instances of I$\wedge$ and I$\vee$, it suffices to use A^* as the antecedent of the transformed premisses. In the case of I$\rightarrow$, I use the t-sequent A^*,v. Consider now an instance of an E-rule, e.g.

$$\frac{M \Longrightarrow v \qquad w, M \Longrightarrow d}{v \rightarrow w, M \Longrightarrow d}$$

with $A = v \rightarrow w, M$. Then the t-sequence A^* can be written as a concatenation p, B^* with the principal formula p (here $p = v \rightarrow w$). If p was not parametric in this instance then I use B^* for premisses without a left side formula, and I use w, B^* for premisses with a left side formula w, i.e.

$$\frac{B^* \Longrightarrow v \qquad w, B^* \Longrightarrow d}{v \rightarrow w, B^* \Longrightarrow d} .$$

If p was parametric, however, then I cannot afford to loose it this way since it may be required as a side formula further upwards. In this case I use the premisses p, B^* and w, p, B^* respectively with the conclusion $p, p, B^* \Longrightarrow d$. I then insert an instance of (RC) and obtain $p, B^* \Longrightarrow d$, e.g.

$$\frac{\dfrac{v \rightarrow w, B^* \Longrightarrow v \qquad w, v \rightarrow w, B^* \Longrightarrow d}{v \rightarrow w, v \rightarrow w, B^* \Longrightarrow d}}{v \rightarrow w, B^* \Longrightarrow d} .$$

An example, showing that the use of (RC) cannot be avoided, is offered by the K_tP-derivation of $v_3 \Longrightarrow y$ where $v_1 = (x \rightarrow y) \rightarrow x$, $v_2 = v_1 \rightarrow x$, $v_3 = v_2 \rightarrow y$:

$$\frac{\dfrac{\dfrac{\dfrac{\dfrac{\dfrac{\dfrac{v_1, x \Longrightarrow x}{x \Longrightarrow v_2} \qquad x, y \Longrightarrow y}{x, v_3 \Longrightarrow y}}{v_3 \Longrightarrow x \rightarrow y} \qquad x, v_3 \Longrightarrow x}{v_1, v_3 \Longrightarrow x}}{v_3 \Longrightarrow v_2} \qquad y, v_3 \Longrightarrow y}{v_3, v_3 \Longrightarrow y}}{v_3 \Longrightarrow y} .$$

The uses of (RC) in a K_tP-derivation may be quite accidental. Obviously, it would suffice to save them up to the very last moment, introducing principal formulas with ever growing multiplicities, until a highly non-canonical expansion of the endsequent is subjected to a succession of contractions.

I now define a further calculus, the *extremal* calculus K_uP, which employs the same t-sequents that are used by K_tP. The only structural rule of K_uP is the permutation rule (RP), and of its logical rules the I-rules are the same as for K_tP. The E-rules of K_uP arise from those of K_tP by forcing a principal occurence in the antecedent to appear also in the premisses:

$$(E\wedge l)\quad \frac{v, v\wedge w, M \Longrightarrow u}{v\wedge w, M \Longrightarrow u} \qquad\qquad (E\wedge r)\quad \frac{w, v\wedge w, M \Longrightarrow u}{v\wedge w, M \Longrightarrow u}$$

$$(E\vee)\quad \frac{v, v\vee w, M \Longrightarrow u \qquad w, v\vee w, M \Longrightarrow u}{v\vee w, M \Longrightarrow u}$$

$$(E\rightarrow)\quad \frac{v\rightarrow w, M \Longrightarrow v \qquad w, v\rightarrow w, M \Longrightarrow u}{v\rightarrow w, M \Longrightarrow u} .$$

K_uP-derivations are defined to always start from generalized axioms. For E-rules there now are principal occurrences (of the principal formula) also in the premisses, and so I change the definition of corresponding occurrences:

for $(E\wedge)$, $(E\vee)$ the occurrence $i+1$ in the antecedent of a premiss corresponds to the occurrence i in the conclusion,

for $(E\rightarrow)$ for $i>0$ the occurrence $i+1$ in the antecedent of the right, premiss corresponds to the occurrence i in the conclusion, and the occurrence i in the antecedent of the the left premiss corresponds to the occurrence i in the conclusion.

This has the effect that for E-rules also the principal occurrences correspond though they are not parametric.

As in the case of K_sP, a derivation of a premiss of the weakening rule (W) becomes a derivation, of the same length, of the corresponding conclusion if the weakening formula is adjoined to each of its sequents (and if appropriate instances of (RP) are performed). More generally, a given rule is called *admissible* for some calculus if derivation(s) of its premiss(es) can, within that calculus, be extended to a derivation of its conclusion. In this sense, the rule (W) is admissible for K_uP. Further, the reflexivity axioms again have derivations in K_uP.

LEMMA 2 The contraction rule (RC) is admissible for K_uP.

There are two operators, transforming K_tP-derivations and K_uP-derivations into each other, which preserve endsequents and lengths.

As for the first statement, let $v,v,M \Longrightarrow u$ be the endsequent of a K_uP-derivation D. Working upwards from the endsequent, I see that the antecedent of every sequent of D contains v with multiplicity at least 2, and that each instance of a logical rule remains such instance if this multiplicity is lowered by 1 both in the conclusion and in the premiss(es). Performing this removal on all sequents of D, I obtain a K_uP-derivation of $v,M \Longrightarrow u$.

The operator $\mathbf{O}$ acting on K_tP-derivations D I define by recursion on the length of D ; its action is trivial if D is of length 0. There is a unique *last* instance of a logical rule (R_t) in D, below which only instances of (RC) can follow; let D' and D" be the subderivations of D leading to the premisses P, Q of this instance. Weakening D' and D" with the principal formula of that rule, I obtain K_tP-derivations H', H" of shorter length such that $\mathbf{O}$(H'), $\mathbf{O}$(H") are available with endsequents P*, Q*. Applying the rule (R_u), which in K_uP corresponds to (R_t), the sequents P*, Q* now give the same conclusion S which P, Q gave under (R_t), and so $\mathbf{O}$(H'), $\mathbf{O}$(H") produce a K_uP-derivation of S. Below S only instances of (RC) lead to the endsequence of D, and as (RC) is admissible, I can remove them and obtain the K_uP-derivation $\mathbf{O}$(D) of this endsequent.

The operator $\mathbf{O}$ acting on K_uP-derivations D I define by recursion on the length of D ; its action is trivial if D is of length 0. Let D', D" be the subderivations of D leading to the premisses P, Q of the last rule (R_u) applied in D. By inductive assumption, I can transform them into K_tP-derivations of P, Q. Applying the rule (R_t), which in K_tP corresponds to (R_u), the derivations $\mathbf{O}$(D'), $\mathbf{O}$(D") produce a K_tP-derivation of a sequent S*. If (R_t) was not an E-rule then S* is the endsequent of D, and otherwise S* differs from S in that it contains the principal formula of (R_t) with its multiplicity risen by 1. Continuing with (RC), I obtain the K_tP-derivation $\mathbf{O}$(D) of S.

As an example, observe that the K_tP-derivation of $v_3 \Longrightarrow y$ is changed into the K_uP-derivation

$$v_1,x,v_1,v_3 \Longrightarrow x$$
$$x,v_1,v_3 \Longrightarrow v_2 \qquad\qquad y,x,v_1,v_3 \Longrightarrow y$$
$$x,v_1,v_3 \Longrightarrow y$$
$$v_3,v_1 \Longrightarrow x \to y \qquad\qquad x,v_3,v_1 \Longrightarrow x$$
$$v_3,v_1 \Longrightarrow x$$
$$v_3 \Longrightarrow v_2 \qquad\qquad y,v_3 \Longrightarrow y$$
$$v_3 \Longrightarrow y$$

Obviously, the sequents of an extremal derivation contain redundant information. But K_uP has the convenient property that a formula b, occurring in the antecedent of a conclusion of a logical rule, occurs with the same multiplicity and at the same places on the left sides of the premiss(es).

It follows immediately from the form of their rules that the three calculi K_sP, K_tP and K_uP have the

Subformula Property: In any derivation of an endsequent $M \Longrightarrow u$ there occur, in its sequents, only such formulas which are in M,u or are subformulas of formulas in M,u

and have the property of

Directness: In any derivation of an endsequent $M \Longrightarrow u$ only such logical rules are used which refer to operations (connectives) occurring in the formulas of M,u .

They also have the property of

Decidability : There is an algorithm which, given any sequent $s: M \Longrightarrow u$, determines in finitely many steps whether it occurs as endsequent of a derivation or not.

It will be no restriction if I show this for the calculus K_tP, because derivability in one of my calculi implies that in the others. Deciding whether the sequent s is derivable is easy if all of its formulas are variables: in that case the only rules employed in a derivation D of s can be structural ones, D does not ramify, and if it starts from an axiom $x \Longrightarrow x$ then the variable x remains present on the right as well as on the left. A sequent $M \Longrightarrow u$, therefore, consisting of variables only, has a derivation if, and only if, u occurs in M .

I shall say that a sequent $s: M \Longrightarrow u$ is *small* if the formulas in M occur there with a multiplicity at most 2. Let $t: N \Longrightarrow u$ be not small and let s be such that M arises from N by lowering the multiplicities of certain formulas. If t can be derived then so can s by performing suitable contractions; if s can be derived then so can t by performing suitable weakenings. Thus it will suffice to decide the derivability of small sequents.

LEMMA 3 Every derivation D of small sequent can be transformed into a derivation containing only small sequents.

Because given D, I rarefy each of its sequents $M \Longrightarrow u$ to a small sequent $M^+ \Longrightarrow u$ by keeping, for each of the formulas v in M, the first two occurrences of v and omitting the later ones. This changes D into an object D^+ of which I shall show that it can be expanded into a derivation $D^\ddagger$, and since the small endsequent of D remains unchanged, $D^\ddagger$ will be the desired derivation. Observe first that axioms, being small already, remain unchanged.

Consider an instance of (W); it is omitted if v occurs already twice in M, and otherwise it remains in effect. An instance of (RC) remains in effect since it concerns only the first two occurrences of v. Consider next an instance of (RP) with a transposition π changing M into $\pi(M)$. If the transposed occurrences both were kept then it remains in effect with a transposition π^+ of $\pi(M)^+ = \pi^+(M^+)$; if the transposed occurrences both were removed then it can be omitted. If the sequence M has the form $M_0vM_1wM_3$ and if $\pi(M)$ has the form $M_0wM_1vM_3$, and if v is kept and w is removed, then $M^+ = M_0{}^+vM_1{}^+M_3{}^+$ and $\pi(M)^+ = M_0{}^+M_1{}^+vM_3{}^+$; thus if $M_1{}^+$ has length n then n instances of (RP), transposing v with the members of $M_1{}^+$, will lead from $M^+ \Longrightarrow u$ to $\pi(M)^+ \Longrightarrow u$. An analogous method handles the case that w is kept and v is removed.

Instances of the I-rules remain such instances under rarefication since in (I$\rightarrow$) the first occurrence of v in the premiss is not omitted. For instances of the E-rules, the rarefied M^+ in premisses, say $(w,M)^+ = w,M^+$, may contain the principal formula r already twice; in these cases another application of (RC) to M^+ (preceded and succeeded by instances of (RP)) will remove one copy of r leading to $M^+{}_r$, and then the E-rule will give the desired result, e.g. in case of (E$\rightarrow$)

$$\begin{array}{cc} M^+ \Longrightarrow v & w,M^+ \Longrightarrow u \\ M^+{}_r \Longrightarrow v & w,M^+{}_r \Longrightarrow u \\ \hline \multicolumn{2}{c}{v\rightarrow w,M^+{}_r \Longrightarrow u} \end{array} ,$$

since $(r,M)^+ = r,M^+{}_r$.

It follows from Lemma 3 that it suffices to check whether a small sequent $s: M \Longrightarrow u$ has a derivation consisting of small sequents. By the subformula property, these small sequents must be formed with subformulas of formulas occurring in s. The set K_0 of these subformulas is finite and its size k_0 is bounded by $1+2^u+\Sigma\langle 1+2^{|m|} \mid m \text{ occurs in } M\rangle$ where $|m|$ is the complexity of the formula m. Thus it remains

to form the set K_1 of all small sequents consisting of formulas in K_0,

to determine for each t in K_1 the sequents in K_1 which can serve as premisses of an instance of a rule leading to t,

to check whether these finitely many dependencies can be put together such as to form a derivation of s.

The number of steps required for these tasks is finite. Of course, it also is enormous, but finite being finite, a decision algorithm has been found.

[Observe that for $t: N \Longrightarrow w$ in K_1 there at most 4 sequents in K_1 which can serve as premisses of an instance of a logical rule leading to t. If N contains n formulas, there are at most $c_n = (2n!)/2^n$ sequents in K_1 from which t can be derived by the rules (RP) and (RC) alone, and there are at most $d_n = \Sigma<\binom{n}{k}\cdot c_k \mid 0<k\leq n>$ sequents from which t can be derived by (W), (RP) and (RC) alone. The number c_{10} has already 16 decimal digits.]

A more economical decision algorithm will be discussed in Chapter 7.

4. Inversion Operators

In this section I consider the calculi K_tP or K_uP. Given a derivation D, I shall define, for every occurrence i in the endsequent of D, the *predecessors* of i together with the sequents in D containing such predecessors. I begin by saying that the occurrence in the endsequent is a predecessor of itself, and I proceed in the tree underlying D by recursion on the height of nodes. So consider a sequent s in D which does not lay on a maximal node and thus is conclusion of an instance of a rule. If s does not contain a predecessor of i then neither shall its premisses. If s does contain an occurrence j which is predecessor of i then any occurrence in a premiss which corresponds to j shall also be a predecessor of i.

The predecessors of an occurrence all carry the same formula. Because in K_tP corresponding pairs are always parametric; in K_uP they also may both be principal. The subtree of D, formed from sequents containing predecessors of the occurrence i, I call the *predecessor tree* or *parameter tree* P of i. For K_uP the parameter tree of an occurrence in the antecedent of the endsequent is always D itself. For a right occurrence i the parameter tree at an instance of (E→) continues into the right but not into the left branch.

What happens at the maximal nodes e of the parameter tree P? If i is the right occurrence of u in $M \Longrightarrow u$ and if u is a variable then e carries an axiom, and if u is composite then e carries the conclusion of an I-rule introducing u. For this right occurrence, therefore, all logical rules applied within the parameter tree must be E-rules.

Consider now the particular case that u is $p \to q$. If e carries a sequent $N_e \Longrightarrow p \to q$ then it will have an upper neighbour e' carrying $p, N_e \Longrightarrow q$. Replace, for every h in P, the sequent $N_h \Longrightarrow p \to q$ at h by $p, N_h \Longrightarrow q$. Then e and e' carry the same sequent. At nodes h below e, the replacement at corresponding parametric occurrences preserves the instances of rules, and the addition of p in the antecedent preserves them (after appropriate permutations) as well. Omitting the step from e' to e, I obtain a derivation of the sequent $p, M \Longrightarrow q$. Thus I have proved the *inversion rule:*

(JI→) There is an operator **JI→** transforming any derivation D of $M \Longrightarrow p \to q$ into a derivation **JI→**(D) of $p, M \Longrightarrow q$ which has at most the lenght of length of D .

Consider next the case $u = p \wedge q$. If e carries a sequent $N_e \Longrightarrow p \wedge q$ then it will have two upper neighbours e', e" carrying $N_e \Longrightarrow p$ and $N_e \Longrightarrow q$. Replace, for every h in P, the sequent $N_h \Longrightarrow p\wedge q$ at h by $N_h \Longrightarrow p$. Then e and e' carry the same sequent as e', and the instances of the E–rules performed in P remain such instances. Consequently, I obtain a derivation of the sequent $M \Longrightarrow p$ and have proved the *inversion rule:*

(JI∧) There are two operators **JI∧L** and **JI∧R** transforming any derivation D of $M \Longrightarrow p\wedge q$ into derivations **JI∧L**(D) of $M \Longrightarrow p$ and **JI∧R**(D) of $M \Longrightarrow q$ which have at most the length of D .

Consider finally the case $u = p\vee q$. If e carries $N_e \Longrightarrow p\vee q$ then it will have an upper neighbour e' carrying either $N_e \Longrightarrow p$ or $N_e \Longrightarrow q$; if there are two maximal nodes e, f in P, the one covered by $N_e \Longrightarrow p$ and the other by $N_f \Longrightarrow q$ with $p \neq q$, then no substitution will work. But it will work as before if there is *only one* maximal node in P, i.e. if P does not ramify. An instance of (E→) in P does not cause a ramification since only the right branch leading to it will continue into P; hence ramifications can only occur if P contains an instance of E∨. A a sufficient condition to exclude this possibility is that M consists of *Harrop* formulas in the following sense: every variable is Harrop; if v and w are Harrop then v∧w is Harrop; if w is Harrop then v→w is Harrop. Thus I have proved the *inversion rule:*

(JI∨) There is an operator **JI∨** transforming any derivation D of $M \Longrightarrow p\vee q$ where M consists of Harrop formulas, into a derivation **JI∨**(D) of either $M \Longrightarrow p$ or $M \Longrightarrow q$ which has at most the length of D .

I now turn to an occurrence i in the antecedent of the endsequent of D. For the time being, I restrict myself to K_uP such that P coincides with D. Let b be the formula occurring at i.

Consider the case $b = p\to q$. Replace, for every node h in T, the sequent $N_h \Longrightarrow c$ at h by $q, N_h{}^\dagger{}_b \Longrightarrow c$ where $N_h{}^\dagger{}_b$ arises from N_h by removing the unique predecessor of the occurrence i. Working in K_uP with an occurrence in the antecedent, a predecessor may be well be principal occurrence of an (E→)–instance, and in that case remove further the entire subderivation leading to the left upper neighbour of h. The resulting object then is a subtree of T with the endsequent $q, M^\dagger{}_b \Longrightarrow u$. Instances of rules, not using a predecessor as principal occurrence, remain such instances. Instances of E→ with a predecessor as principal occurrence

$$\frac{p\to q, N_h' \Longrightarrow p \qquad q, p\to q, N_h' \Longrightarrow c}{p\to q\,, N_h' \Longrightarrow c} \qquad \text{become} \qquad \frac{q, q, N_h' \Longrightarrow c}{q, N_h' \Longrightarrow c}$$

and thus are instances of (RC). Applying the operator which removes these instances of the rule (RC), admissible according to Lemma 2, I have proved the *inversion rule*

(JE→L) There is an operator **JE→L** transforming any derivation D of $p\to q, M \Longrightarrow u$ into a derivation **JE→L**(D) of $q, M \Longrightarrow u$ which has at most the length of D .

Consider the case $b = p \wedge q$. Replace, for every node h in T, the sequent $N_h \Longrightarrow c$ at h by $p,q,N_h{}^{\dagger}{}_b \Longrightarrow c$ where $N_h{}^{\dagger}{}_b$ arises from N_h by removing the unique predecessor of the occurrence of b. Instances of rules, not using a predecessor as principal occurrence, remain such instances. Instances of E$\wedge$ with a predecessor as principal occurrence, say

$$\frac{p,p\wedge q,N_h' \Longrightarrow c}{p\wedge q,N_h' \Longrightarrow c} \quad \text{become} \quad \frac{p,q,p,N_h' \Longrightarrow c}{p,q,N_h' \Longrightarrow c}$$

and thus instances of (RC) together with (RP). As the analogous construction can be performed for q, I have proved the *inversion rule*

(JE$\wedge$) There is an operator **JE**$\wedge$ transforming any derivation D of $p\wedge q,M \Longrightarrow u$ into a derivation **JE**$\wedge$(D) of $p,q,M \Longrightarrow u$ which has at most the length of D .

Consider finally the case that $b = p\vee q$. Replace, for every node h in T, the sequent $N_h \Longrightarrow c$ at h by $p,N_h{}^{\dagger}{}_b \Longrightarrow c$ where $N_h{}^{\dagger}{}_b$ arises from N_h by removing the the unique predecessor of the occurrence of b. Also, if $N_h \Longrightarrow c$ is the conclusion of a E$\vee$ rule with principal formula b, then remove the entire subderivation leading to the right upper neighbour of h. In the resulting object the instances of rules, not using a predecessor as principal occurrence, remain such instances. Instances of E$\vee$ with a predecessor as principal occurrence

$$\frac{p,p\vee q,N_h' \Longrightarrow p \qquad q,p\vee q,N_h' \Longrightarrow c}{p\vee q\,,N_h' \Longrightarrow c} \quad \text{become} \quad \frac{p,p,N_h' \Longrightarrow c}{p,N_h' \Longrightarrow c}$$

and thus instances of (RC) together with (RP). I so have proved the *inversion rule*

(JE$\vee$) There are two operators **JE**$\vee$**L** and **JE**$\vee$**R** transforming any derivation D of $p\vee q,M \Longrightarrow u$ into derivations **JE**$\vee$**L**(D) of $p,M \Longrightarrow u$ and **JE**$\vee$**R**(D) of $q,M \Longrightarrow u$ which have at most the length of D .

So far, the inversion rules (JE–) have been proved only for K_uP. But now the operators, presented in Lemma 1 and Lemma 2, permit to translate between derivations of the three calculi, and so the inversion rules hold true also for them. It will be a useful exercise for the reader to give explicit proofs the case of K_sP in the situations when the parameter tree of b is a proper subtree of T.

5. Tableaux

The question, whether a sequent has a derivation, can be decided by the technique described in section 3. In this section, I shall briefly discuss a slightly different method.

Assume that I am given two (disjoint) copies of the set of propositional formulas which I distinguish by prefixing the ones with the letter P and the others with the letter Q ; I shall call them *signed* (propositional) formulas. The composite signed formulas I classify as *unramified* and *ramified,* and to each of them I assign two *components* as follows:

unramified			ramified		
$Q\,v\wedge w$:	$Q\,v$	$Q\,w$	$P\,v\wedge w$:	$P\,v$	$P\,w$
$P\,v\vee w$:	$P\,v$	$P\,w$	$Q\,v\vee w$:	$Q\,v$	$Q\,w$
$P\,v\rightarrow w$:	$Q\,v$	$P\,w$	$Q\,v\rightarrow w$:	$P\,v$	$Q\,w$

Also, I shall have occasion to consider trees and functions t assigning signed formulas to their nodes. In that case, a node f is called P-predecessor of a node e if t(f) is the highest P-signed formula less or equal to t(e): t(f) is P-signed, $f \leq e$, and $f<g\leq e$ implies that t(g) is not P-signed. Observe that f then is unique and that e is its own P-predecessor if t(e) is P-signed.

Further, speaking of trees T, a *root piece* R of T shall be a set of nodes of T such that (i) R is linearly ordered by the order of T and (ii) if $r \epsilon R$ then $r<e$ for every node e of T not in R. The root of a tree always forms a root piece.

A *tableau* shall consist of four data: a finite 2-ary tree T, a root piece R of T, and two functions t and ε such that

t assigns signed formulas to the nodes of T , in particular:

t assigns a P-signed formula to the maximal node of the root piece R and assigns Q-signed formulas to the other nodes of R .

The function ε assigns reference nodes to certain nodes e of T whose value under t determines how T and t proceed above e:

ε is defined for the nodes which are non-maximal in T and which, if they belong to the root piece R , are the (one) maximal node of R .

$\varepsilon(e)$ is a node such that $\varepsilon(e)\leq e$, $t(\varepsilon(e))$ is composite, and there holds:

(t1) If $t(\varepsilon(e))$ is unramified, or is ramified with identical components, then e has one upper neighbour e' only, and t(e') is one of the components of $t(\varepsilon(e))$. In addition, if $t(\varepsilon(e))$ is $P\,v\rightarrow w$ then also e' has one upper neighbour e'' only, and $t(e') = Qv$, $t(e'') = Pw$, $\varepsilon(e') = \varepsilon(e)$.

(t2) If $t(\varepsilon(e))$ is ramified with distinct components then e has two upper neighbours e', e'', and t(e'), t(e'') are the two components of $t(\varepsilon(e))$.

(t3) If $t(\varepsilon(e))$ is P-signed then $\varepsilon(e)$ is the P-predecessor of e.

The nodes and branches of the tree underlying a tableau I shall also call nodes and branches of the tableau itself; the formula $t(\varepsilon(e))$ is the formula *dissolved* at the node e . A tableau with root piece R uniquely determines an t-sequent $M \Longrightarrow v$: Pv is the one P-signed formula assigned by t to a node of R, and M is a sequence of the formulas w such that t assigns Qw to the nodes of R; the tableau then is said to be a *tableau for* $M \Longrightarrow v$.

A branch B of a tableau is called *closed* if it contains oppositely signed variables and, in case the P-signed one is the lower, it is the P-predecessor of the upper: there are nodes f, e of B and there is a variable x such that $t(f) = Px$, $t(e) = Qx$, and if $f<e$ then f is P-predecessor of e. A tableau is called *closed* if each of its branches is so. The following is an example of a closed tableau:

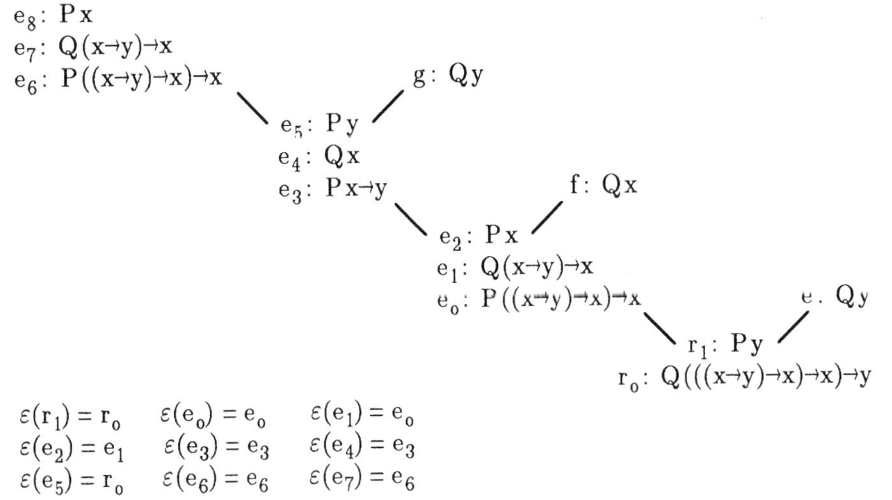

$\varepsilon(r_1) = r_0$ $\varepsilon(e_0) = e_0$ $\varepsilon(e_1) = e_0$
$\varepsilon(e_2) = e_1$ $\varepsilon(e_3) = e_3$ $\varepsilon(e_4) = e_3$
$\varepsilon(e_5) = r_0$ $\varepsilon(e_6) = e_6$ $\varepsilon(e_7) = e_6$

where the root piece consists of r_0, r_1. Observe that the leftmost branch contains oppositely signed variables already at e_4, e_2, but they do not close this branch since e_2 is not P-predecessor of e_4. This shows that, in order to obtain a closed branch, it cannot be avoided to dissolve a Q-signed formula more than once.

Clearly, the above tableau is closely connected to the example of a K_tP- and K_uP-derivation discussed following Lemma 1 and Lemma 2. That this is no accident will be clear from the following

LEMMA 4 There are operators effecting, for any t-sequent s, transformations of the closed tableaux for s into K_uP-derivations of s and vice versa.

Let T, t, ε be a closed tableau; let T_0 be the subtree of T starting at the maximal node of the root piece R of T; it will be no restriction to assume that the branches of T end at the nodes which are the first to make them closed. Let T_1 arise from T_0 by (a) duplicating the branch beginning with e' if $t(\varepsilon(e))$ in (t1) is ramified with identical components, and (b) omitting e' between e and e'' in t(1) if $t(\varepsilon(e))$ is $Pv\rightarrow w$. To every node e of T_1 I assign an t-sequent $d(e)$: $M \Longrightarrow v$ where M consists of the formulas w such that $t(f) = Qw$ for some f in T_0 with $f \leq e$, and where v is such that $t(f) = Pv$ for the P-predecessor of e in T_0; clearly, my tableau is tableau *for* the sequent s assigned to the root of T_0. To every non-maximal node e of T_1 I assign a rule, namely (E∧), (I∨), (I→), (I∧), (E∨), (E→) if $t(\varepsilon(e))$ is $Qp\wedge q$, $Pp\vee q$, $Pp\rightarrow q$, $Pp\wedge q$, $Qp\vee q$, $Qp\rightarrow q$ respectively. It remains to convince oneself that T_1 and its assignments of sequents and rules give rise to a K_uP-deduction.

If e in T_1 is maximal then d(e) is a generalized axiom, because the branch ending with e is closed, and if t(e) is a Q-signed variable then the corresponding P-signed variable lies on the P-predecessor of e. Assume now that e is not maximal, $d(e)$: $M \Longrightarrow v$. Assume that the composite formula $t(\varepsilon(e))$ is a Q-signed formula Qw; applying the permutation rule (RP) if w is not the first member of M, d(e) may be written as $w, M_w \Longrightarrow v$. The formula w is one of $p\wedge q$, $p\vee q$, $p\rightarrow q$; in the first case there is only one upper neighbour e' of e, and by (t1) the sequent d(e') is $p, p\wedge q, M_w \Longrightarrow v$ or $q, p\wedge q, M_w \Longrightarrow v$; thus it is an instance of (E∧l) or (E∧r) which leads from d(e') to d(e). The other two cases are ramified, the components t(e'), t(e'') at the upper neighbours e', e'' of e are Qp, Qq or Pp, Qq respectively, and so it is instances of (E∨) and (E→) which lead from d(e'), d(e'') to d(e). Assume next that $t(\varepsilon(e))$ is a P-signed formula whence by (t3) the node $\varepsilon(e)$ is the P-predecessor f of e; by definition of d(e) then t(f) is Pv, and the formula v is one of $p\wedge q$, $p\vee q$, $p\rightarrow q$. The first case is ramified, and by (t2) the sequents d(e'), d(e'') at the upper neighbours e', e'' of e are $M \Longrightarrow p$, $M \Longrightarrow q$, and so an instance of (I∧) leads from d(e'), d(e'') to d(e). The other two cases are unramified, the components t(e'') at the upper neighbour e'' of e in T_1 are Pp or Pq and Qp or Pq respectively, and so it is instances of (I∨) and (I→) which lead from d(e'') to d(e). - Inserting appropriate instances of the permutation rule (RP), I arrive at an K_uP-deduction.

Conversely, consider the situation that there is a K_uP-derivation of an end-sequent $M \Longrightarrow v$; contracting on its underlying tree the pairs of succeeding nodes which carry instances of the permutation rule (RP), I obtain a tree T_1. From T_1 I obtain a tree T_0 by (a) cutting out one of the branches starting at the premisses of a ramifying principal formula with identical side formulas, and (b) inserting between each neighbouring pair e and e'' a new node e_0 if d(e''), d(e) form an instance of (I→). From T_0 I obtain a tree T by putting below T_0 a linearly ordered set R_0 of new nodes, one for each member of M, which together with the root of T_0 becomes the root piece R of T. On R I define t by sending its nodes to the Q-signed formulas Qw for w in M, and the root of T_1 I send to Pv. The definition of the functions ε and t on T now proceeds by recursion on the height of nodes in T_1; assume that e in T_1 is not maximal and such that

(1e) ε has been defined for all nodes f of T, not in R, such that $f<e$,

(2e) t has been defined for all nodes f of T such that $f \leq e$,

(3e) If d(e) is a sequent $N \Longrightarrow u$ then, for every w in N, there is a node f in T, $f \leq e$, which under t is mapped to Qw, and the P-predecessor f_p of e is mapped to Pu under t.

Having omitted instances of (RP), the sequent d(e) is conclusion $N \Longrightarrow u$ with a principal formula w which, depending on whether it is in N or is u, corresponds to a signed formula Qw or Pu ; depending on whether that is unramified or not, there are one or two premisses at upper neighbours e' or e'' of e in T_1. I define $\varepsilon(e)$ to be the node f such that, by (3e), t(f) is is the appropriately signed principal formula w, and if w is u then $\varepsilon(e) = f_p$. If w is $Pv \to w$ with the premiss at e'' then I define $t(e_0) = Qv$, $\varepsilon(e_0) = e$ for the inserted e_0, and t(e'') = Pw . Otherwise I define t(e') and, in the ramified case with distinct components, t(e'') to be the appropriately signed side formulas. Thus (1e'), (2e') and (1e''), (2e'') follow from (1e) and (2e). If w is $Pv \to w$ as above then e'' is its own P-predecessor and e is that of e_0; if w is $Qv \to w$ with t(e') = Pv and t(e'') = Qw then e' is its own P-predecessor and e is that of e''. In the remaining two cases (I−), the nodes e', and e'' if present, are their own P-predecessors; in the remaining two cases (E−) the P-predecessors remain those of e. Thus in any case (3e') and (3e'') follow from (3e).

It follows that t is defined for all nodes of T and that ε is defined for all nodes in T which are not maximal. Consider now a branch of T ending with a maximal node e''. Then e'' is also maximal in T_1, and d(e'') will be a generalized axiom $x,N \Longrightarrow x$, while for the lower neighbour e of e'' the sequent d(e): $M \Longrightarrow u$ was no generalized axiom yet. Consequently, either u is not x or M is not x,N , i.e. for the rule leading from d(e'') to d(e) either x on the right is a side formula whence t(e'') = Px , or x on the left is a side formula whence t(e'') = Qx . In the first case, Px is not below Qx ; in the second case, (3e) applied to d(e): $M \Longrightarrow x$ shows that t(f) = Px for the P-predecessor f of e which, therefore, remains P-predecessor of e'' as well. This concludes the proof of Lemma 4.

The *Peirce formula* $((x \to y) \to x) \to x$ is an example of a formula v for which there is no closed tableau for $\blacktriangle \Longrightarrow v$. Because let T be tableau for $\blacktriangle \Longrightarrow v$ and observe first that dissolutions of $P((x \to y) \to x) \to x$ can only lead to $Q(x \to y) \to x$, Px, $Px \to y$, Qx, Py; hence no branch of T can close for y. A branch containing Px will refer there to the root $P((x \to y) \to x) \to x$; in that case, no other P-signed formula must occur below Px , and in particular Px cannot occur twice on a branch. A branch containing Qx will refer there to some $Px \to y$ or $Q(x \to y) \to x$ below. That T cannot be closed will follow from the two observations:

(a) If a branch B of T closes with Px above some Qx then T contains a new branch which cannot be closed .

(b) If a branch of T closes (for the first time) with Qx above Px then T contains a new branch which cannot be closed.

Consider first the case (a). Then no closing Qx can refer to some $Px\dashv y$, as for Px the reference to the root must be possible. Hence Qx will refer to some $Q(x\dashv y)\dashv x$, giving rise to a ramification from B into a new branch B' with a $Px\dashv y$ at the side of Qx . B' cannot be closed as it does not contain a Px: Px did not occur in B below the maximal node, and no Px referring to the root can appear above $Px\dashv y$ in B'.

Consider next the case (b). That T closes for the first time with Qx implies that Qx does not occur in B below the maximal nore. Now Qx cannot refer to some $Px\dashv y$ above the closing Px because that would violate closure. But neither can Qx refer to some $Px\dashv y$ below Px , as for Px the reference to the root must be possible. Thus Qx must refer to some $Q(x\dashv y)\dashv x$, giving rise to a ramification from B into a new branch B' with a $Px\dashv y$ at the side of Qx . B' does not contain a Qx below $Q(x\dashv y)\dashv x$, hencing a closing Qx of B' would have to appear above the $Px\dashv y$. On B', no Px can appear above the $Px\dashv y$ as it has to refer to the root; no Px below $Px\dashv y$ (hence already on B) can effect closure with a Qx above $Px\dashv y$. Thus B' cannot be closed.

Calculi of sequents (*Sequenzen*) were invented and thoroughly investigated by GENTZEN 34. Tableaux were invented by BETH, cf. 59 and 62, though used with a different notation.

References

E.W.Beth: The Foundation of Mathematics. 1959

E.W.Beth: Formal Methods. Dordrecht 1962

G.Gentzen: Untersuchungen über das logische Schliessen I,II. Math.Z. **39** (1934/35), 176–210 and 405–431

Chapter 2. Cuts

The *transitivity rule* or *cut rule*

$$(\mathrm{CUT}_1) \qquad \frac{M \Longrightarrow v \qquad v,N \Longrightarrow u}{M,N \Longrightarrow u}$$

expresses a property which deductions can reasonably be expected to have. Adding it to the rules of a given calculus, however, will certainly destroy both the subformula property and the property of directness, and it will introduce a moment of inconstructiveness: being informed that $M,N \Longrightarrow u$ has been derived with this rule, we cannot guess which formula v may have been used for that.

Fortunately, (CUT_1) can be shown to be *admissible* for KP: there are algorithms which, given derivations of its premisses, produce a derivation of its conclusion. In this Chapter, I shall present a first such algorithm.

There are variants of (CUT_1), namely the *symmetric* cut rule

$$(\mathrm{CUT}_0) \qquad \frac{M \Longrightarrow v \qquad v,M \Longrightarrow u}{M \Longrightarrow u}$$

and the *radical* cut rule

$$(\mathrm{CUT}_2) \qquad \frac{M \Longrightarrow v \qquad v,N \Longrightarrow u}{M,N^{\dagger} \Longrightarrow u}$$

where $N^{\dagger}$ is $N-\{v\}$ in the case of s-sequents formed with sets, and otherwise $N^{\dagger}$ is the sequence obtained from N by removing *all* occurrences of v. They are considered mainly for technical reasons: (CUT_0) permits to avoid the contraction rule in certain situations, while (CUT_2) is required when instances of the cut rule shall be exchanged with contractions. Contrasting (CUT_1) with (CUT_2) it is sometimes called the *simple* cut rule.

As all my calculi admit the weakening rule, the admissibility of (CUT_0) implies that of (CUT_1), and as they admit the contraction rule, the admissibility of (CUT_1) implies that of (CUT_0). The admissibility of (CUT_1) implies that of (CUT_2) since $v,N \Longrightarrow u$ may first be contracted to $v,N^{\dagger} \Longrightarrow u$. The admissibility of (CUT_2) implies that of (CUT_1) since missing occurrences of v in the conclusion may be restored by weakenings. – In the following, (CUT) will refer to an arbitrary one of (CUT_i), $i<3$.

That (CUT) is admissible for KP will be shown by considering the calculi KPC, obtained from KP by adjoining the rule (CUT), and constructing a *cut elimination operator* which, acting on a KPC-derivation, removes all instances of (CUT) and so transforms it into a KP-derivation; in this connection, KP-derivations are also called *cut free.* The precise definition of KPC is this:

K_sPC_i and K_tPC_i, for $i \le 2$, is the calculus obtained from K_sP, K_tP by adjoining (CUT_i),

K_uPC_0 is the calculus obtained from K_uP by adjoining (CUT_0),

K_uPC_i, for $1 \le i \le 2$, is the calculus obtained from K_uP by adjoining both (CUT_i) *and* the rule (RC).

All these calculi admit the rule (W) if, in the case of (CUT_2), weakenings are lifted into the left premiss. The calculi K_sPC, K_tPC and K_uPC_0 admit the rule (RC). [The rule (RC) is not admissible in the calculus obtained from K_uP by adjoining only (CUT_1) or (CUT_2). Because consider an instance of (CUT_i) in which both premisses contain in their antecedents a formula p used earlier as a principal formula of a logical rule. If p is not the formula v, the conclusion of (CUT_i) must contain *two* copies of p. Thus second copies of a formula below such the instance of (CUT_i) cannot simply be omitted as it had been possible for K_uP.]

The formula v in (CUT) is always called the *cut formula,* and the number $|v|+1$, $|v|$ the complexity of v, is called the *cut degree* of the particular instance; that instance then is called a $(|v|+1)$-*cut*. The occurrences of u and of the formulas in M, N (or $N^\dagger$) are *parametric,* and there are neither principal nor side formulas. For (CUT_1), in the case of s-sequents, if v was not in the set N of the right premiss, it will not be in the set N of the conclusion. In the case of t-sequents, the multiplicity of v is lowered by 1 when changing from N,v into N.

If, in the right premiss of (CUT_1) or (CUT_0), the formula v occurs in N then the conclusion arises from that premiss by weakenings and contractions. Because a weakening of $v,N \Longrightarrow u$ with M gives $M,N,v \Longrightarrow u$ from where a contraction leads to the conclusion $M,N \Longrightarrow u$.

LEMMA 1 If one of the premisses of (CUT) is a reflexivity axiom then the conclusion arises from one of the premisses by weakenings and contractions.

If $v,N \Longrightarrow u$ is a reflexivity axiom then either $u = v$ or u occurs in N. If $u = v$ then the first premiss becomes $M \Longrightarrow u$, and from that I obtain the conclusion by weakening. If $u \neq v$ and u occurs in N then also $N^\dagger \Longrightarrow u$ is a reflexivity axiom, and from that I obtain the conclusion by weakening. If the second premiss is not a reflexivity axiom then $u \neq v$, and now the first premiss $M \Longrightarrow v$ will be a reflexivity axiom such that v occurs in M. From the second premiss I obtain $v,N^\dagger \Longrightarrow u$ by contractions, and weakening this with the elements of M different from v I obtain $M,N^\dagger \Longrightarrow u$.

In the definition of depths of nodes and of lengths of KPC-derivations, I shall count instances of (CUT) as I did count instances of logical rules; thus

(CUT) is *not* viewed as a structural rule. For a KPC–derivation D the *cut degree* of D shall be the supremum of cut degrees of instances of (CUT) in D; thus KPC–derivations of cut degree 0 are KP–derivations.

The translations between K_sP-, K_tP- and K_uP-derivations remain in effect for the corresponding calculi KPC; also, they preserve the lengths of the derivations. Thus every cut elimination operator for *one* of these calculi KPC gives rise to operators for the *others:* translate a given D into a derivation G for the particular calculus for which an elimination operator has been found, transform G into a cut free G*, and then translate G* back into a derivation of the given calculus. Similarly, it will suffice to eliminate either simple or radical cuts, since the admissibility of the one kind implies that of the other.

However, in order to actually set up an elimination algorithm, the choice of calculus and type of cut will *not* be arbitrary since there *is* an interaction between the kind of cuts and the type of rules.

1. Cut Elimination with Exchange Operators

For this section I shall make the *Permanent Assumption* :

KPC *is* K_sPC_i for $i \leq 2$ or
KPC *is* K_tPC_2 or
KPC *is* K_uPC_0 or K_uPC_2 .

The arguments, for which these specific choices matter, will be indicated at the appropriate locations.

Let $e_i(k,n)$ be the function defined by $e_i(0,n) = n$ and $e_i(k+1,n) = i^{e_i(k,n)}$.

I shall show that there is an operator **A** acting on KPC–derivations D such that there holds the

THEOREM 1 **A**(D) is a KP–derivation with the same endsequent as D. If n is the length and k is the cut degree of D then the length of **A**(D) is at most $e_3(k,n)$.

The operator **A** will be defined from an operator $\mathbf{A}_1$ acting on KPC–derivations D such that

(A1) $\mathbf{A}_1(D)$ is a derivation with the same endsequent as D; if n is the length and k is the cut degree of D then $\mathbf{A}_1(D)$ is at most of cut degree k–1 and of length 3^n .

Because given $\mathbf{A}_1$, **A** can be defined by recursion on its cut degree k as the k–fold iteration of $\mathbf{A}_1$.

A derivation D is said to *essentially* end with an instance of a non-structural rule (R) if that instance is followed only by instances of structural rules (which in K_uPC can only be permutations).

A derivation D is called k-*extreme* if it essentially ends with a k-cut, and if all other cuts occurring in D are of degrees below k; it is called *extreme* if it is k-extreme for some suitable k.

Consider an instance of (CUT) in a derivation D, and let a and b be the depths of the two premisses of this cut. Then it is called an (a,b)-*cut* in D, and (a,b) is its *weight* in D. Cuts of weight (0,b) or (a,0) then are those which have a (generalized) axiom as premiss. I shall write $(c,d) \lesssim (a,b)$ if $c<a$ and $d \leq b$ or if $c \leq a$ and $d<b$. This defines a well founded order relation on weights, and I can use induction on weights for proofs, and recursion on weights for definitions.

I define $\mathbf{A}_1$ by recursion on the length d of D. If $d=0$ then I set $\mathbf{A}_1(D) = D$. If $d>0$ then let (R) be the non-structural rule essentially ending D. Let D', and in case of a two premiss rule also D", be the subderivations of D leading to the premiss(es) of that (R)-instance. Since they have a length shorter than D, both $\mathbf{A}_1(D')$ and $\mathbf{A}_1(D'')$ are defined.

Let $D^\ddagger$ be the derivation obtained from D if D' is replaced by $\mathbf{A}_1(D')$ and D" is replaced by $\mathbf{A}_1(D'')$.

If (R) is not a k-cut then I set $\mathbf{A}_1(D) = D^\ddagger$. If (R) is a k-cut then I set $\mathbf{A}_1(D) = \mathbf{A}_0(D^\ddagger)$ where $\mathbf{A}_0$ is an operator which I will define below and of which I only need that it has the property

(A0) $\mathbf{A}_0(D) = D$ if D is not extreme. If D is k-extreme then $\mathbf{A}_0(D)$ is a derivation with the same endsequent as D and of a cut degree at most k-1. In that case, the length of $\mathbf{A}_0(D)$ is at most a+2b where D ends with an (a,b)-cut.

It remains to verify the property (A1). Since together with D also D', D" have at most the cut degree k, both $\mathbf{A}_1(D')$ and $\mathbf{A}_1(D'')$ have cut degrees at most k-1. If (R) is a k-cut then $D^\ddagger$ is k-extreme, hence $\mathbf{A}_0(D^\ddagger)$ has a cut degree at most k-1. If (R) is not so then $D^\ddagger$ has a cut degree at most k-1.

If D has length n then D', D" have at most length n-1. Hence $\mathbf{A}_1(D')$, $\mathbf{A}_1(D'')$ have at most length 3^{n-1}. If (R) is a k-cut then in $D^\ddagger$ its weight (a,b) is such that a and b are bounded by 3^{n-1}. Thus the length of $\mathbf{A}_1(D) = \mathbf{A}_0(D^\ddagger)$ is at most $3^{n-1}+2 \cdot 3^{n-1} = 3^n$. If $\mathbf{A}_1(D) = D^\ddagger$ then its length is at most $1+3^{n-1}$.

I define $\mathbf{A}_0(D)$ for an extreme D by recursion on the weight (a,b) of the cut essentially ending D. If $a \cdot b = 0$, i.e. $a = 0$ or $b = 0$, then one of the premisses is a generalized axiom, hence so is the endsequent of D by Lemma 1; let $\mathbf{A}_0(D)$ be the derivation of length 0 leading to it. If $a \cdot b \neq 0$ then $\mathbf{A}_0$ is defined from an operator **E** which I will define below and of which I only need that it has the property

(AE) **E**(D) = D if D does not essentially end with a cut or with an (a,b)-cut with $a \cdot b = 0$. Otherwise **E**(D) still has the same endsequent and the same cut degree as D; the length of **E**(D) is at most 2 plus the length of D.

If D is k-extreme and essentially ends with an (a,b)-cut with $a \cdot b \neq 0$ then **E**(D) contains one or two k-cuts with weights (c,d), and the subderivations E', E" ending with them are k-extreme (hence none of them occurs above the other). There holds

(e_0) $length(\mathbf{E}(D)) \leq i + \max(length(E'), length(E''))$ with $i \leq 2$
(e_1) $c \leq a-1,\ d \leq b$ or $c \leq a,\ d \leq b-1$

and if $i = 2$ in holds (e_0) then in (e_1) the second case occurs.

Now D satisfies the hypotheses of the second part of (AE), and so there are the k-extreme subderivations E', E", of **E**(D), ending with k-cuts which are (c,d)-cuts with $c+d < a+b$. Hence $\mathbf{A}_0(E')$, $\mathbf{A}_0(E'')$, are defined; let $\mathbf{A}_0(D)$ be the derivation obtained from **E**(D) if E', E" are replaced by the $\mathbf{A}_0(E')$, $\mathbf{A}_0(E'')$. This concludes the construction of $\mathbf{A}_0$ from **E**, and it remains to verify the property (A0). The subderivations E', E" end with the k-cuts in **E**(D), and when they are replaced by $\mathbf{A}_0(E')$, $\mathbf{A}_0(E'')$ then still no k-cuts will occur below in $\mathbf{A}_0(D)$. Hence together with $\mathbf{A}_0(E')$, $\mathbf{A}_0(E'')$ also $\mathbf{A}_0(D)$ has at most cut degree k−1. The lengths of $\mathbf{A}_0(E')$, $\mathbf{A}_0(E'')$ are at most c+2d and thus bounded by

$$a + 2b - 1 \qquad \text{or} \qquad a + 2b - 2$$

in the two cases of (e_1). If $i = 1$ in (e_0) then

$$\begin{aligned} length(\mathbf{A}_0(D)) &\leq 1 + \max(length\,\mathbf{A}_0(D'), length\,\mathbf{A}_0(D'')) \\ &\leq 1 + a + 2b - 1 = a + 2b \end{aligned}$$

and if $i = 2$ then

$$\begin{aligned} length(\mathbf{A}_0(D)) &\leq 2 + \max(length\,\mathbf{A}_0(D'), length\,\mathbf{A}_0(D'')) \\ &\leq 2 + a + 2b - 2 = a + 2b \ . \end{aligned}$$

It remains to define the *exchange operator* **E** for derivations D ending with an (a,b)-cut with $a \cdot b \neq 0$. **E** acts in two steps upon the bottom part of D, and I shall depict this action graphically, marking the nodes where the chosen bottom part starts by numbers 1, 2, ... ; the replacing part then, written to the right of (or below) the old one, is to be completed by making use of subderivations of D leading to the sequents at the marked nodes. In all diagrams no instances of (RP) will be indicated. All diagrams will be written for the worst case, i.e. with parametric principal formulas such as in $K_u P$; the simplifications possible in non parametric cases are left to the reader. All diagrams will be written for radical cuts, but unless explicitly noted may be read for simple cuts as well. In the case of symmetric cuts, M usually will have to be written for N, but the premisses of the new symmetric cuts may require additional weakenings in advance.

E is constructed from two operators $\mathbf{E}_0$ and $\mathbf{E}_1$. $\mathbf{E}_0(D)$ is D if D is *good* in the sense that none of the premisses of the cut essentially ending D arises under the rules (W) or (RC). Thus $\mathbf{E}_0(D) = D$ for K_uPC_0. Assume now that D is not good. Then $\mathbf{E}_0(D)$

has the same endsequent, the same length and the same cut degree as D

consists of a good subderivation H_0 which essentially ends with a cut with the same cut formula and the same weight as the cut ending D.

I begin with an auxiliary operator $\mathbf{E}_0^0$ which acts on D by performing the following exchanges at the cut essentially ending D:

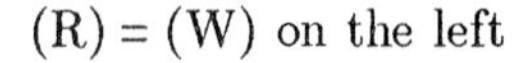
(R) = (W) on the left

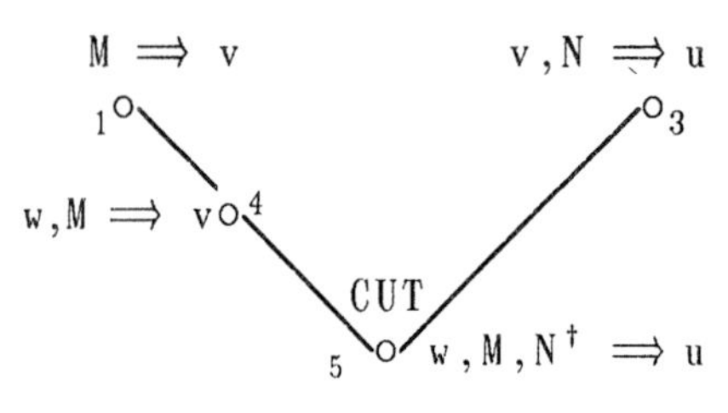

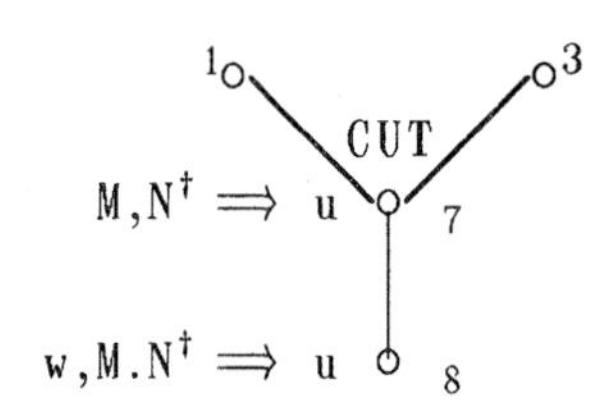

(R) = (W) on the right

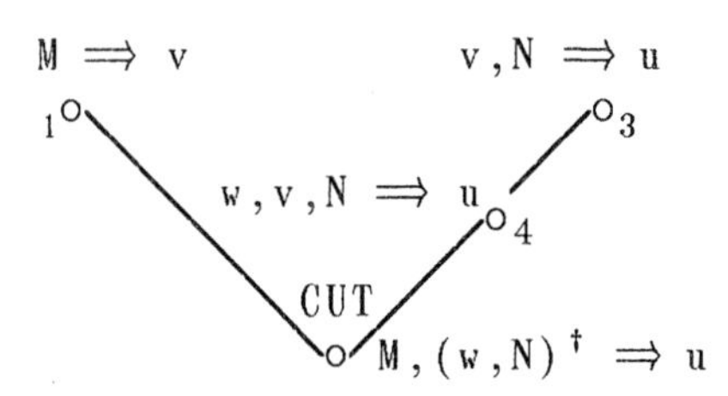

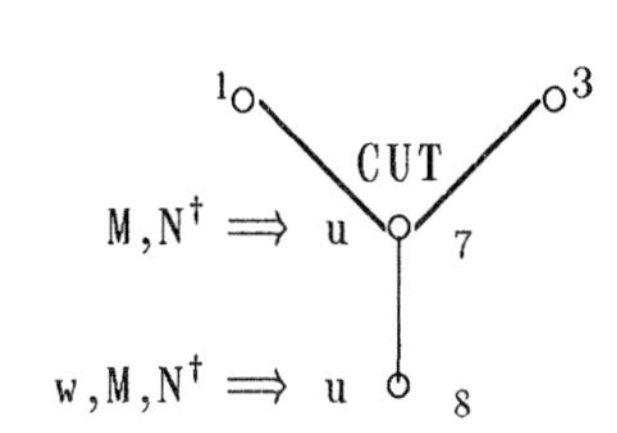

If w = v then 8 is 7.

(R) = (RC) on the left

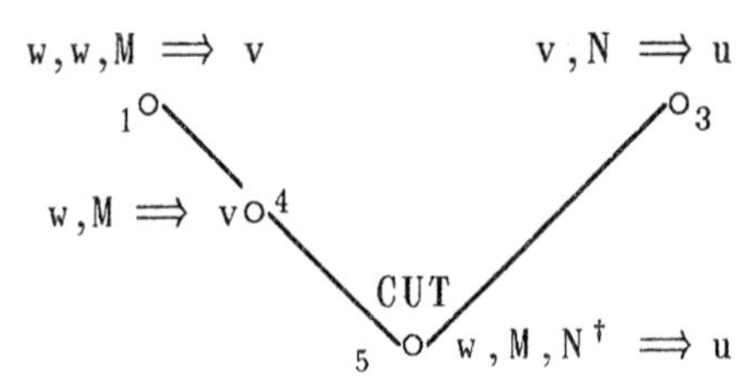

1 3 CUT

w , w , M , N† ⇒ u 7

w , M , N† ⇒ u 8

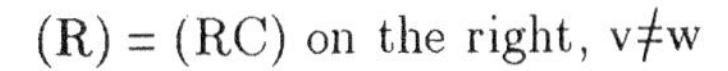

(R) = (RC) on the right, v$\neq$w

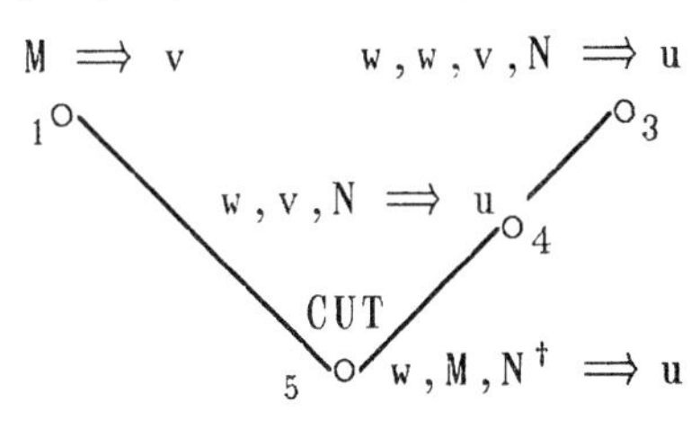

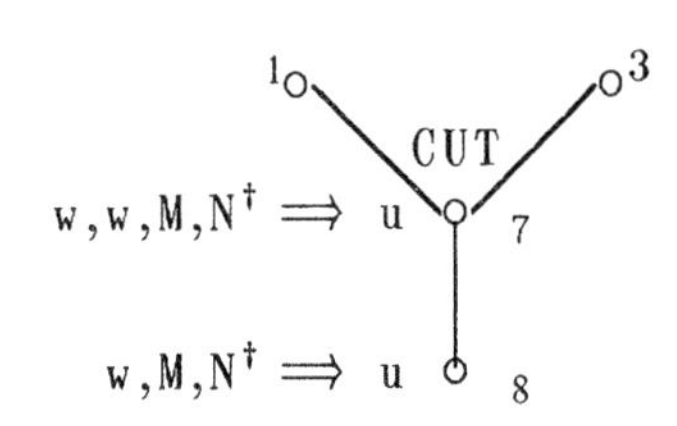

(R) = (RC) on the right, v = w

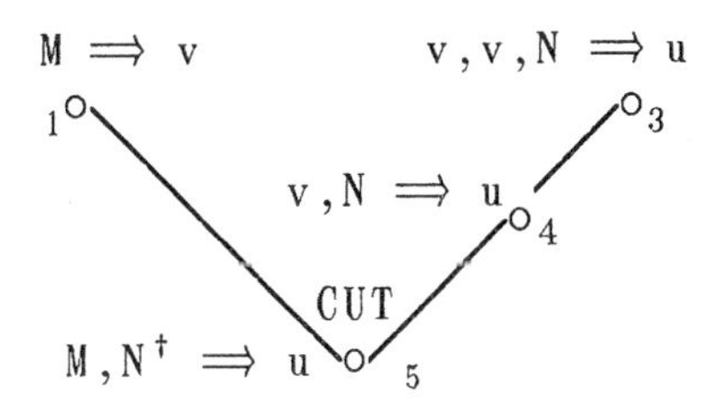

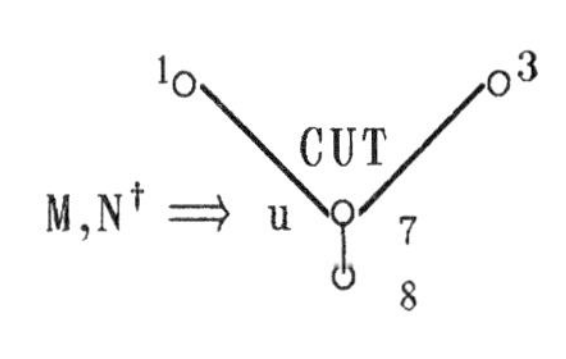

This is the case where (CUT) must be radical. Because consider a simple cut and assume that v does not occur in M or N. Then 5 carries M,N $\Longrightarrow$ u and 7 carries M,v,N $\Longrightarrow$ u ; here v cannot be removed by contraction.

⟦ For the calculi to be discussed in Chapters **4** other rules have to be taken into account. The reader may omit this discussion at present but should return to it when verifying that the present constructions remain in effect in the framework of Chapter **4**.

(R) = AINΔ or (R) = AIN on the left

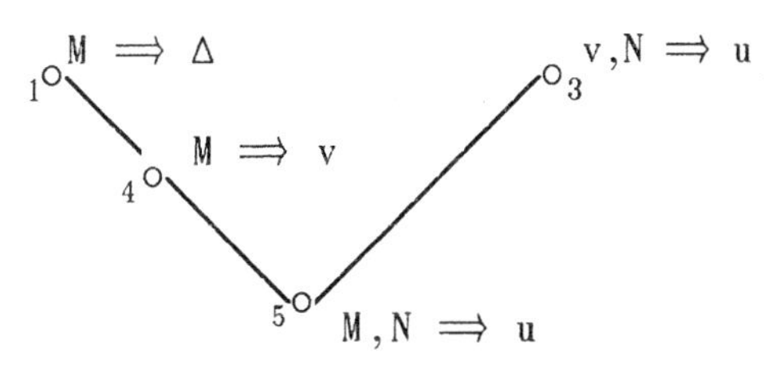

M $\Longrightarrow$ Δ 1

M,N $\Longrightarrow$ Δ 7

M,N $\Longrightarrow$ u 8 .

(R) = AINΔ or (R) = AIN on the right

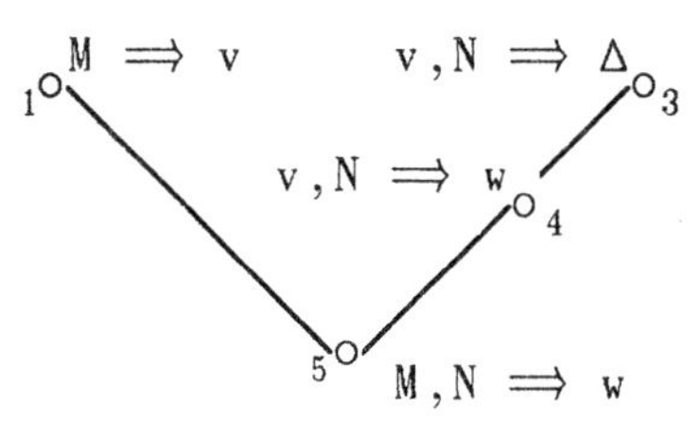

1 3 CUT

M,N† $\Longrightarrow$ Δ 7

M,N† $\Longrightarrow$ w 8 ⟧

It follows from a$\cdot$b>0 that there are nodes e_1, e_r, still of depths a, b, above the premisses of the (a,b)-cut essentially ending D, and these carry sequents obtained either by (other) cuts or by logical rules. Let $m_l(D)$, $m_r(D)$ be the number of instances of structural rules leading from e_l, e_r to those premisses. I set $D_0 = D$ and $D_{i+1} = \mathbf{E}_0^0(D_i)$ if $m_l(D_i)+m_r(D_i) > 0$, otherwise $D_{i+1} = D_i$. Since $\mathbf{E}_0^0$ decreases m_l or m_r for the cut essentially ending D, I can set $\mathbf{E}_0(D) = D_s$ with $s = m_l+m_r$. Observe that if D was k-extreme then so is $\mathbf{E}_0(D)$.

The operator $\mathbf{E}_1$ will be defined for sequents D essentially ending with a good (a,b)–cut with $a \cdot b \neq 0$. For such D then the operator **E** is defined as $\mathbf{E}_1(D)$. If $\mathbf{E}_0(D)$ is not D then $\mathbf{E}(D)$ shall be the derivation obtained by replacing H_0 in $\mathbf{E}_0(D)$ with $\mathbf{E}_1(H_0)$.

Thus it remains to define $\mathbf{E}_1$ for its arguments D, and with v as the cut formula of the cut ending D, I distinguish the three cases

E1. The left premiss of (CUT) is conclusion of a rule (R) which is either a cut or a logical rule for which v is not its principal formula.

E2. The first case does not hold, and the right premiss of (CUT) is conclusion of rule (R) wich is either a cut or a logical rule such that no occurrence of v is its principal formula.

E3. Both premisses of (CUT) are conclusions of logical rules (R) with v as principal formulas.

Then the bottom part of D is exchanged as follows.

Case E1

$(R) = E\wedge$

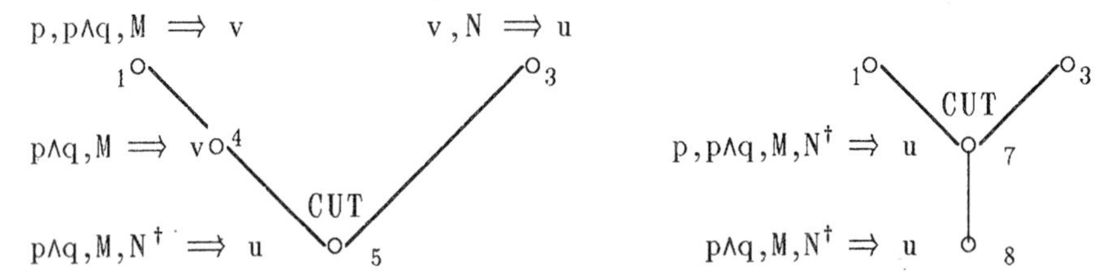

For symmetric cuts, 3 on the left diagram carries $v,p\wedge q,M \Longrightarrow u$ and the weakened sequent $v,p,p\wedge q,M \Longrightarrow u$ on the right diagram.

Here, as in all following subcases of Case 1, the original (a,b)–cut from nodes 4 and 3 to 5 (a the depth of 4) is shifted upwards to an (a–1, b)–cut from 1 and 3 to 7. If the depth of 5 is determined by 4 then D and $\mathbf{E}_1(D)$ have the same length; if it is determined by 3 then the length of $\mathbf{E}_1(D)$ is 1 plus the length of D. If D is k–extreme then so is the subderivation E' ending at 7.

(R) = E$\vee$

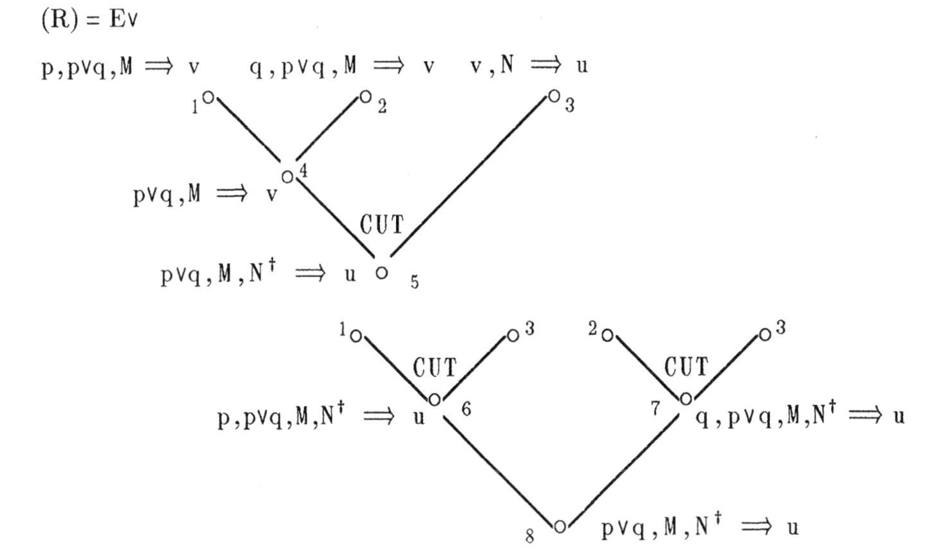

Here the (a,b)–cut from 4 and 3 to 5 is shifted upwards to two (c,b)–cuts from 1 and 3 to 6 and from 2 and 3 to 7, both with $c \leq a-1$ since 1 and 2 both have depths below that of 4. Again, if the depth of 5 is determined by 4 then D and $\mathbf{E}_1(D)$ have the same length; if it is determined by 3 then the length of $\mathbf{E}_1(D)$ is 1 plus the length of D.

(R) = E$\rightarrow$

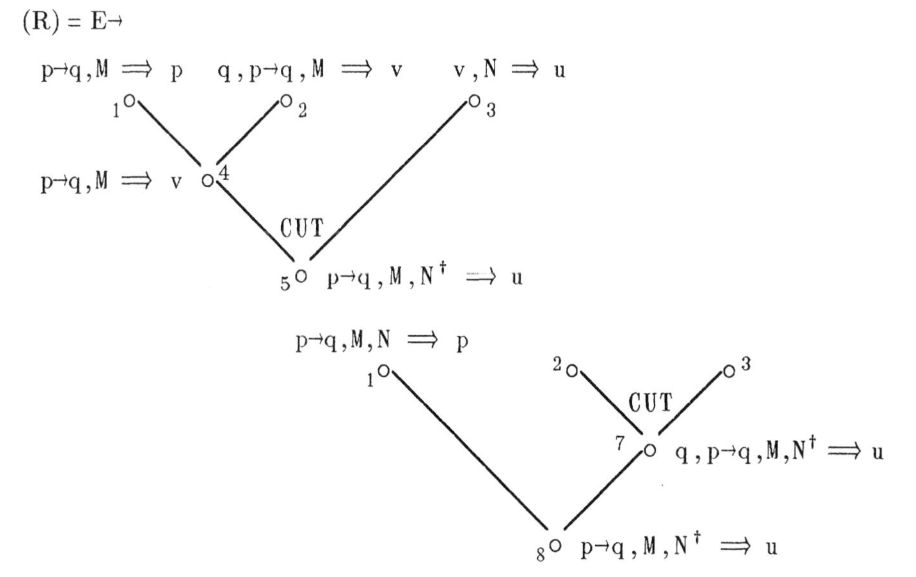

For symmetric cuts, 3 carries $v, p \rightarrow q, M \Longrightarrow u$ on the left and the weakened sequent $v, q, p \rightarrow q, M \Longrightarrow u$ on the right.

(R) is a cut with cut formula w, v≠w

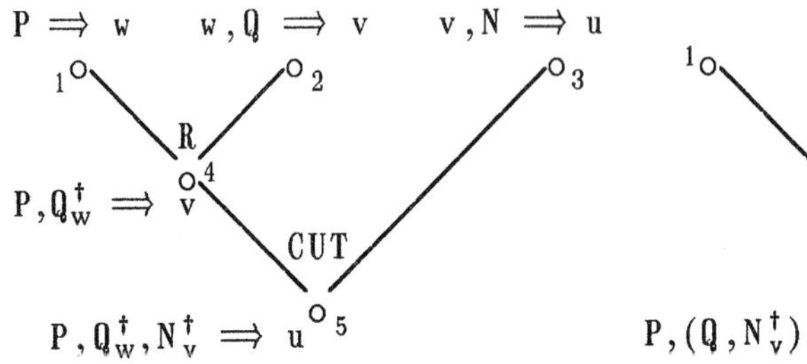

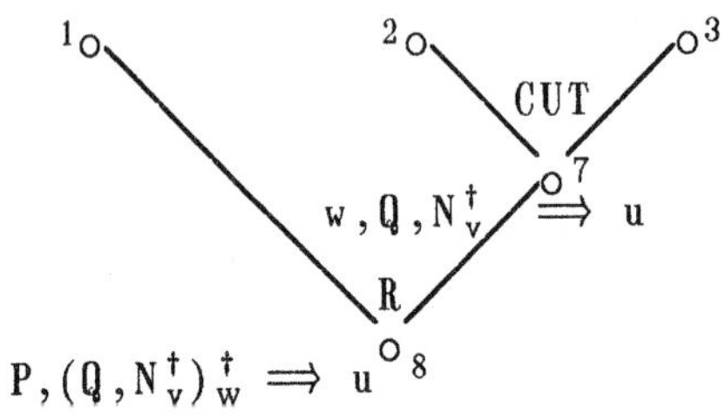

At 8 then w may have to be restored in $N^{\dagger}{}_{v}$. For symmetric cuts, 3 carries v,P,$Q^{\dagger}_{w}$ ⟹ u on the left and v,w,P,$Q^{\dagger}_{w}$ ⟹ u on the right.

If D is k-extreme then the cut (R) at 4 and 8 has cut degree k−1. Thus the subderivation ending at 7 remains k-extreme.

(R) is a cut with cut formula w, v = w

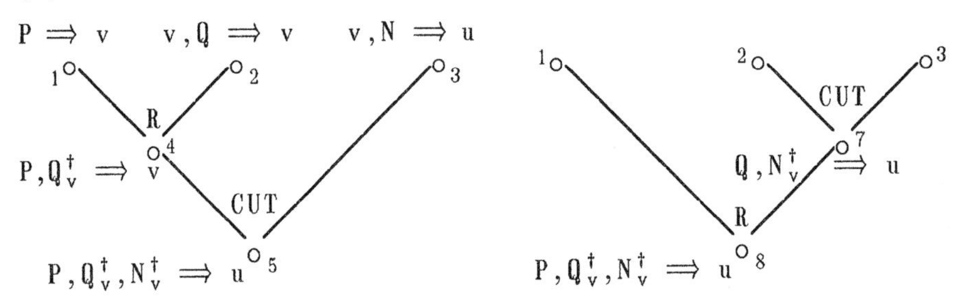

[For the calculi to be discussed in Chapters 4 and 5 other rules have to be taken into account.

(R) = K_0

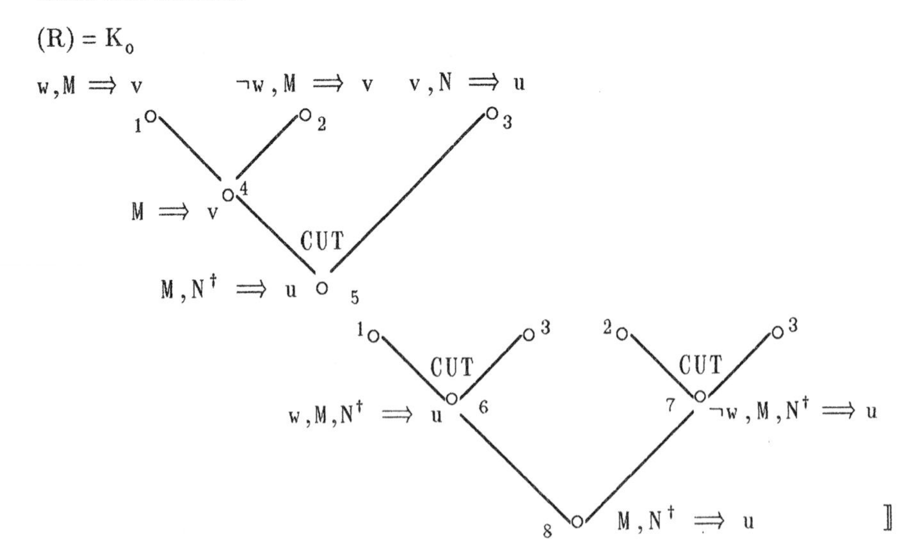

]

Case E2

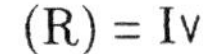

(R) = I∨

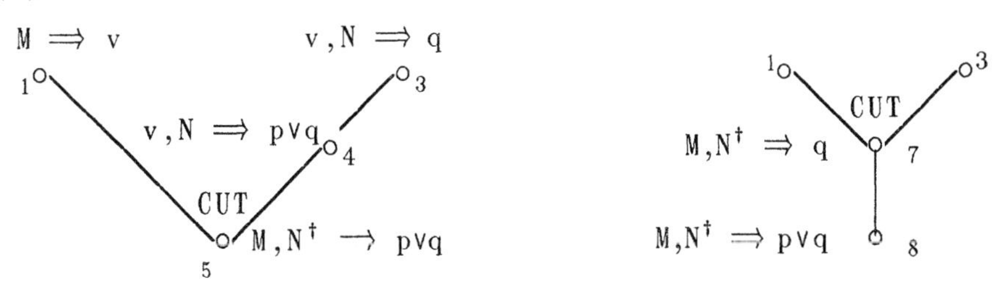

Here, as in all following subcases of Case 2, the (a,b)-cut from 1 and 4 to 5 (b the depth of 4) is shifted upwards to an (a, b−1)-cut from 1 and 3 to 7. If the depth of 5 is determined by 4 then D and $\mathbf{E}_1(D)$ have the same length; if it is determined by 1 then the length of $\mathbf{E}_1(D)$ is 1 plus the length of D. If D is k-extreme then so is the subderivation E' ending at 7.

(R) = I→

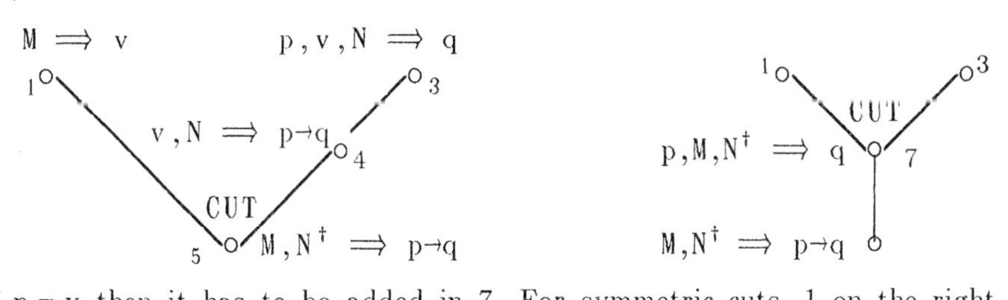

If p = v then it has to be added in 7. For symmetric cuts, 1 on the right carries p, M ⟹ v .

(R) = E∧

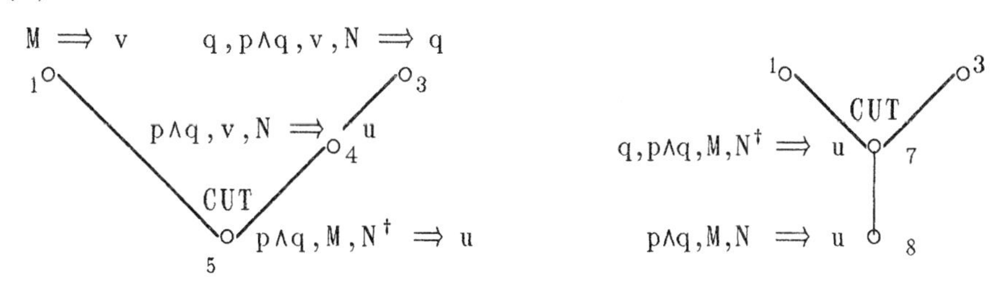

If q = v then 8 is 7. For symmetric cuts, 1 carries p∧q,M ⟹ u on the left and q,p∧q, M ⟹ v on the right.

(R) = I∧

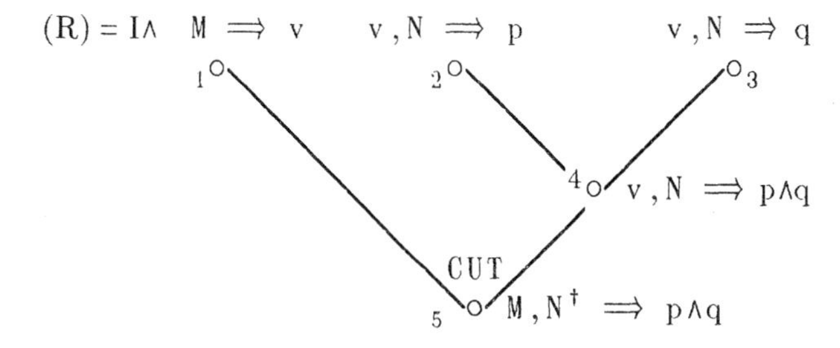

1
2
1
3
CUT
CUT
M,N† ⇒ p
6
M,N† ⇒ q
7
M,N† ⇒ p∧q
8

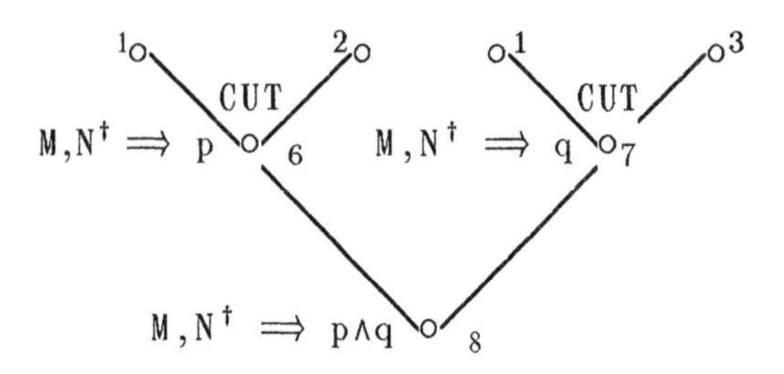

Here the (a,b)-cut from 1 and 4 to 5 is shifted to two (a,d)-cuts from 1 and 2 to 6 and from 1 and 3 to 7 with $d \leq b-1$ since 2 and 3 both have depths below that of 4. Again, if the depth of 5 is determined by 4 then D and $\mathbf{E}_1(D)$ have the same length; if it is determined by 1 then the length of $\mathbf{E}_1(D)$ is 1 plus the length of D.

(R) = E∨

M ⇒ v
p, p∨q, v, N ⇒ u
q, p∨q, v, N ⇒ u
1
2
3
4
p∨q, v, N ⇒ u
CUT
5
p∨q, M, N† ⇒ u

1
2
1
3
CUT
CUT
p,p∨q,M,N† ⇒ u
6
q, p∨q, M, N† ⇒ u
7
8
p∨q, M, N† ⇒ u

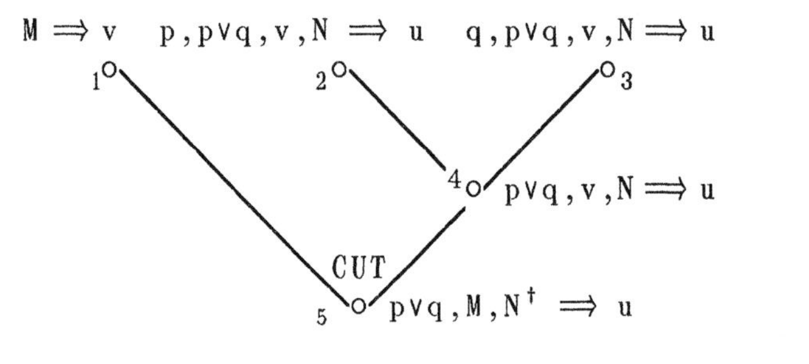

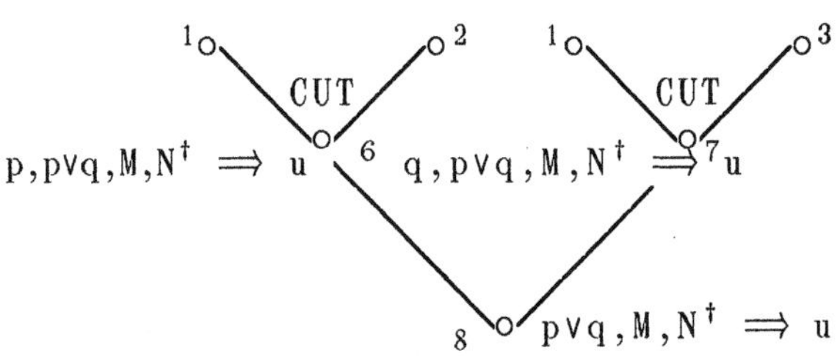

For symmetric cuts, the left 1 has p∨q,M ⇒ u , the right 1 has q,p∨q,M ⇒ v .

(R) = E→

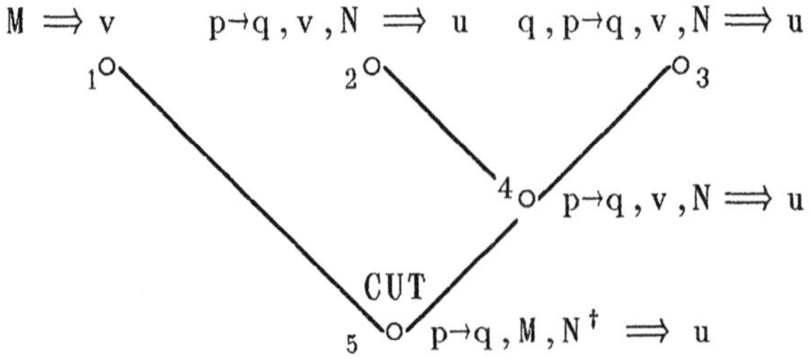

M ⇒ v
p→q, v, N ⇒ u
q, p→q, v, N ⇒ u
1
2
3
4
p→q, v, N ⇒ u
CUT
5
p→q, M, N† ⇒ u

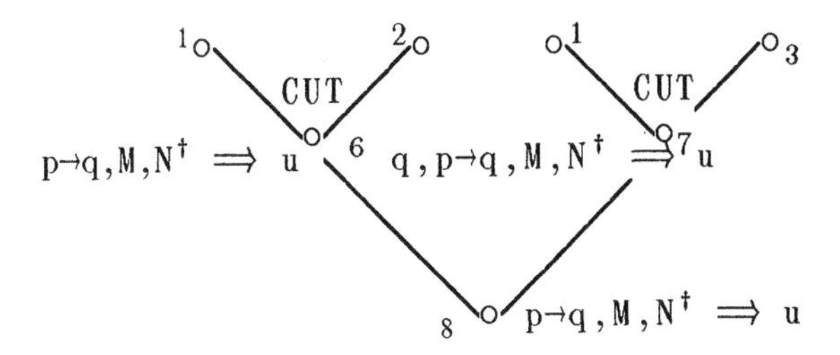

If q = v then 8 is 7. For symmetric cuts, 1 carries p→q, M ⟹ u on the left and middle copies, and q,p→q, M ⟹ v on the right copy.

(R) is a cut

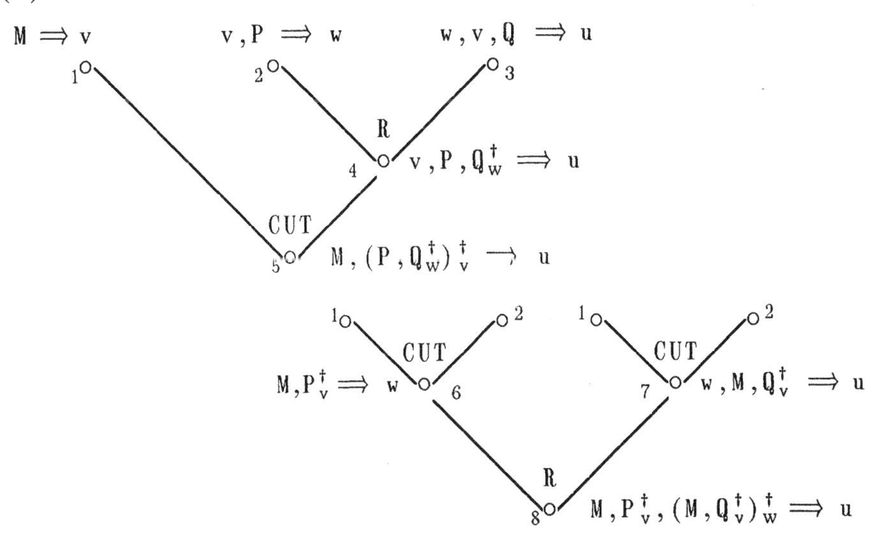

At 8 the sequence $M^{\dagger}{}_{w}$ has to be removed by contractions. For symmetric cuts, 1 on the utter right carries M,w ⟹ v . If D is k-extreme then the cut (R) at 4 and 8 has cut degree k−1. Thus the subderivation ending at 7 remains k-extreme.

[[For the calculi to be discussed in Chapters **4** and **5** more rules have to be taken into account:

(R) = IΔ¬ or I¬

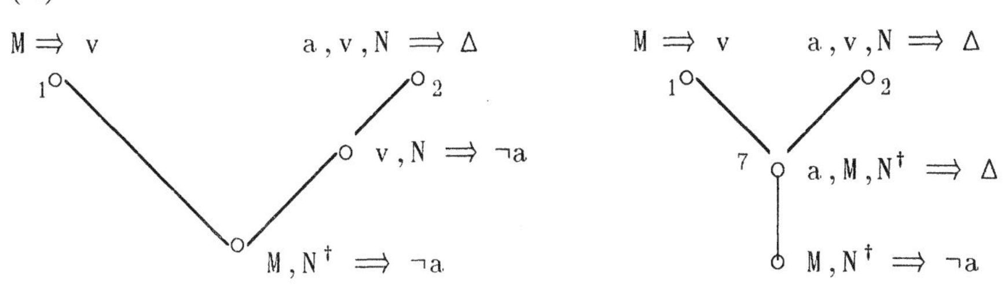

If a = v then this formula may have to be added at 7 by a weakening.

(R) = E$\Delta\neg$ or E$\neg$

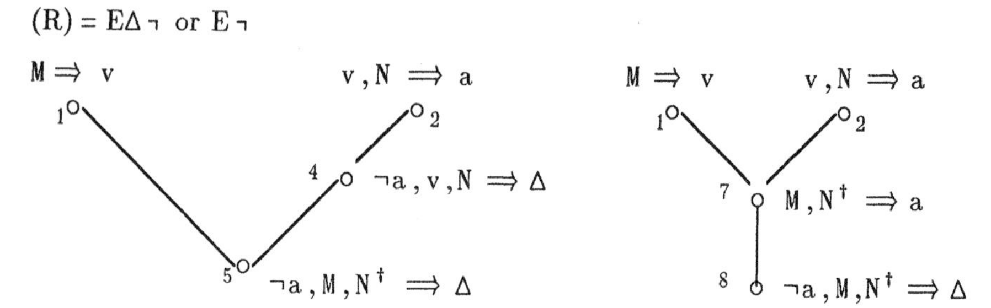

For K_uPC it may became necessary to weaken with $\neg a$ above 7 if v = $\neg a$.

(R) = K_0

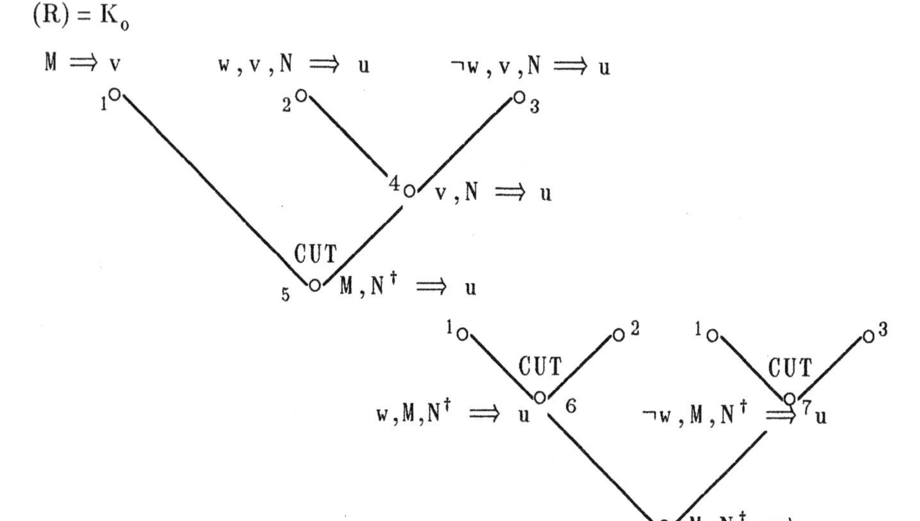

If w = v then the sequent at 6 can be taken in place of that at 8.

Case E3

Here I distinguish the three possibilities for the connective used to form the cut formula v. I shall write (CIT) for an instances of (CUT) in which the the cut formulas are subformulas of v.

v = p$\wedge$q .

$M \Rightarrow p$ $\quad$ $M \Rightarrow q$ $\quad$ $p, p\wedge q, N \Rightarrow u$

1 $\quad$ 2 $\quad$ 3

4 a $\quad$ 4 b

$M \Rightarrow p\wedge q$ $\quad$ $p\wedge q, N \Rightarrow u$

CUT

5 $\quad$ $M, N^{\dagger}_{v} \Rightarrow u$

is changed into

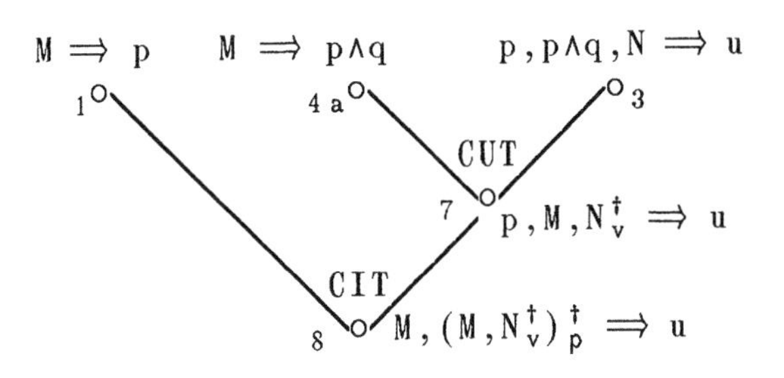

The sequence $M^{\dagger}{}_{p}$ may be removed by contractions, and p may have to be reintroduced in order to obtain $N^{\dagger}{}_{v}$. For symmetric cuts, in the right diagram take 4a as M,p ⟹ p∧q .

v = p∨q .

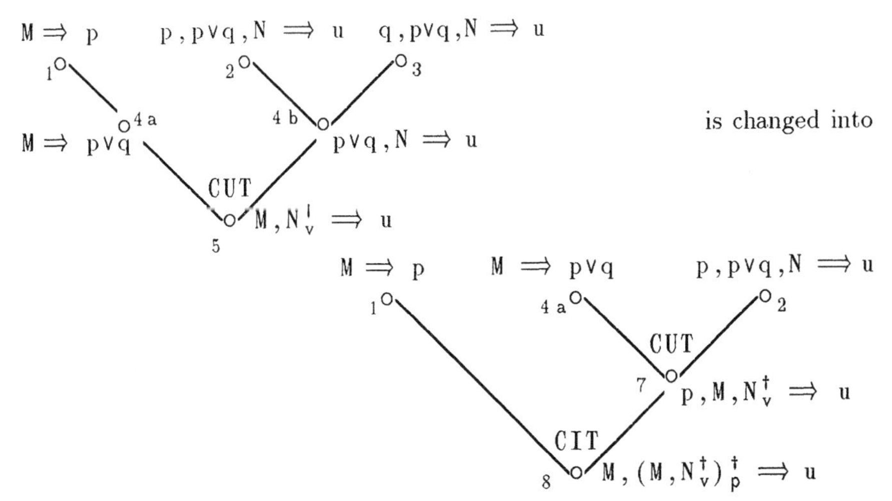

The sequence $M^{\dagger}{}_{p}$ may be removed by contractions, and p may have to be reintroduced in order to obtain $N^{\dagger}{}_{v}$. For symmetric cuts, in the right diagram take 4a as M,p ⟹ p∨q .

v = p→q .

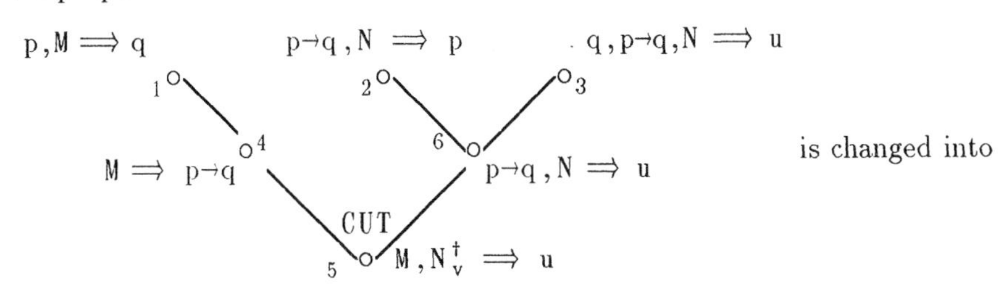

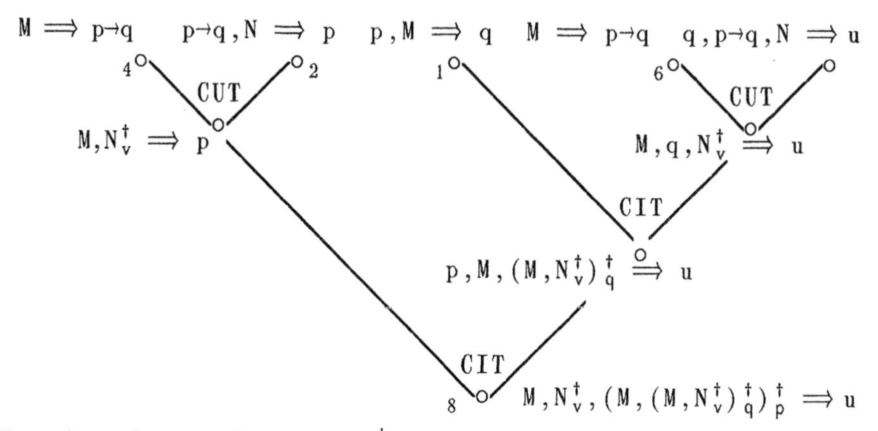

Here the endsequent becomes $M,N^{\dagger}{}_{v} \Longrightarrow u$ after contractions. For symmetric cuts in the lower diagram 4 has to be taken as $q,M \Longrightarrow p{\rightarrow}q$.

⟦For the calculi to be discussed in Chapter **4** another connective ¬ with the rules (IΔ¬), (EΔ¬) or (I¬), (E¬) has to be taken into account:

$v = \neg p$:

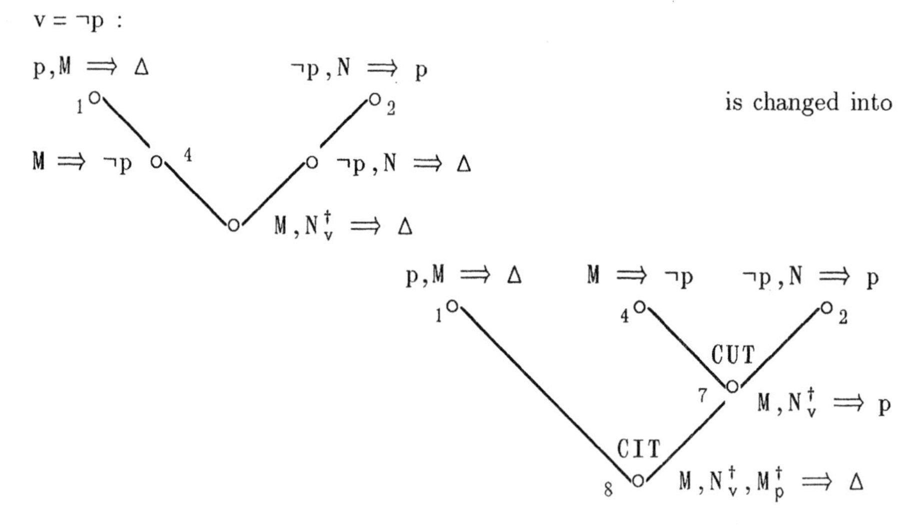

where $M^{\dagger}_{p}$ may be removed by contractions. ⟧

This concludes the definition of the exchanges in the bottom part of D under the operator $\mathbf{E}_1$ and, thus, the definition of $\mathbf{E}_1$ and $\mathbf{E}$ themselves. As for the property (AE), I have already mentioned during the definitions above that in the Cases E1 and E2 the length of $\mathbf{E}_1(D)$ grows at most by 1 and that in Case E1 the first and in Case E2 the second alternative of (e_1) holds. In Case E3 for $v = p \wedge q$ the length grows at most by 1, and the (a,b)-cut is moved to an (a, b−1)-cut; for $v = p \vee q$ the length grows at most by 1, and the (a,b)-cut is moved to an (a,d)-cut with $d \leq b-1$. For $v = p{\rightarrow}q$ the length grows at most by 2, namely if the depth of 5 was determined by 4, and the (a,b)-cut is moved to (a,d)-cuts with $d \leq b-1$ - which, as requested, is the second alternative of (e_1).

This concludes the verification that all my operators have the properties stated, and it thus ends the proof of Theorem 1.

In K_tP the rule (W) can be omitted if generalized axioms are used. During the definition of $\mathbf{E}_1$ there were several cases in which a formula, removed by radical cuts, had to be restored at node 8. Thus in K_tPC_2 the rule (W) cannot be avoided by using generalized axioms. Also, there were several cases in which parts of the antecedents of sequents first were duplicated and then had to be removed by contractions. The number of these contractions is certainly bounded by the number of formulas in the sequent at the original node 5, but the latter number does not depend in a simple manner on the size of the endsequent of D. For this reason, it is difficult to compute the growth of the length of the underlying tree of D under the operator **E** and the operators derived from it.

Acting on a derivation D, the operator **A** stepwise reduces the cut degree k by applying $\mathbf{A}_1$. The operator $\mathbf{A}_1$ acts upon D by first acting on shorter subderivations and then, having arrived at an extremal situation, by applying $\mathbf{A}_0$. In the action upon shorter subderivations of D, therefore, $\mathbf{A}_0$ comes to act first, beginning with the shortest extremal subderivations of D. Hence it is the k-cuts in *maximal* position in D which are attacked first. These cuts are shifted upwards under **E** into k-cuts of lesser weight and, by definition of $\mathbf{A}_0$, the shifts are repeated until they arrive at k-cuts of weight (a,b) with $a \cdot b = 0$, i.e. those for which one premiss in an axiom.

Also, under $\mathbf{E}_1$ a cut to which Case E1 or Case E2 applies, and with a left premiss $M \Longrightarrow v$, is always moved in such a manner that the shifted cut again has a premiss of the form $M' \Longrightarrow v$. If v is composite then it cannot occur in an axiom, and so after repeated moves either the other premiss of the shifted cut will become an axiom, or v will be dissolved in that the Case E3 occurs.

The definition of the operator **E** can, obviously, be generalized such that **E** acts on a pair, consisting of a derivation D and an instance of (CUT) at a particular position in D. In this way, exchanges can also be performed on cuts not necessarily in maximal position. It follows from the above that in order to determine $\mathbf{A}_1(D)$ I need to start very far from the endsequent of D, namely at the k-nodes in maximal position, and it might be tempting to look for another approach, starting near the endsequent at the k-cuts in minimal position. But if there are several k-cuts located on the same branch, shifting the lowest one will finally result in a situation in which two or more k-cuts succeed upon each other. Shifting the order within such a block of k-cuts will do no good, and I should try to shift upwards the entire block above a logical rule leading to the premiss of the highest member of my block. The following example, in which the depth of 3 shall be larger than that of 5, shows that this attempt will, in general, result in a situation worse than that I started from:

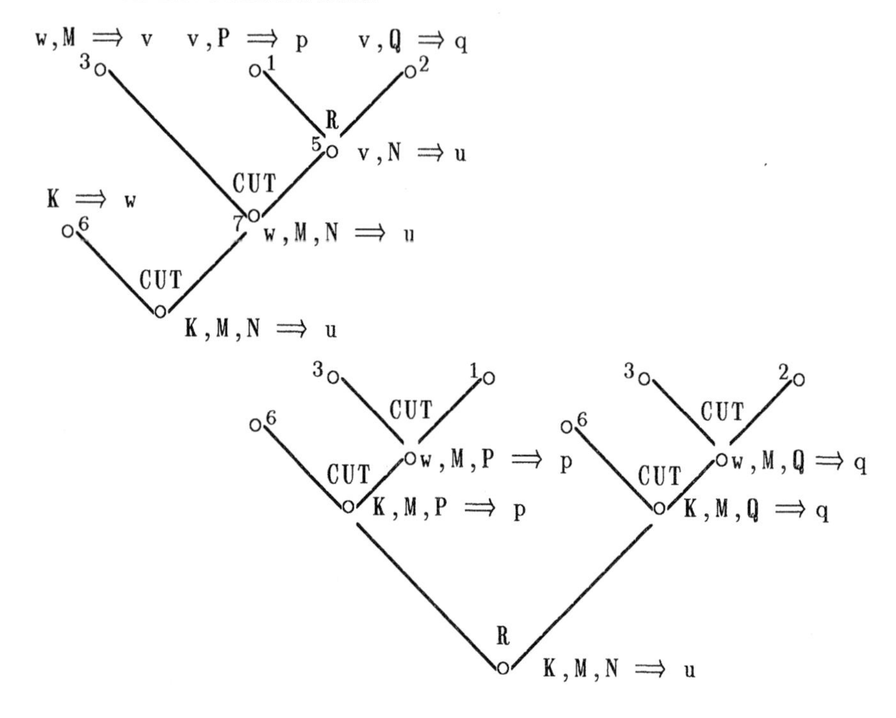

2. Arithmetization

The notions about recursive functions used in this Section can be found in Chapter **5** of Book 3 of these Lectures or in FELSCHER 93 .

In Chapter **1.5** I discussed how to code, or arithmetize, formulas by numbers, and it is not hard to see that such arithmetization can be carried out also for sequences and sequents of formulas as well as for trees and for derivations of K_tP and K_uP; an example for the arithmetization of another calculus can be found in FELSCHER 93. If D is a derivation, then $\ulcorner D \urcorner$ shall be its code, and the same notation shall be used for the formation of codes of other objects. Since the basic coding and decoding functions are *elementary* recursive functions, I can define elementary functions all of which will have the value 0 unless the following circumstances hold:

SEQ(x) = 1 if x = $\ulcorner s \urcorner$ for a sequent s
MEM(x,k) = y+1 if SEQ(x) = 1 and y = $\ulcorner v \urcorner$ is the k-th member of that sequent
DEDU(x) = 1 if x = $\ulcorner D \urcorner$ for a derivation D
CDEDU(x) = 1 if x = $\ulcorner D \urcorner$ for a cutfree derivation D,

$LE(x) = n+1$ if $x = \ulcorner D \urcorner$ and n is the length of D,
$LE_0(x) = n+1$ if $x = \ulcorner D \urcorner$ and n is the length of the tree underlying D,
$CDEG(x) = n+1$ if $x = \ulcorner D \urcorner$ and n is the cut degree of D
$ENDS(x) = y$ if $x = \ulcorner D \urcorner$ and y is code of the endsequent of D,
$RULE(x) = r$ if $x = \ulcorner D \urcorner$ and r is (a number for) the last rule applied in D,
$BRA_l(x) = y$ if $x = \ulcorner D \urcorner$ and $LE(x) > 1$ and y is code of the subderivation of the left or the only premiss of D's endsequent
$BRA_r(x) = y$ analogously in case there is a second right premiss
$JOIN_1(x,z,r) = u$ if $u = \ulcorner D \urcorner$, $x = BRA_l(u)$, $z = ENDS(x)$, $RULE(u) = r$ and this is a 1-premiss rule
$JOIN_2(x,y,z,r) = u$ if $u = \ulcorner D \urcorner$, $x = BRA_l(u)$, $y = BRA_r(u)$, $z = ENDS(x)$, $RULE(u) = r$ and this is a 2-premiss rule.

Making use of these functions, I can explicitly define an elementary function EX_1 such that $x = \ulcorner D \urcorner$ implies $EX_1(x) = \ulcorner \mathbf{E}_1(D) \urcorner$.

In the same manner, I can elementarily describe the exchanges performed by $\mathbf{E}_0^0$. But $\mathbf{E}_0$, if it is not the identity, is defined by recursion (and with help of the numbers $m_s(D) = LE_0(BRA_s(x)) - LE(BRA_s(x))$ for $s = l, r$). Hence the functions EX_0 and EX such that $x = \ulcorner D \urcorner$ implies $EX_0(x) = \ulcorner \mathbf{E}_0(D) \urcorner$ and $EX(x) = \ulcorner \mathbf{E}(D) \urcorner$ are primitive recursive for K_sPC and K_tPC and are elementary for K_uPC.

A primitive recursive function A_0 such that $x = \ulcorner D \urcorner$ implies $A_0(x) = \ulcorner \mathbf{A}_0(D) \urcorner$ is defined as $A_0(x) = A_0^0(x, LE(BRA_l(x)) \dot- 1, LE(BRA_r(x)) \dot- 1)$ from a function $A_0^0(x,y,z)$ which is defined by primitive course of values recursion imitating the definition of $\mathbf{A}_0$. A primitive recursive function A_1 such that $x = \ulcorner D \urcorner$ implies $A_1(x) = \ulcorner \mathbf{A}_1(D) \urcorner$ is defined as $A_1(x) = A_1^0(x, LE(x) \dot- 1)$ from a function $A_1^0(x,y)$ which is defined by primitive course of values recursion imitating the definition of $\mathbf{A}_1$. A primitive recursive function A such that $x = \ulcorner D \urcorner$ implies $A(x) = \ulcorner \mathbf{A}(D) \urcorner$ is defined as $A^o(x, CDEG(x) \dot- 1)$ where $A^o(x, n)$ is the n-th iteration of A_1.

In this way, the operator **A** has a primitive recursive arithmetization.

Reference

W.Felscher: Berechenbarkeit. Rekursive und programmierbare Funktionen. Berlin 1993

Chapter 3. Continuous Cut Elimination

1. Explicit Retracing as a Motivation

For the calculi K_sPC and K_tPC the Case 3 in the definition of the exchange operator $\mathbf{E}_1$ can be simplified drastically if the cut formula v happens not to occur as parameter in the logical rules leading to the premisses of the cut: in this situation, the three given subderivations are replaced by

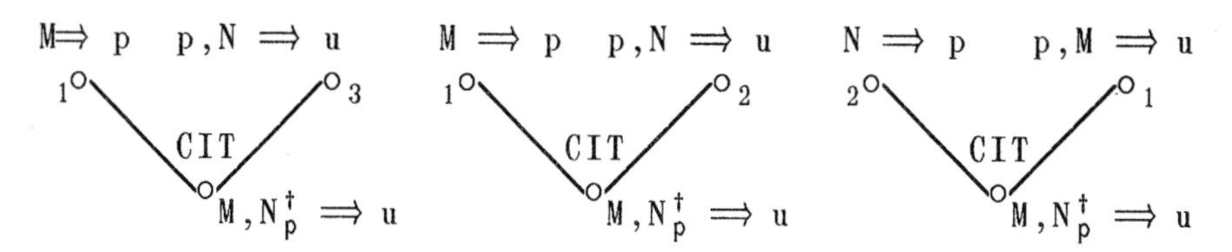

and the k-cut is actually removed. While such an event is but accidental, a different approach to the operator $\mathbf{A}_0$ will always produce such situations, and I shall discuss this in the next two Lemmata as a motivation for later developments.

Throughout this Chapter the *Permanent Assumption* shall be that that KPC is K_uPC_0; the case of K_uPC_1 then is a simple generalization. I begin by defining an operator $\mathbf{F}_u$ acting on pairs H, J of KPC-derivations such that there holds the

LEMMA 1 If H ends with $M \Rightarrow v$, and
if v is a variable or is principal formula of a rule leading to this endsequent, and
if J ends with $v, M \Rightarrow u$, and
if all cuts in H and J have cut degrees below the degree of v,

then $\mathbf{F}_u(H,J)$ is a KPC-derivation of $M \Rightarrow u$, all cuts in $\mathbf{F}_u(H,J)$ have degrees below the degree of v , and

$$\mathit{length}(\mathbf{F}_u(H,J)) \leq \mathit{length}(H) + 2 \cdot \mathit{length}(J) .$$

At each of its nodes e, the derivation J carries a sequent J(e) of the form $B,v,C \Rightarrow a$ where v appears at the unique predecessor of the occurrence of v in J's endsequent $v,M \Rightarrow u$. Replacing, for every e, J(e) by $B,C \Rightarrow a$, I obtain an object J* formed by a tree with sequents at its nodes. Observe that all members of M occur in B,C .

I now distinguish the two cases that the formula v is a variable or is composite.

If v is a variable, then there may have been maximal nodes e such that J(e) is a generalized axiom $v,A \Longrightarrow v$, changed in J^* to $A \Longrightarrow v$. As all members of M occur in A, there is a unique sequence A/M consisting of the members of A *not* in M. For every such e, I define H_e by (1) taking a copy of H, obtained by writing A/M to the *right* of the antecedents of all its sequents, and then (2) performing the permutation which changes its end-sequent $M,A/M \Longrightarrow v$ into $A \Longrightarrow v$. I define $\mathbf{F}_u(H,J)$ from J by placing H_e on top of every such e and identifying its last node with e. Clearly, $\mathbf{F}_u(H,J)$ is a KPC-derivation with the desired properties, and so $length(\mathbf{F}_u(H,J)) \leq length(H) + length(J)$.

Assume that v is composite. Let k be the lowest node of H, let k_0, k_1 be its one or two upper neighbours, and let H_{k0} and H_{k1} be the subderivations of H leading to k_0, k_1. Let n be the length of H.

I construct, for every e in J, a derivation J_e^* of the sequent $J^*(e)$. The construction proceeds by recursion on the depth $|e|$ of e in J, and during the construction I prove that $length(J_e^*) \leq n + 2\cdot|e|$.

If e is of depth 0 then J(e) is a generalized axiom $B,v,C \Longrightarrow a$ and $J^*(e)$ is $B,C \Longrightarrow a$. Since v is composite, it is different from a, and so $J^*(e)$ is a generalized axiom which I take as the derivation J_e^* of length 0.

Assume that J(e) arises from $J(e_0)$, $J(e_1)$ under a cut or under a logical rule (R) for which the omitted occurrence of v is not principal. Then $J^*(e)$ arises in the same manner from $J^*(e_0)$, $J^*(e_1)$: the omitted occurrence of v was already in the conclusion J(e) and, therefore, neither one of a cut formula nor a of a side formula, and by hypothesis it was not that of the principal formula either. Hence I continue the derivations J_{e0}^*, J_{e1}^* of $J^*(e_0)$, $J^*(e_1)$ by (R) and obtain the derivation J_e^*. As $|e_0| < |e|$, $|e_1| < |e|$, there follows

$$length(J_e^*) \leq 1 + max(n+2\cdot|e_0|, n+2\cdot|e_1|) \leq 1+n+2\cdot max(|e_0|, |e_1|)$$

which is $n + 2\cdot|e|$. Assume that J(e) arises from J(e'), J(e'') under a logical rule for which the omitted occurrence of v is principal. If $v = p\wedge q$ and

$$J(e_0)\colon p,v,A \Longrightarrow a \qquad J^*(e_0)\colon p,A \Longrightarrow a$$
$$J(e)\colon v,A \Longrightarrow a \qquad J^*(e)\colon A \Longrightarrow a$$

then I define $H_{e\,k0}^*$ by taking a copy of H_{k0} with $H_{e\,k0}^*(k_0)\colon M, A/M \Longrightarrow p$, obtained by writing A/M to the antecedents of all sequents in H_{k0}, and then continue with the permutation producing $A \Longrightarrow p$. I define J_e^* by

$$\dfrac{\dfrac{H_{e\,k0}^*}{A \Longrightarrow p} \qquad \dfrac{J_{e\,0}^*}{p,A \Longrightarrow a}}{A \Longrightarrow a} \quad .$$

Here H_{k0} and $H_{e\,k0}^*$ have at most length n−1, whence

$$length(J^*_e) \leq 1 + max(n-1,\ n+2\cdot|e_0|)) = 1 + n + 2\cdot|e_0|) = n+2\cdot|e|\ .$$

If $v = p \vee q$ and

$J(e_0)$: $p,v,A \Longrightarrow a$ $\qquad$ $J^*(e_0)$: $p,A \Longrightarrow a$
$J(e_1)$: $q,v,A \Longrightarrow a$ $\qquad$ $J^*(e_1)$: $q,A \Longrightarrow a$
$J(e)$: $v,A \Longrightarrow a$ $\qquad$ $J^*(e)$: $A \Longrightarrow a$

then I define $H^*_{e\,ko}$ by taking a copy of H_{ko}, say with $H^*_{e\,ko}(k_o)$: $M,\ A/M \Longrightarrow p$, obtained by writing A/M to the antecedents of all sequents in H_{ko}, and then continue with the permutation producing $A \Longrightarrow p$. I define J^*_e by

$$\dfrac{\dfrac{H^*_{e\,ko}}{A \Longrightarrow p} \qquad \dfrac{J^*_{e\,o}}{p,A \Longrightarrow a}}{A \Longrightarrow a}\ .$$

Again, there holds $length(J^*_e) \leq 1 + max(n-1,\ n + 2\cdot|e_0|)) \leq n + 2\cdot|e|)$. If $v = p \rightarrow q$ and

$J(e_0)$: $v,A \Longrightarrow p$ $\qquad$ $J^*(e_0)$: $A \Longrightarrow p$
$J(e_1)$: $q,v,A \Longrightarrow a$ $\qquad$ $J^*(e_1)$: $q,A \Longrightarrow a$
$J(e)$: $v,A \Longrightarrow a$ $\qquad$ $J^*(e)$: $A \Longrightarrow a$

then I define $H^*_{e\,ko}$ by taking a copy of H_{ko} with $H^*_{e\,ko}(k_o)$: $p,M,A/M \Longrightarrow q$, obtained by writing A/M to the antecedents of all sequents in H_{ko}, and then continue with the permutation producing $p,A \Longrightarrow q$. I define J^*_e by

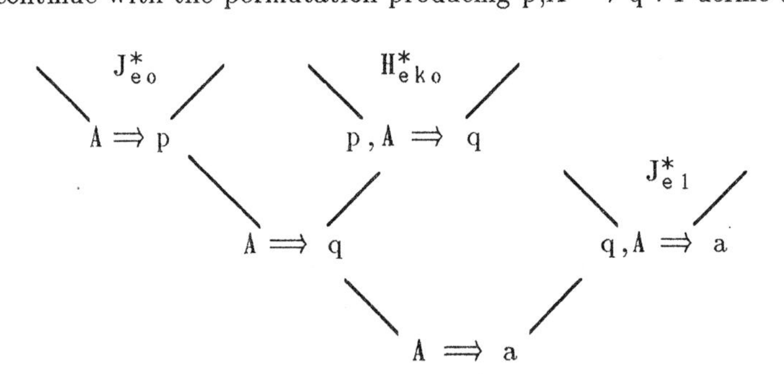

This time, I find

$$\begin{aligned} length(J^*_e) &\leq 1 + max(1 + max(n + 2\cdot|e_0|, n-1),\ n + 2\cdot|e_1|) \\ &= 1 + max(1 + n + 2\cdot|e_0|,\ n + 2\cdot|e_1|) \\ &\leq n + 2 + 2\cdot max(|e_0|, |e_1|) = n + 2\cdot|e|\ . \end{aligned}$$

It follows, that J^*_e is defined for every node e of J and derives $J^*(e)$. In particular, for the minimal node g of J, J^*_g derives $J^*(g)$ which is $M \Longrightarrow u$. Thus I define $\mathbf{F}_u(H,J) = J^*_g$, and it will have at most the length $n + 2\cdot|g|$. This concludes the proof of Lemma 1 .

⟦ The connective ¬ to be studied in Chapter **4** requires to consider

$$J(e_0) : \neg p, A \Longrightarrow p \qquad J^*(e_0) : A \Longrightarrow p$$
$$J(e) \; : \neg p, A \Longrightarrow \Delta \qquad J^*(e) \; : A \Longrightarrow \Delta$$

which is handled by

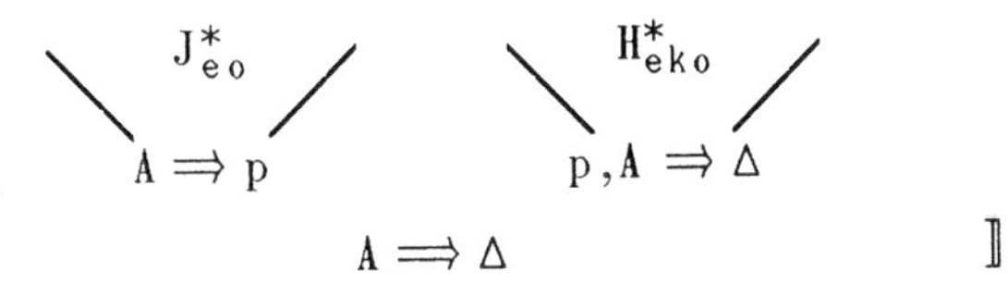

⟧

As a replacement for $\mathbf{A}_0$ I define an operator $\mathbf{F}_0$ acting on KPC–derivations D such that there holds the

LEMMA 2 If D is a KPC–derivation then $\mathbf{F}_0(D)$ is a KPC–derivation of the same endsequent, and if D is k–extreme then all cuts in $\mathbf{F}_0(D)$ have degrees below k, and there holds $length(\mathbf{F}_0(D)) \leq 3 \cdot (length(D) - 1)$.

If D is not k–extreme then I set $\mathbf{F}_0(D) = D$. It remains to consider the case that D ends with a cut with premisses $M \Longrightarrow v$ and $v, M \Longrightarrow u$ and with subderivations H and J leading to them. If v is a variable then I apply Lemma 1 and set $\mathbf{F}_0(D) = \mathbf{F}_u(H,J)$. If v is not a variable then let P be the predecessor tree in H of the right occurrence of v in $M \Longrightarrow v$ which I define as before, stipulating that in an instance of (CUT_0) the right occurrence in the conclusion corresponds to the right occurrence in the right premiss.

For every node e of P, the sequent H(e) is of the form $B \Longrightarrow v$. Replacing, for every such e, H(e) by $B \Longrightarrow u$, I obtain an object H^* formed by the tree P with sequents at its nodes.

For nodes k of H, let H_k be the subderivation ending with H(k).

As v is not a variable, the maximal nodes k of P are such that v is introduced into $H(k): M_k \Longrightarrow v$ as principal formula of a logical rule. Hence H_k and the derivation J_k, obtained from J by weakening with the sequence M_k/M of formulas in M_k not in M, are of the form such that $\mathbf{F}_u$ can be applied, and $\mathbf{F}_u(H_k, J_k)$ then ends with $M_k \Longrightarrow u$, i.e. with $H^*(k)$.

Let D^* arise from H^* by placing, on top of every maximal k of P, the derivation $\mathbf{F}_u(H_k, J_k)$, and then identify k with its last node. Then D^* is a KPC–derivation. This is clear above every maximal k since D^* there acts as $\mathbf{F}_u(H_k, J_k)$. If e in D^* is below a maximal k of P then e is in P and has one or two upper neighbours e_0, e_1 in P. The rule, leading from $H(e_0)$, $H(e_1)$ to H(e), remains unaffected when changing the parametric occurrence of v to u; hence it also leads from $D^*(e_0) = H^*(e_0)$, $D^*(e_1) = H^*(e_1)$ to $D^*(e) = H^*(e)$.

I now set $\mathbf{F}_0(D) = D^*$. It remains to establish the bound for $length(\mathbf{F}_0(D))$. The length of H_k is the depth $|k|_H$ of f in H. Since $length(\mathbf{F}_u(H_k,J_k)) \leq length(H_k) + 2\cdot length(J)$, in D^* the nodes k maximal in P have a depth $|k|_{D^*}$ bounded by $|k|_H + 2\cdot length(J)$. The length of D^* is that of a longest branch in D^*; let k be maximal in P on this branch and let n_k be the number of nodes below k. The length of H is that of a longest branch in H, whence $n_k+|k|_H \leq length(H)$. Thus

$$\begin{aligned} length(D^*) &= n_k + |k|_{D^*} \\ &\leq n_k+|k|_H + 2\cdot length(J) \\ &\leq length(H) + 2\cdot length(J) \\ &\leq 3\cdot(length(D) - 1) \ . \end{aligned}$$

This concludes the proof of Lemma 2. Now $\mathbf{F}_0$ can take the place of $\mathbf{A}_0$ in the proof of Theorem 1.

There is a second, differently organized way to construct D^*. Observe that, for every node e of J, the sequent J(e) is of the form $B,v,C \Longrightarrow a$; let J_r be the subderivation ending with J(r). Replacing every J(e) by $B,C \Longrightarrow a$ I obtain an object J^* and objects J_e^*. If e is a maximal node of J then $J^*(e)$: $B,C \Longrightarrow a$ can only cease to be a generalized axiom if v is a variable (and a is v), and if such e occurs then I define again $F_0(D)$ as $F_u(H,J)$.

So it remains to consider a composite v and the *critical* nodes r at which J_r ends with a rule having v as principal occurrence in J(r): $v,C_r \Longrightarrow a_r$. If the root e_J of J itself is critical then I use the above construction of $\mathbf{F}_0$ and obtain $\mathbf{F}_0(D)$. Otherwise I consider the *minimal* critical nodes r and form, for each of them, the derivation H^r, obtained from H by weakening with the sequence C_r/M of formulas in C_r not in M. Then H^r and J_r continued by a cut with v give a derivation D_r of $J^*(r)$: $C_r \Longrightarrow a_r$, and so with my former construction of $\mathbf{F}_0$ also $\mathbf{F}_0(D_r)$ ends with $J^*(r)$. Replacing J_r^* by $\mathbf{F}_0(D_r)$ in J^* (and identifying r with its last node) I obtain a derivation D^* which I define to be $\mathbf{F}_0(D)$. – It will be this second description of $\mathbf{F}_0$ which will be put into the form of an algorithm in the following section.

The operators $\mathbf{F}_u$ and $\mathbf{F}_0$ can be extended to the calculi K_sPC and K_tPC and to the other forms of cuts. A first complication in the case, say, of K_tPC is caused by the fact that logical rules now produce *new* occurrences of the principal formula, making it necessary to define also the parameter tree P of the occurrence of v in $v,M \Longrightarrow u$ and to restrict the formation of J^*, performed above on all of J, now to the nodes of that subtree of J. A second complication arises in the case of K_tPC where the presence of the contraction rule necessitates the use of radical cuts. Radical cuts, however, remove all occurrences of v from sequents $v,A \Longrightarrow a$, even if they are needed later in order to use v as a side formula.

2. The Reduction Operator R_0

The elimination algorithms discussed so far proceeded by recursion on the *length* of a derivation D and proceeded *downwards* in the tree of D, starting from the k-cuts in maximal position. There are situations more general than the ones considered presently (which will occur in Book 3) in which one studies a more general sort of derivations, and for them it is *not* possible to assign a finite number as their length. MINTS 75 has invented an elimination algorithm which uses recursion on the *height* of nodes and proceeds by ascending the tree of D *upwards*. The machinery of this algorithm, however, can already be explained here, and it may even become clearer if no additional concepts distract the reader's attention.

I continue with the *Permanent Assumption* that KPC is K_uPC_0. I shall use the permutation rule (RP) for (1) transpositions and (2) for the *identical* permutation; in the latter case it may be viewed as a *repetition* rule (RR). This special rule will be *essential* for the operator $\mathbf{R}_1$ to be studied in the next section.

Already in my earlier constructions, for instance during the proof of Lemma 1, I changed given derivations into objects which, after some additional observations, turned out be derivations as well. In the following definitions I shall introduce names for these (intermediary) objects as well as some concepts which organize the work required to finally recognize some of them as derivations.

A *rule datum* shall be an object consisting of (1) the name of a rule (R), (2a) a formula, to be used as the principal formula for an instance of (R) in case (R) is a logical rule, or (2b) a formula to be used as cut formula in case R is (CUT_0), or (2c) two place numbers, to be exchanged in case (R) is the transposition rule.

A tree T, together with a function d assigning sequents to its nodes, I call a D-*object*. A D-object, together with a partial function r assigning rule data to certain nodes e of T, I temporarily call a *prederivation* in a calculus if

> the domain of r is descending in T: if $k<e$ and $e \in def(r)$ then $k \in def(r)$,
>
> d is *locally correct* for r: if e has one upper neighbour e_0 and if e_0 is in def(r) then the sequents d(e), $d(e_0)$ form an instance of the rule named in r(e) with the accompanying data; if e has two upper neighbours e_0 and e_1 and if e_0, e_1 are in def(r) then the sequents d(e), $d(e_0)$, $d(e_1)$ form an instance of the rule in r(e) with the accompanying data.

A derivation $<T,d,r>$ then is a prederivation such that r is defined on all non maximal nodes of T.

There is one more technical preparation which I should mention in advance. The operator $\mathbf{R}_0$ will, essentially, do what the operator $\mathbf{F}_0$ did in Lemma 2, and once it is available it will be applied to simplify other derivations. Now all my constructions will be recursive in the *height* of nodes and will produce, ascending from below, D-objects which, when completed, shall become derivations. Applying $\mathbf{R}_0$ to some (as yet uncompleted) subderivations starting at a node k of a (pre)derivation D, I so am led to consider the *relative* heights above k in these subderivations when studying the action of $\mathbf{R}_0$ upon them. Rather than relativizing heights between different derivations, it will be simpler to use a *universal* domain of nodes from which all my derivation trees are built - and it also will relieve me of the task to *choose* new nodes during the repeated stages of my construction.

But if I recall that trees may be represented as sets of sequences of natural numbers then the set ω^∞ of all these sequences offers itself as a universal domain of nodes. And if, as it is the case for my present derivations, I only consider 2-ry trees, then it suffices to take the *universal* tree 2^∞ of finite sequences of 0 and 1: its root is the empty sequence $< >$, and for a node $e = <e_0, \ldots, e_{n-1}>$ the two upper neighbours are the sequence $e*0 = <e_0, \ldots, e_{n-1}, 0>$ and $e*1 = <e_0, \ldots, e_{n-1}, 1>$. Being a sequence, every node e has a length $\|e\|$ which, viewed in the tree, is its *height.* The order relation between nodes then means that $e<g$ if e is an initial part of g: $e = g \upharpoonright \|e\|$. [In the same way, every node e in ω^∞ has the ω many upper neighbours $e*i$ for $i \epsilon \omega$.]

A 2-ry tree T in general then is a subset of 2^∞ containing a fixed node k as its root and such that, for every e in T, there holds $k \leq e$ and, for every g, $k \leq g \leq e$ implies $g \epsilon T$. Observe that the root k will, in general, have a positive length, such that the height of a node e in T is the difference $\|e\| - \|k\|$. Further, it will simplify my notations below if I assume that the trees underlying derivations are subtrees of 2^∞ in such a way that, for the rules ($E\wedge l$), ($I\vee l$), ($I\rightarrow$) and ($E\wedge r$), ($I\vee r$), if e carries the conclusion then $e*0$, respectively $e*1$, carries the premiss.

Having set the stage, I now shall construct the operator $\mathbf{R}_0$ which will have three arguments: two (sub-) derivations H and J and the root k_T of the derivation which it shall produce:

LEMMA 3 If H ends with $h(k_H)$: $M \Longrightarrow v$ and
if J ends with $j(k_J)$: $v, M \Longrightarrow u$

then $\mathbf{R}_0(H, J, k_T)$ is a KPC-derivation of $M \Longrightarrow u$ with root k_T in which all cuts have cut formulas which either are proper subformulas of v or are cut formulas of cuts occurring in H or in J.

Thus $\mathbf{R}_0$ shall have the same result as had the operator $\mathbf{F}_0$ in Lemma 1.

Let $H = <T_H, h, r_H>$, $J = <T_J, j, r_J>$ be given; I abbreviate $\mathbf{R}_0(H, J, k_T)$ as $<T, f, r>$. Beginning at k_T, $<T, f, r>$ will be constructed by recursion on the height of nodes e in 2^∞ with $k_T \leq e$ such that the numbers $\|e\| - \|k_T\|$ are the *stages* of my construction. This construction will be described as a *program* with three subprograms, called *Cases* 0, 1 and 2. Case 0 will run until termination; a run of Case 1 may terminate, may branch into Case 0 or may be interrupted by runs of Case 2; a run of Case 2 will always end with a resumption of Case 1. The programs use *bookkeeping* functions η, φ, ψ, χ, δ_J, δ_H and if, at an upper neighbour e*i of a node e, a function, say ψ, is defined as $\psi(e*i) = \psi(e)$ then I shall say that the value of ψ has been *kept* when going from e to e*i.

Beginning at k_T, I enter Case 1 and copy J while omitting the predecessors of the formula v indicated in the endsequent. For each of the new nodes e, I denote as $\varphi(e)$ the *reference* node in J for which $j(\varphi(e))$ is copied to e. If v is a variable then the only occasion requiring attention is the arrival at a maximal node e for which $j(\varphi(e))$ happens to be an axiom $v, C_r \Longrightarrow v$. In that Case 1b I rewrite $C_r \Longrightarrow v$ as $M, C_r{}^o \Longrightarrow v$ where $C_r{}^o$ consists of the formulas in C_r which are not predecessors of formulas in M. I then enter Case 0 and copy all sequents from H with antecedents weakened by $C_r{}^o$.

If v is not a variable then I may arrive at the Case 1ab of a node e carrying $C_r \Longrightarrow a_r$ such that, for $r = \varphi(e)$, v is principal occurrence in $j(r): v, C_r \Longrightarrow a_r$. At the following node e*0 I enter Case 2 and save $r = \varphi(e)$ as $\psi(e*0)$; I also save the seqence $C_r{}^o$ of formulas as $\delta_J(e*0)$, and rewrite $C_r \Longrightarrow a_r$ as $M, C_r{}^o \Longrightarrow a_r$. I then begin to copy H with antecedents weakened by $C_r{}^o$ and the formula v on the right replaced by a_r; for each of the newly arising nodes g I now denote as $\varphi(g)$ the *reference* node in H for which $h(\varphi(g))$ is copied to g, and I also continue ψ and δ_J by keeping their values. Making use of these definitions, the sequent at g is $h(\varphi(g))$ with v replaced by a_r and weakened by $\delta_J(g) = C_r{}^o$, and these data come from $j(\psi(g))$. This copying process preserves almost all rules as long as v is not principal; only instances of E→ or CUT will have left premisses in H in which v is not on the right, and the left branches above these premisses I again copy from H as Case 0 by only weakening them with $C_r{}^o$. Since v now is composite, I must arrive at nodes g carrying $A_k, C_r{}^o \Longrightarrow a_r$ for which $\varphi(g) = k$ in H has $h(k): A_k \Longrightarrow v$ with a principal occurrence of v; this now are the Cases 2b which are treated essentially as in Lemma 1.

For instance, if $v = q_0 \wedge q_1$ then $\psi(g) = \varphi(e) = r$ in J had an upper neighbour $r*i_0$ carrying $q_{i(o)}, v, C_r \Longrightarrow a_r$, and $\varphi(g) = k$ has an upper neigbour $k*i_0$ in H carrying $A_k \Longrightarrow q_{i(o)}$. I now add an upper neighbour g*0 of g with $\varphi(g*0) = k*i_0$ carrying $A_k, C_r{}^o \Longrightarrow q_{i(o)}$, and I also add an upper neighbour g*1 carrying $q_{i(o)}, A_k, C_r{}^o \Longrightarrow a_r$; from g*0, g*1 a cut with $q_{i(o)}$ leads to $A_k, C_r{}^o \Longrightarrow a_r$ at g. The node k I save as the first member of a sequence $\chi(g*1)$. Ascending from g*0 I again copy from H as Case 0 the branch leaving from $\varphi(g*0)$ by weakening with $\delta_J(g) = C_r{}^o$ from $\psi(g) = r$. Above

$g*1$ I add a node $g*1*0$ and change $q_{i(o)}, A_k, C_r{}^o \Longrightarrow a_r$ by a permutation to $q_{i(o)}, C_r, A_k{}^o \Longrightarrow a_r$ where $A_k{}^o$ consists of the formulas in A_k which are not predecessors of formulas in M. The reference node $\varphi(g*1*0)$ shall be $r*i_0$ in J, and for the intermediary $g*0$ I define a *delayed* reference node $<\S, r*i_0>$. The set $A_k{}^o$ of formulas I save as $\delta_H(g*1*0)$ Ascending from $g*1*0$ I re-enter the Case 1 and, beginning with the reference node, copy from J with v omitted and with antecedents weakened by $\delta_H(g*1) = A_k{}^o$; for all these newly arising nodes e I continue χ, δ_J, δ_H by keeping their values. Making use of these definitions, the sequent at e is $j(\varphi(e))$ with v removed and weakened by $\delta_H(g) = A_k{}^o$, and these data come from h(k) for the (one and) last member k of $\chi(e)$. - The case of the other connectives is analogous, though for $v = p \rightarrow q$ two cuts have to be performed and the reference node in J has to be delayed twice.

While the pursuit of branches in the Cases 0 does not need further attention, the copying process from J in Case 1 may, of course, arrive at new nodes e carrying $C_s, \delta_H(e) \Longrightarrow a_s$ such that for $s = \varphi(e)$ in $j(s): v, C_s \Longrightarrow a_s$ has v as principal occurrence. Here I will have to start a new run of Case 2, this time with $\psi(e*0) = s$ and $\delta_J(e) = C_s{}^o, \delta_H(e)$. Weakened with $\delta_J(e)$, H will be copied until the right occurrence of v becomes principal at nodes m, and then I proceed as above, and at the appropriate new nodes $e*1*0$ I enlarge, by the sets $A_m{}^o$, the values of χ, kept alive so far from the last passing of Case 1. All this may have to be repeated, and every time Case 1 is left for a run through Case 2 I have to re-define $\psi(e)$, $\delta_J(e)$ during this run, the former to be available when this run ends at some m and the latter to be used for the weakening also in the newly arising Cases 0. A node g from a run of Case 2 then carries $h(\varphi(g))$ with v replaced by a_r and weakened by $C_s{}^o, \delta_H(g)$ where $C_s{}^o$ comes from $j(\psi(g))$ and $\delta_H(g)$ is the concatenation of the $A_k{}^o$ for all k in $\chi(g)$. Every time a run of Case 2 is left and Case 1 is re-entered I have to append χ and to enlarge $\delta_H(e)$. A node g from a run of Case 1 then carries $j(\varphi(g))$ with v omitted and weakened by the union $\delta_H(g)$ of the $A_k{}^o$ for all k in $\chi(g)$.

I now present the algorithm producing the result of $\mathbf{R}_0$; it will be noticed that χ is not actually needed in its definition and is introduced only for later application. At stage 0, T shall be the tree consisting of k_T only, and I define $f(k_T): M \Longrightarrow u$, $\eta(k_T) = 1$, $\varphi(k_T) = k_J$; I define $\psi(k_T) = k_T$; $\delta_J(k_T)$ and $\delta_H(k_T)$ shall be the empty sets and $\chi(k_T)$ shall be the empty sequence of formulas. Assume that at stage n there have been defined

the nodes e of T with $\|e\| - \|k_T\| \leq n$,
the functions f, η, φ, ψ, χ, δ_J, δ_H at these nodes,
the function r at the nodes e with $\|e\| - \|k_T\| < n$.

At stage n+1, I consider the e with $\|e\| - \|k_T\| = n$ and, if appropriate, extend T by including their upper neighbours $e*0$, $e*1$. I define the values of all functions distinct from r at $e*0$, $e*1$, and I then also define the value of r at e. I list the following cases in which such extensions will occur:

0. $\eta(e) = 0$, $\varphi(e)$ is non maximal in H.

⟦Preceded by Cases 1c, 2ab, 2ba, 2bb, 2bd, 2cb⟧

$\varphi(e)$ has upper neighbours $\varphi(e)*i$.

T is extended by the e*i, and I set

$\eta(e*i) = 0$, $\varphi(e*i) = \varphi(e)*i$, $r(e) = r_H(\varphi(e))$,
$f(e*i)$ is $h(\varphi(e)*i)$ weakened by $\delta_J(e)$,
the values of ψ, χ, δ_J, δ_H are kept.

⟦Succeded by Case 0⟧

1. $\eta(e) = 1$.

⟦Preceded by the Start or by Cases 1a, 1aa, 2ab, 2ca⟧

$\varphi(e)$ is in J

1a. $\varphi(e)$ is non maximal

1aa. $r_J(\varphi(e))$ is not a logical rule with the predecessor of v as principal formula.

$\varphi(e)$ has upper neighbours $\varphi(e)*i$.

T is extended by the e*i, and I set

$\eta(e*i) = 1$, $\varphi(e*i) = \varphi(e)*i$, $r(e) = r_J(\varphi(e))$,
$f(e*i)$ is $j(\varphi(e)*i)$ weakened by $\delta_H(e)$ and with v omitted,
the values of ψ, χ, δ_J, δ_H are kept.

⟦$j(\varphi(e))$: $w, B, v, C \Longrightarrow u$, $j(\varphi(e)*i)$: $[w_i], w, B, v, C \Longrightarrow u_i$,
$f(e)$: $w, B, C, \delta_H(e) \Longrightarrow u$, $f(e*i)$: $[w_i], w, B, C, \delta_H(e) \Longrightarrow u_i$.⟧

⟦Succeded by Cases 1⟧

1ab. $r_J(\varphi(e))$ is a logical rule with the predecessor of v as principal formula.

$j(\varphi(e))$: $v, C_r \Longrightarrow a_r$.

T is extended by e*0, and I set

$\eta(e*0) = 2$, $\varphi(e*0) = k_H$, $\delta_J(e*0) = C_r{}^o, \delta_H(e)$,
$r(e) =$ (PR) for the permutation sending $C_r, \delta_H(e)$ into $M, C_r{}^o, \delta_H(e)$,
$f(e*0)$ is $h(k_H)$ weakened by $\delta_J(e*0)$ and with v replaced by a_r ,
$\psi(e*0) = \varphi(e)$, the values of χ, δ_H are kept.

⟦$f(e)$: $C_r, \delta_H(e) \Longrightarrow a_r$, $f(e*0)$: $M, C_r{}^o, \delta_H(e) \Longrightarrow a_r$ ⟧

⟦Succeded by Cases 2⟧

1b. $\varphi(e)$ is maximal in J and $j(\varphi(e))$: $v, C_r \Longrightarrow v$.

$[j(\varphi(e))$: $v, C_r \Longrightarrow v$.]

T is extended by e*0, and I set

$\eta(e^*0) = 0$, $\varphi(e^*0) = k_H$, $\delta_J(e^*0) = C_r{}^o, \delta_H(e)$,
r(e) = (PR) for the permutation sending $C_r, \delta_H(e)$ into $M, C_r{}^o, \delta_H(e)$,
$f(e^*0)$ is $h(k_H)$ weakened by $\delta_J(e^*0)$,
$\psi(e^*0) = \varphi(e)$, the values of χ, δ_H are kept.

⟦$f(e)$: $C_r, \delta_H(e) \Longrightarrow v$, $f(e^*0)$: $M, C_r{}^o, \delta_H(e) \Longrightarrow v$ ⟧

⟦Succeded by Case 0⟧

2. $\eta(e) = 2$

⟦Preceded by Cases 1ab, 2a, 2bc, 2cb⟧

$\psi(e) = r$ is in T_J , $\varphi(e) = k$ is in T_H ,
$j(\psi(e))$: $v, C_r \Longrightarrow a_r$.

2a. $r_H(\varphi(e))$ is not a logical rule with the predecessor of v as principal formula.

2aa. $r_H(\varphi(e))$ is not E→ or (CUT) .

$\varphi(e)$ has upper neighbours $\varphi(e)^*i$.

T is extended by the e*i, and I set

$\eta(e^*i) = 2$, $\varphi(e^*i) = \varphi(e)^*i$, $r(e) = r_H(\varphi(e))$,
$f(e^*i)$ is $h(\varphi(e)^*i)$ weakened by $\delta_J(e)$ and with v replaced by a_r ,
the values of ψ, χ, δ_J, δ_H are kept.

⟦$h(\varphi(e))$: $A_k \Longrightarrow v$, $h(\varphi(e)^*i)$: $A_i \Longrightarrow v$, $\delta_J(e) = C_r{}^o, \delta_H(e)$,
$f(e)$: $A_k, \delta_J(e) \Longrightarrow a_r$, $f(e^*i)$: $A_i, \delta_J(e) \Longrightarrow a_r$ ⟧

⟦Succeded by Cases 2⟧

2ab. $r_H(\varphi(e))$ is E→ or (CUT) .

T is extended by e*0, e*1, and I set

$\eta(e^*0) = 0$, $\eta(e^*1) = 2$, $\varphi(e^*i) = \varphi(e)^*i$, $r(e) = r_H(\varphi(e))$,
$f(e^*i)$ is $h(\varphi(e)^*i)$ weakened by $\delta_J(e)$ and, for e*1, with v replaced by a_r, the values of ψ, χ, δ_J, δ_H are kept .

⟦$h(\varphi(e)^*i)$: $A_i \Longrightarrow u_i$ with $u_1 = v$, $\delta_J(e) = C_r{}^o, \delta_H(e)$,
$f(e^*0)$: $A_o, \delta_J(e) \Longrightarrow u_o$, $f(e^*1)$: $A_1, \delta_J(e) \Longrightarrow a_r$ ⟧

⟦Succeded by Case 0 at e*0 and by Cases 2 at e*1⟧

2b. $r_H(\varphi(e))$ is a logical rule with the predecessor of v as principal formula.

2ba. $v = q_0 \wedge q_1$.

Let i_0 be such that $\varphi(e)*i_0$ exists in T_J ,
$j(\psi(e)*i_0): q_{i(o)}, v, C_r \Longrightarrow a_r$, $h(\varphi(e)*i_0): A_K \Longrightarrow q_{i(o)}$.

T is extended by e*0, e*1, and I set

$\eta(e*0) = 0$, $\eta(e*1) = 2$,
$\varphi(e*0) = \varphi(e)*i_0$, $\varphi(e*1) = <\S, \psi(e)*i_0>$, $r(e) = <\mathrm{CUT}, q_{i(o)}>$,
$\chi(e*0) = \chi(e)$, $\chi(e*1)$ is $\varphi(e)$ appended to $\varphi(e)$,
$f(e*0)$ is $h(\varphi(e)*i_0)$ weakened by $\delta_J(e)$,
$f(e*1)$ is obtained from $j(\psi(e)*i_0)$ by omitting v, weakening C_r to $A_K^o, C_r, \delta_H(e)$ and permuting this first to $A_K^o, M, C_r^o, \delta_H(e)$ and then to $A_K, \delta_J(e)$ employing $\delta_J(e) = C_r^o, \delta_H(e)$,
the values of ψ, δ_J, δ_H are kept .

[[$f(e): A_K, \delta_J(e) \Longrightarrow a_r$,
$f(e*0): A_K, \delta_J(e) \Longrightarrow q_{i(o)}$, $f(e*1): q_{i(o)}, A_K, \delta_J(e) \Longrightarrow a_r$]]

[[Succeded by Case 0 at e*0 and by Case 2ca at e*1]]

2bb. $v = q_0 \vee q_1$.

Let i_0 be such that $\varphi(e)*i_0$ exists in T_H ;
$h(\varphi(e)*i_0): A_K \Longrightarrow q_{i(o)}$, $j(\psi(e)*i_0): q_{i(o)}, v, C_r \Longrightarrow a_r$.

T is extended by e*0, e*1, and η, φ, χ, f, ψ, δ_J, δ_H are defined as above in 2aa.

[[$f(e*0): A_K, \delta_J(e) \Longrightarrow q_{i(o)}$, $f(e*1): q_{i(o)}, A_K, \delta_J(e) \Longrightarrow a_r$]]

[[Succeded by Case 0 at e*0 and by Case 2ca at e*1]]

2bc. $v = p \rightarrow q$.

$j(\psi(e)*0): v, C_r \Longrightarrow p$, $j(\psi(e)*1): q, v, C_r \Longrightarrow a_r$,
$h(\varphi(e)*0): p, A_K \Longrightarrow q$.

T is extended by e*0, e*1, and I set

$\eta(e*0) = \eta(e*1) = 2$,
$\varphi(e*0) = <\S\S, \varphi(e)*0, \psi(e)*0>$, $\varphi(e*1) = <\S, \psi(e)*1>$,
$r(e) = <\mathrm{CUT}, q>$,
$\chi(e*0) = \chi(e)$, $\chi(e*1)$ is $\varphi(e)$ appended by $\chi(e)$,
$f(e*0)$ is $h(\varphi(e)*0)$ weakened by $\delta_J(e)$ and with p omitted,
$f(e*1)$ is obtained from $j(\psi(e)*0)$ by omitting v, weakening C_r to $A_K^o, C_r, \delta_H(e)$ and permuting this first to $A_K^o, M, C_r^o, \delta_H(e)$ and then to $A_K, \delta_J(e)$ employing $\delta_J(e) = C_r^o, \delta_H(e)$,
the values of ψ, δ_J, δ_H are kept .

[[$f(e): A_K, \delta_J(e) \Longrightarrow a_r$,
$f(e*0): A_K, \delta_J(e) \Longrightarrow q$, $f(e*1): q, A_K, C_r, \delta_J(e) \Longrightarrow a_r$]]

[[Succeded by Case 2cb at e*0 and by Case 2ca at e*1]]

[2bd. $v = \neg p$

$h(\varphi(e)*0)\colon p, A_k \Longrightarrow \Delta$, $j(\psi(e)*0)\colon v, C_r \Longrightarrow p$.

T is extended by e*0, e*1 and I set

$\eta(e*0) = 2$, $\eta(e*1) = 0$,
$\varphi(e*0) = <\S, \psi(e)*0>$, $\varphi(e*1) = \varphi(e)*0$, $r(e) = <\text{CUT}, p>$,
$\chi(e*0)$ is $\chi(e)$ appended by $\varphi(e)$, $\chi(e*1) = \chi(e)$,
$f(e*0)$ is obtained from $j(\psi(e)*0)$ by omitting v, weakening C_r to $A_k{}^o, C_r, \delta_H(e)$ and permuting this first to $A_k{}^o, M, C_r{}^o, \delta_H(e)$ and then to $A_k, \delta_J(e)$ employing $\delta_J(e) = C_r{}^o, \delta_H(e)$,
$f(e*1)$ is $h(\varphi(e)*0)$ weakened by $\delta_J(e)$,
the values of ψ, δ_J, δ_H are kept .

⟦$f(e)\colon A_k, \delta_J(e) \Longrightarrow \Delta$,
$f(e*0)\colon A_k, \delta_J(e) \Longrightarrow p$, $f(e*1)\colon p, A_k, \delta_J(e) \Longrightarrow \Delta$ ⟧

⟦Succeded by Case 0 at e*1 and by Case 2ca at e*0 ⟧]

2c. $\varphi(e)$ is a delayed node .

2ca. $\varphi(e) = <\S, \psi(e)*i>$.

⟦Preceded by Cases 2ba – 2bd, 2cb ⟧

$j(\psi(e)*i)\colon [q_i], v, C_r \Longrightarrow d$,
k the last member of $\chi(e)$ and $h(k)\colon A_k \Longrightarrow v$,

$d = a_r$ for 2ba, 2bb, 2bc, $d = p$ for 2bd, 2cb, $q_i = q$ for 2bc,
no q_i for 2bd, 2cb .

T is extended by e*0, and I set

$\eta(e*0) = 1$, $\psi(e*0) = e*0$, $\varphi(e*0) = \psi(e)*i$,
$\delta_J(e*0) = \delta_J(e)$, $\delta_H(e*0) = \delta_H(e), A_k{}^o$,
$r(e) = (\text{PR})$ for the permutation sending $C_r, \delta_H(e), A_k{}^o$ first into $A_k{}^o, M, C_r{}^o, \delta_H(e)$ and then into $A_k, C_r{}^o, \delta_H(e) = A_k, \delta_J(e)$.
$f(e*0)$ is $j(\psi(e)*i)$ weakened by $\delta_H(e*0)$ and with v omitted.
the values of ψ, χ are kept .

⟦$f(e)\colon [q_i], A_k, \delta_J(e) \Longrightarrow d$, $f(e*0)\colon [q_i], C_r, \delta_H(e*0) \Longrightarrow d$ ⟧

⟦Succeded by Case 1⟧

2cb. $\varphi(e) = <\S\S, k*0, \psi(e)*0>$,

⟦Preceded by Case 2bc⟧

$h(k*0)\colon p, A_k \Longrightarrow q$, $j(\psi(e)*0)\colon v, C_r \Longrightarrow p$.

T is extended by e*0, e*1, and I set

$\eta(e*0) = 2$, $\eta(e*1) = 0$, $\psi(e*i) = \psi(e)$,
$\varphi(e*0) = <\S, \psi(e)*0>$, $\varphi(e*1) = k*0$, $r(e) = <\text{CUT}, p>$,

$f(e*0)$ is obtained from $j(\psi(e)*0)$ by omitting v, weakening C_r to $A_k{}^o, C_r, \delta_H(e)$ and permuting this first to $A_k{}^o, M, C_r{}^o, \delta_H(e)$ and then to $A_k, \delta_J(e)$ employing $\delta_J(e) = C_r{}^o, \delta_H(e)$,
$f(e*1)$ is $h(k*0)$ weakened by $\delta_J(e)$,
the values of ψ, χ, δ_J, δ_H are kept .

⟦$f(e)$: $A_k, \delta_J(e) \Longrightarrow q$,
$f(e*0)$: $A_k, \delta_J(e) \Longrightarrow p$, $f(e*1)$: $p, A_k, \delta_J(e) \Longrightarrow q$ ⟧

⟦Succeded by Case 2ca at e*0 and by Case 0 at e*1 ⟧

In all other cases, T is not extended. This concludes the step from stage n to stage n+1 of the recursive construction of $\mathbf{R}_0(H,J,k_T)$.

Since T_H and T_J were finite trees, they will be exhausted after a finite number of stages and so the construction terminates.

There remains the proof, proceeding by induction on heights (stages), that $<T, f, r>$ actually *is* a prederivation. In view of the remarks, placed in double brackets within the various subcases, it is clear that f is locally correct. $<T, f, r>$ is a derivation since f is defined for all nodes of T and r is defined for all non-maximal nodes. That it ends with the sequent $M \Longrightarrow u$ was enforced already at stage 0. It also follows from the construction that all cuts occurring in $<T, f, r>$ were either copied from H or J, or arise in the Cases 2b or 2cb , and the latter have as cut formulas proper subformulas of v. This concludes the proof of Lemma 3 .

From the construction of T there follow some observations which will become useful immediately. Let me set $\| <\S, a> \| = \|a\|$, $\| <\S\S, b, a> \| = \max(\|a\|, \|b\|)$.

LEMMA 4 If $\|k_J\| \leq \|k_T\|$, $\|k_H\| \leq \|k_T\|$ then for every e in T :

(a) $\|\psi(e)\| \leq \|e\|$,
(b) $\|\varphi(e)\| \leq \|e\|$,
(c) if k occurs in $\chi(e)$ then $\|k\| \leq \|e\|$,

The proof is by simultaneous induction on $\|e\| - \|k_T\|$. Then (a) follows immediately from the Cases 1ab, 1b which are the only ones changing the values of ψ. Statement (b) follows for $e = k_T$ from $\|k_J\| \leq \|k_T\|$, and I prove it for e*i. In Cases 0, 1aa, 2a, and in the undelayed Cases 2b, the node $\varphi(e*i)$ is the upper neighbour $\varphi(e)*i$ of $\varphi(e)$ whence $\|\varphi(e)\| \leq \|e\|$ implies $\|\varphi(e*i)\| = \|\varphi(e)\|+1 \leq \|e\|+1 = \|e*i\|$. In Cases 1ab, 1b the statement follows from $\|\varphi(e*0)\| = k_H \leq k_T \leq \|e*0\|$. In the delayed Cases 2b I have $\|\psi(e)\| \leq \|e\|$ whence $\|\varphi(e)*i\| = \|<\S, \psi(e)*j>\| = \|\psi(e)*j\| \leq \|e*i\|$, and $\|\varphi(e)\| \leq \|e\|$ implies $\max(\|\varphi(e)*0\|, \|\psi(e)*0\|) \leq \|e*0\|$ whence $\|<\S\S, \varphi(e)*0, \psi(e)*0>\| \leq \|e*0\|$. In Case 2ca the inductive hypothesis gives $\|\varphi(e)\| = \|\psi(e)*i\| \leq \|e\|$ whence $\|\varphi(e)*0\| = \|\psi(e)*i\| \leq \|e*0\|$. In Case 2cb the inductive hypothesis gives $\|\varphi(e)\| = \max(\|k{*}0\|, \|\psi(e){*}0\|) \leq \|e\|$ whence $\|\varphi(e){*}0\| = \|\psi(e){*}0\| \leq \|e{*}0\|$ and $\|\varphi(e){*}1\| = \|k{*}0\| < \|e{*}0\|$. – Statement (c) follows immediately from the cases 2b which are the only ones enlarging the values of χ.

LEMMA 5 For any e, f in T

(a) If $e<f$ and $\eta(e) = 0$ then $\eta(f) = 0$ and $\varphi(e)<\varphi(f)$.
(b) If $e<f$ and $\eta(e) = \eta(f) = 1$ then $\varphi(e)<\varphi(f)$.
(c) If $e<f$ and $\eta(e) = \eta(f) = 2$ and $\psi(e) = \psi(f)$ and if $\varphi(e)$, $\varphi(f)$ are not delayed then $\varphi(e)<\varphi(f)$.

Here (a) follows immediately from Case 0. I prove (b) by induction on $s = \|f\| - \|e\|$; for $s = 1$ only Case 1aa can occur which implies (b). If $s>1$ and if $\eta(g) = 1$ for the finitely many g with $e<g<f$ then (b) follows again. Otherwise, there exists a lowest of these g such that $\eta(g) = 1$ and $\eta(g')\neq 1$ for the following g'; observe that $\varphi(e)<\varphi(g)$. Then Case 1ab must occur with $g' = g{*}0$, $\eta(g') = 2$, $\varphi(g') = k_H$, $\psi(g') = \varphi(g) = r$. Consider now the largest h with $g' \leq h \leq f$ such that $g' \leq g'' \leq h$ implies $\eta(g'') = 2$, $\psi(g'') = \psi(g')$, and let h' be the following node below f. Thus the step from h to h' must either change $\eta(h')$ to 0 or 1, or change $\psi(h')$. This excludes Case 2aa for h, and (a) excludes the cases 2ab, 2b, 2cb which might result in $\eta(h') = 0$, because then Case 0 would imply $\eta(f) = 0$ for any f with $h' \leq f$. Thus Case 2ca will hold with $h' = h{*}0$, $\eta(h') = 1$, $\varphi(h') = r{*}i$ whence $\varphi(e)<\varphi(g) = r< \varphi(h')$. Since $\|h'\| - \|e\| < s$, the inductive hypothesis implies $\varphi(h') < \varphi(f)$ whence also $\varphi(e) < \varphi(f)$.

The statement (c) is clear if e, f are such that also $\eta(g) = 2$, $\psi(e) = \psi(g)$ for the finitely many g with $e \leq g \leq f$. Otherwise, there would be the last such g, say h, whence for $h' = h{*}0$ there holds $\eta(h') = 1$, $\psi(e)<\varphi(h')$ as just described. Also, there would be the first d' such that $h' < d' \leq f$ and $\eta(d'') = 2$, $\psi(d'') = \psi(f)$ for all d'' with $d' \leq d'' \leq f$. Then d' must arise as $d{*}0$ from some d with $h' \leq d$ under Case 1ab whence $\eta(d) = 1$, $\psi(f) = \psi(d') = \varphi(d)$. By (b) this would imply $\psi(e)<\varphi(h') \leq \varphi(d) = \psi(f)$.

Let me now, for any pair $D = \langle d, r\rangle$ of functions on the universal tree 2^∞ [or ω^∞], consider D as the function with values $D(e) = \langle d(e), r(e)\rangle$ for nodes e of that tree. I denote by $O(n)$ the set of all nodes e such that $\|e\|<n$. For any n, I denote by $D\restriction n$ the restriction of D to $O(n)$. Then $\mathbf{R}_0$ has the *continuity property* :

LEMMA 6 Assume that $\|k_J\| \leq \|k_T\|$, $\|k_H\| \leq \|k_T\|$. If $H\restriction n = H'\restriction n$ and $J\restriction n = J'\restriction n$ then $\mathbf{R}_0(H, J, k_J)\restriction n = \mathbf{R}_0(H', J', k_J)\restriction n$

The algorithm defining $\mathbf{R}_0(H, J, k_J)$ determines the values of the functions f and r at upper neighbours e*i from the values of h, r_H, j, r_J, δ_H at nodes which are at most upper neighbours of $\varphi(e)$. But $\|e*i\| \leq n$ implies $\|\varphi(e)\| \leq \|e\| < n$, $\|\varphi(e)*i\| \leq n$ by Lemma 4. - I should add that Lemma 6 may be generalized to the case where the root k_T is chosen arbitrarily. If c is the larger of the numbers $\|k_H\|-\|k_T\|$ and $\|k_J\|-\|k_T\|$, then Lemma 4(a) takes the form $\|\varphi(e)\| \leq \|e\|+c$, $\|k\| \leq \|e\|+c$; Lemma 6 says that $H\restriction n+c = H'\restriction n+c$ and $J\restriction n+c = J'\restriction n+c$ imply $\mathbf{R}_0(H, J, k_J)\restriction n = \mathbf{R}_0(H', J', k_J)\restriction n$.

[The construction of $\mathbf{R}_0(H,J,k_T)$ remains in effect for *infinitary* derivations the appearance of which is actually the raison d'être for continuous cut elimination procedures. Infinite derivations occur if there are rules with ω many premisses generalizing 2-ary conjunctions and disjunctions: to certain sequences $<q_i \mid i\epsilon\omega>$ of formulas there are assigned formulas $Q_0<q_i \mid i\epsilon\omega>$ and $Q_1<q_i \mid i\epsilon\omega>$, and for these as principal formulas, and the q_i as side formulas, the rules for $\wedge$ and $\vee$ are extended to rules (IQ_0), (EQ_0), (IQ_1), (EQ_1) of which (IQ_0) and (EQ_1) require an ω-sequence of premisses. Instead of from 2^∞, the underlying trees of such derivations must be taken from ω^∞ whence $e = <e_0, \ldots, e_{n-1}>$ has upper neighbours $e*i$ for all i in ω, and with this modification all of the previous definitions, in particular that of $\mathbf{R}_0(H,J,k_T)$, remain in effect with obvious generalizations of Case 2ba, 2bb. Of course, if H or J are infinite then the construction may not terminate after a finite number of stages, and the new tree T becomes infinite as well.

A tree shall be called *cowell founded* if it does not contain an *ascending* ω-sequence, i.e. a sequence of nodes d_i with $d_i<d_{i+1}$ for every i.

LEMMA 7 If H and J are cowell founded then so is the tree T underlying $\mathbf{R}_0(H,J,k_T)$.

Assume that there exists an ascending ω-sequence $<d_i \mid i\epsilon\omega>$ in T which determines the sequence $<h_i \mid i\epsilon\omega>$ of its reference nodes $h_i = \varphi(d_i)$. If there would be a j such that $\eta(d_j) = 0$ then by Lemma 5(a) the sequence of the h_i with $i \geq j$ would violate the cowell foundedness of H. If there would be a j such that $\eta(d_j) = 2$ for $i \geq j$ then by Lemma 5(c) the sequence of these h_i would again violate the cowell foundedness of H. Hence there is a $d_j = e$ with $\eta(e) = 1$, and for every d_j with $\eta(d_j) = 1$ there is an h with $j<h$ and $\eta(d_h) = 1$. Thus I obtain a subsequence $<e_i \mid i\epsilon\omega>$ with $\eta(e_i) = 1$ for every i, and by Lemma 5(b) the sequence of the corresponding reference nodes violates the cowell foundedness of J.]

The construction of $\mathbf{R}_0(H,J,k_T)$ can easily be put into an arithmetical form. The nodes $e = <e_0, \ldots, e_{n-1}>$ of the universal tree 2^∞ should not be coded by the numbers they represent as binary digits since there arises the problem of leading zeros; rather they may be viewed as the numbers represented by $<e_0+1, \ldots, e_{n-1}+1>$ to the base 3; the empty node $< >$ can be coded as the number 0. The height $\|e\|$ of e then becomes the number $1+{}_3\log(e)$, and for any two nodes e and g in 2^∞, if $\|e\| < \|g\|$ then e *as a number* is smaller than g *as a number.* [In the case of ω^∞ nodes can be coded e.g. as exponents of prime factor decompositions.]

Arithmetizing this approach, a derivation $<T,d,r>$ may then be re-defined as a pair D of functions d, r, defined for all natural numbers such that

> d and r have values different from 0 precisely outside of T, and d is locally correct for r.

Here, again, I assume that formulas and sequents (as values of d), and rule data (as values of r), have been coded as numbers in some arithmetical manner using elementary recursive functions. Roots k_D of derivations $D = <d, r>$ shall always be taken such as to have a value different from 0. Once k_T is given, I define $f(k_T): M \Longrightarrow u$, $\eta(k_T) = 1$ and $\varphi(k_T) = k_H$ as before, define $r_0(k_T) = 0$, and finally define $f(e) = \eta(e) = \varphi(e) = r_0(e) = 0$ for all other e such that $\|e\| \leq \|k_T\|$. In view of the rôle played by the values of η, however, I must assume that the numbers 0 and 1 do not occur as nodes of H (otherwise H can be replaced by multiplying all its nodes by 3).

The definition of the functions f, r_0, η, φ, ψ, χ, δ_J, δ_H then proceeds by a simultaneous course-of-values recursion. The values at e*i are defined as above in the cases listed, and in all other cases they are set to be 0. The value of r_0, at the left upper neighbour $e{*}i_T$ in T, I define as that one which formerly was set to be r(e); once the definition is completed, I define r by setting its value at e to be that of r_0 at $e{*}i_T$. The recursive definition of f then has the form

$$f(e{*}0) = F(h(\varphi(e)), j(\varphi(e)), r_H(\varphi(e)), r_J(\varphi(e)), \chi(\varphi(e)), \delta_H(\varphi(e))) ,$$

and analogously for e*1 and for the other functions. The recursion function F describes the decoding of the sequents $h(\varphi(e))$, $h(\varphi(e){*}0)$, $h(\varphi(e){*}1)$, ... , the distinction of the cases 0 to 2 above, and then the coding of the sequent f(e) as described for these cases. In this way, the operator $\mathbf{R}_0$ has a primitive recursive arithmetization.

[An infinitary derivation $D = <d, r>$ is called *elementarily bounded* if the arithmetizised function d is *elementary* recursive as a function of the length $\|e\|$ of nodes e ; this may be expressed as saying that it is primitive recursive and that there exists a number b_d such that $e_2(b_d,-)$ is a bound of d, i.e. for all nodes e:

$$d(e) \leq e_2(b_d, \|e\|) .$$

It then follows that $\mathbf{R}_0$ is *elementary* in the sense that if H and J are elementarily bounded then so is $\mathbf{R}_0(H,J,k_T) = <f, r>$. For the following discussion let b be such that both the function h and j are bounded by $e_2(b,-)$.

It was used already for Lemma 6 that the algorithm defining $\mathbf{R}_0(H,J,k_J)$ determines the values of its functions at nodes e*i from the values of h, r_H, j, r_J , δ_H at smaller nodes. In Case 2aa, for instance, f(e*0) is obtained from $h(\varphi(e)*i_0)$ by weakening with $C_{\psi(e)}{}^o, \delta_H(e)$ where $\delta_H(e)$ consists of the sets $A_k{}^o$ for k in $\chi(e)$. Arithmetically, this requires to concatenate the antecedent of $h(\varphi(e)*i_0)$ with the sequence ϑ of all the sequences $C_{\psi(e)}{}^o$ and the $A_k{}^o$ of formulas. The lenght $|\vartheta|$ of ϑ may grow with $\|e\|$, and the usual arithmetical concatenation functions CAT have the property that CAT(ϑ) is elementarily recursive depending on $|\vartheta|$ and max(ϑ). Thus there is a number c such that

$$\text{CAT}(\vartheta) \leq e_2(c, |\vartheta| + \max(\vartheta)) .$$

Now $|\vartheta| = 1+|\chi(e)|$, and since χ is enlarged only in the Cases 2b, an immediate induction shows $|\chi(e)| \leq \|e\|-2$; hence $|\vartheta| < \|e\|$. Also, $C_{\psi(e)}{}^o$ and the $A_k{}^o$ come from values of f at nodes e_p with $\|e_p\| < \|e*i\|$ whence for the arithmetized sequences $C_{\psi(e)}{}^o \leq e_2(b, \|e_p\|)$, $A_k{}^o \leq e_2(b, \|e_p\|)$; hence max($\vartheta$) $\leq e_2(b,n)$ for the maximum n of $|\vartheta|$ and the $\|e_p\|$, $n < \|e*i\|$. Thus

$$\text{CAT}(\vartheta) \leq e_2(c, n + e_2(b,n)) \leq e_2(c+b, n) ,$$

and so f(e*i) is bounded by $e_2(c+b, n)$ where c depends only on CAT and b only on H and J .]

3. The Reduction Operator R_1 and the Elimination Operator

I now shall introduce a *reduction operator* $\mathbf{R}_1$ which acts upon derivations D and reduces the cut degree by at least 1. Recall that the *cut degree cdg*(D) of a derivation D was defined globally to be 0 if D contains no cuts, and to be 1 plus the maximum of all degrees of cuts in D otherwise; this can be made into a recursive definition by setting $cdg(D) = \sup < cdg(D \upharpoonright n) \mid n \epsilon \omega >$. Observe that, in the situation of Lemma 3, the continuation of H and J to a derivation D by a cut with the formula v of degree $|v|$ would result in $cdg(D) = \max(|v|+1, cdg(H), cdg(J))$. The same endsequent produced by $\mathbf{R}_0(H,J,k_T)$ requires at most the cut degree $\max(|v|, cdg(H), cdg(J))$.

For any derivation D, I abbreviate by n(D) the height $\|k_D\|$ (in BT) of the root k_D of D .

Following the ideas of GORDEEV 88 , I define a more general operator $\mathbf{R}_1$ which acts on pairs, formed from natural numbers n and derivations D. $\mathbf{R}_1(n,D)$ shall be a derivation with the same endsequent as D. I define explicitly

(r0) $\mathbf{R}_1(n,D) = D$ if $n \leq n(D)$

(r1) $\mathbf{R}_1(n,D) = D$ if D is a derivation of length 0 , i.e. a single axiom.

I now make use of the function $n \dot{-} m$ defined as n−m for $m \leq n$ and 0 otherwise. If D is of positive length and ends with an instance of a rule (R), then let H, J be the subderivations of D leading to the premisses of that rule. I proceed by recursion on $n \dot{-} n(D)$: if $n \leq n(D)$ then $n \dot{-} n(D) = 0$, and as D has positive length, there holds $n(D) < n(H)$ and $n(D) < n(J)$ whence $n(D) < n$ implies $n \dot{-} n(H) < n \dot{-} n(D)$ and $n \dot{-} n(J) < n \dot{-} n(D)$.

$\mathbf{R}_1(n,D)$ shall be undefined if $\mathbf{R}_1(n,H)$, $\mathbf{R}_1(n,J)$ are *not* derivations ending with the endsequents of H and J respectively. Otherwise I set

(r2) $\mathbf{R}_1(n,D) = D^\ddagger$ if $(R) \neq (CUT)$

where $D^\ddagger$ arises from $\mathbf{R}_1(n,H)$, $\mathbf{R}_1(n,J)$ by an application of a rule (R)

(r3) $\mathbf{R}_1(n,D) = D^\ddagger$ if $(R) = (CUT)$

where $D^\ddagger$ arises from $\mathbf{R}_0(\mathbf{R}_1(n,H), \mathbf{R}_1(n,J), k_T)$, k_T the root of $\mathbf{R}_1(n,J)$, by continuing with an instance of the repetition rule (RR), i.e. (RP) for the identical permutation.

There follows by an obvious induction on $n \dot{-} n(D)$ the first statement of the

LEMMA 8 $\mathbf{R}_1(n,D)$ is a derivation, has the same root and the same endsequent as D, and there holds $cdg(\mathbf{R}_1(D)) \leq cdg(D) \dot{-} 1$.

The last statement is trivial if $n \dot{-} n(D) = 0$. If $n \dot{-} n(D) > 0$, it still is trivial if D is a single axiom, i.e. in the case (r1). Assume now that D ends with an instance of a rule (R). Observe that $n \dot{-} n(H) < n \dot{-} n(D)$ and $n \dot{-} n(J) < n \dot{-} n(D)$. Hence the inductive hypothesis gives

$$cdg(\mathbf{R}_1(n,H)) \leq cdg(H) \dot{-} 1 \quad \text{and} \quad cdg(\mathbf{R}_1(n,J)) \leq cdg(J) \dot{-} 1 .$$

In the case (r2), therefore, there holds

$$\begin{aligned} cdg(\mathbf{R}_1(n,D)) &= \max(cdg(\mathbf{R}_1(n,H)), cdg(\mathbf{R}_1(n,J))) \\ &\leq \max(cdg(H) \dot{-} 1, cdg(J) \dot{-} 1) = cdg(D) \dot{-} 1 . \end{aligned}$$

In the case (r3), (R) is a cut with a cut formula v, and making use of the property of $\mathbf{R}_0$ mentioned above, I find

$$\begin{aligned} cdg(\mathbf{R}_1(n,D)) &= cdg(\mathbf{R}_0(\mathbf{R}_1(n,H), \mathbf{R}_1(n,J), k_T)) \\ &\leq \max(|v|, cdg(\mathbf{R}_1(n,H)), cdg(\mathbf{R}_1(n,J))) \\ &\leq \max(|v|, cdg(H) \dot{-} 1, cdg(J) \dot{-} 1) \\ &= \max(|v|+1, cdg(H), cdg(J)) \dot{-} 1 \\ &= cdg(D) \dot{-} 1 . \end{aligned}$$

This concludes the proof of Lemma 8. Working with a recursive definition of *cdg* instead of the global one, the second statement can be obtained by the same argument in the form $cdg(\mathbf{R}_1(n,D) \restriction n) \leq cdg(D \restriction n) \dot{-} 1$.

In condition (r3), $\mathbf{R}_0(\mathbf{R}_1(n,H), \mathbf{R}_1(n,J), k_T)$ is defined with the root of $\mathbf{R}_1(n,J)$ (which turns out to be that of J) rather than with the root of D. The only reason for this choice is that then the Lemma 4 becomes available.

In order to have $D^\ddagger$ still start at the root of D, the repetitive use of (RR) then becomes necessary. Observe that, by these definitions, there holds

(x) in case (r2): if $\|e\| > \|k_D\|$ then $\mathbf{R}_1(n,D)(e) = \mathbf{R}_1(n,H)(e)$ respectively $\mathbf{R}_1(n,D)(e) = \mathbf{R}_1(n,J)(e)$,

(y) in case (r3): if $\|e\| > \|k_D\|$ then $\mathbf{R}_1(n,D)(e) = \mathbf{R}_0(\mathbf{R}_1(n,H),\mathbf{R}_1(n,J),k_J)(e)$.

I now want to define the operator $\mathbf{R}_1$, acting upon derivations D, by $\mathbf{R}_1(D)(e) = \mathbf{R}_1(\|e\|,D)(e)$. To this end, I first prove the

LEMMA 9 If $\|e\|<n$ then $\mathbf{R}_1(\|e\|,D)(e) = \mathbf{R}_1(n,D)(e)$.

The proof is by induction on $n \dot{-} n(D)$. If $n \dot{-} n(D) = 0$ then $\|e\| \dot{-} n(D) = 0$. So (r0) implies $\mathbf{R}_1(\|e\|,D) = D = \mathbf{R}_1(n,D)$. If $n \dot{-} n(D) > 0$ and if D is a single axiom, then (r1) likewise implies $\mathbf{R}_1(\|e\|,D) = D = \mathbf{R}_1(n,D)$. Assume now that D ends with an instance of a rule (R). Observe that $n \dot{-} n(H) < n \dot{-} n(D)$ and $n \dot{-} n(J) < n \dot{-} n(D)$. Hence the inductive hypothesis gives

(z) if $\|e\|<n$ then $\mathbf{R}_1(\|e\|,H)(e) = \mathbf{R}_1(n,H)(e)$ and $\mathbf{R}_1(\|e\|,J)(e) = \mathbf{R}_1(n,J)(e)$.

In particular, this holds at the roots k_H and k_J whence these derivations have the same endsequents. In the case (r2), these endsequents uniquely determine $\mathbf{R}_1(\|e\|,D)(k_D)$ and $\mathbf{R}_1(n,D)(k_D)$ by an application of (R), and together with (x) this proves (a) for this case. In the case (r3), I rewrite (z) as

$$\mathbf{R}_1(\|e\|,H)\restriction n-1 = \mathbf{R}_1(n,H)\restriction n \quad \text{and} \quad \mathbf{R}_1(\|e\|,J)\restriction n-1 = \mathbf{R}_1(n,J)\restriction n$$

and then obtain from Lemma 4

$$\mathbf{R}_0(\mathbf{R}_1(\|e\|,H),\mathbf{R}_1(\|e\|,J),k_T)\restriction n = \mathbf{R}_0(\mathbf{R}_1(n,H),\mathbf{R}_1(n,J),k_T)\restriction n .$$

Together with (y), this proves (a) also for this case.

As already announced, I now define, for any derivation D, the operator $\mathbf{R}_1$, acting on D, to produce the pair $\mathbf{R}_1(D) = \langle d, r \rangle$ of functions given by $\mathbf{R}_1(D)(e) = \mathbf{R}_1(\|e\|,D)(e)$.

LEMMA 10 For every n : $\mathbf{R}_1(D)\restriction n = \mathbf{R}_1(n,D)\restriction n$.

For if $\|e\|<n$ then $\mathbf{R}_1(D)(e) = \mathbf{R}_1(\|e\|,D)(e) = \mathbf{R}_1(n,D)(e)$ by Lemma 9.

THEOREM 1 The operator $\mathbf{R}_1$, acting on a derivation D, produces a derivation $\mathbf{R}_1(D)$ with the same endsequent and the same root as D and such that $cdg(\mathbf{R}_1(D)) \leq cdg(D) \dot{-} 1$.

In order to see that $\mathbf{R}_1(D)$ is a derivation, is suffices to check that d is locally correct for r at nodes e_0, e. Choosing n such that $\|e\| < \|e_0\| < n$, this follows from Lemma 10 and the local correctness of $\mathbf{R}_1(n,D)$. Also, Lemma 8 and Lemma 10 imply for every n

$$cdg(R_1(D)\upharpoonright n) = cdg(\mathbf{R}_1(n,D)\upharpoonright n) \leq cdg(D\upharpoonright n) \dot{-} 1$$

whence $cdg(R_1(D)) \leq cdg(D) \dot{-} 1$.

Together with $\mathbf{R}_0$ also the operator $\mathbf{R}_1(n,D)$ is primitive recursive [and is elementarily bounded] ; hence also the operator $\mathbf{R}_1(D)$ has these properties.

Finally, I obtain a cut elimination operator $\mathbf{R}_2$ as iteration of $\mathbf{R}_1$: given a derivation D, $\mathbf{R}_1$ iterated *cdg*(D) times will be cut free. Thus also $\mathbf{R}_2$ is primitive recursive [but will, in the case of infinitary derivations, in general not be elementarily bounded anymore] .

References

L.Gordeev: Proof-theoretical Analysis: Weak Systems of Functions and Classes. Ann.Pure Appl.Logic **38** (1988) 1-121

G.E.Mints (Minc): Finite Investigations of Transfinite Derivations. Zap.Nauchn.Semin.LOMI **49** (1975) [Russian] and J.Sov.Math. **10** (1978) 548-596

Chapter 4. Sequent Calculi for Minimal and Intuitionistic Logic

1. Negation in Deductive Situations

At the beginning of Chapter **1**, the positive connectives *and, or* and *if–then* were given a deductive interpretation. Just as these connectives do, also negation refers to a semantical meaning, namely that something *is so* or *is not so*. But which something is it, to which *not* might refer with respect to deductive situations? What shall it mean that a deduction leads from M to *not* v :

(a_1) $\quad M \Longrightarrow \neg v$.

As a first attempt at an answer, it might be suggested to read (a) as

(b) $\quad$ there is no deduction leading from M to v ,

explaining *negation as failure.* Given a calculus of deductive situations such as KP, this may be specialized to say that, in this calculus,

(c) $\quad M \Longrightarrow v$ has no derivation .

Unfortunately, this attempt will meet several obstacles.

(1) Given a calculus such as KP, the claim that there *is* a derivation of some sequent can be secured by *presenting* a derivation. The claim (b), however, is of quite a different character: it requires either to verify for *all* derivations that they do not produce $M \Longrightarrow v$, or it requires the possibility of a decision algorithm which, given a sequent, decides in finitely many steps whether it does or does not have a derivation in my calculus. And while for KP, and for several other calculi, decision algorithms can indeed be found (as was discussed in Chapter **1**), the more interesting logical calculi do *not* admit such decision algorithms (as will be seen in Theorem **3**.**9**.7). [Among them, for instance, are the calculi codifying the set theoretical argumentations about numbers and functions. So if M is a suitable set of axioms for analysis and v is the Riemann hypothesis then, at the time of this being written, it simply is not known whether $M \Longrightarrow v$ has a derivation.] For such calculi, therefore, the use of (c) would make (a_1) depend on the progress of human knowledge.

(2) Even for calculi which do admit a decision algorithm, the reduction of (a_1) to (c) would require cumbersome methodological distinctions. If the connective $\neg$ is included in my language, then still no sequents $M \Longrightarrow \neg v$ are derivable by the rules of KP from the axioms $x \Longrightarrow x$. Now a calculus such as KP is defined by a fixed system of rules. If (c) can be established for KP then this refers to *these* rules, and introducing the conditional rule (a_1) defines a new calculus KP_1 which is more expressive than KP. If it can be established that $M \Longrightarrow v$ is not derivable in KP_1 then I may set up a

next calculus KP_2 with an analogous rule (a_2) in which $M \Longrightarrow \neg v$ is derivable. In this way, I arrive at a hierarchy of calculi KP_i with $KP_0 = KP$ and such that KP_{i+1} extends KP_i by a conditional rule (a_{i+1}).

(3) I continue to assume that the connective $\neg$ belongs to my language. Were I to use only the axioms $x \Longrightarrow x$ then it would be obvious that no sequent $M \Longrightarrow \neg v$ is derivable in KP, and it would be not hard to see that also no sequent $\neg w \Longrightarrow v$ can be derivable. Hence in KP_1 all sequents $M \Longrightarrow \neg\neg v$ and all sequents $\neg w \Longrightarrow \neg v$ would be derivable. In order to avoid such undesirable effects, I should admit *all* sequents $\neg v \Longrightarrow \neg v$ as KP-axioms in which case all sequents $v \Longrightarrow v$ will become KP-derivable. Since KP derives $v \Longrightarrow v$, it is *not* the case that KP does not derive $v \Longrightarrow v$; hence KP_1 does not derive $v \Longrightarrow \neg v$, and so KP_2 derives $v \Longrightarrow \neg\neg v$.

In view of these difficulties, the notion of negation as failure does not seem to be the appropriate one with which to describe negation within a general setting. There is, however, a second approach to the deductive analysis of negation in the form of JOHANSSONs 37 *minimal logic*. This approach consists in introducing the intermediary notion of *refutability* and then reducing negation to relative refutability; it was suggested by JOHANSSON when analyzing KOLMOGOROVs 25 interpretation of intuitionistic negation and was elaborated by CURRY 63.

Refutability does not need to be handled by presenting a list $f_0, f_1, \ldots$ of refutable propositions, or an algorithm generating such, but can be treated axiomatically by using one distinguished refutable proposition written as Δ (and if $f_0, f_1, \ldots$ should be given explicitly then also using a rule

$$\frac{M \Longrightarrow f_i}{M \Longrightarrow \Delta} \qquad \text{for } i = 0,1,\ldots \text{).}$$

Refutability shall be governed by the deductive principle

if $v, M \Longrightarrow w$ and if w is refutable
then v is refutable under the assumptions M .

In a language containing $\neg$, the derivability of $M \Longrightarrow \neg u$ shall be set up such that it is equivalent to u being refutable. So the absolute refutability of Δ amounts to an axiom

(da) $\quad M \Longrightarrow \neg\Delta$

and the deductive principle can be captured in the rule

$$\text{(db)} \qquad \frac{v, M \Longrightarrow w \qquad M \Longrightarrow \neg w}{M \Longrightarrow \neg v} \quad .$$

In particular, if w is Δ then the right premiss is (da) and thus there is the particular rule

$$\frac{v, M \Longrightarrow \Delta}{M \Longrightarrow \neg v} \ .$$

A third approach to the deductive analysis of negation is KOLMOGOROVs 25 interpretation of *intuitionistic logic;* it consists in introducing the intermediary notion of *absurdity,* governed by the principle of *ex absurdo quodlibet,* and then reducing negation to relative absurdity. It uses again a distinguished absurd proposition, again written as Δ , and *ex absurdo quodlibet* is captured in the rule

$$\frac{M \Longrightarrow \Delta}{M \Longrightarrow v} \ .$$

2. The Calculi KM_0 of Minimal Logic and KJ_0 of Intuitionistic Logic

I consider formulas from a propositional language Fm(Δ) with the connectives $\wedge$, $\vee$ and $\rightarrow$ and a constant (0-ary connective) Δ. Thus Fm(Δ) may be viewed as a propositional language as before in which now a particular variable has been chosen as Δ. Sequents, namely s- and t-sequents, I define as before; let the index x be one of s, t or, when appropriate, also u.

I define the calculi K_xM_0 with the same rules as are used for K_xP and with the axioms $x \Longrightarrow x$ and $\Delta \Longrightarrow \Delta$ (or their generalized forms). I define the calculi K_xJ_0 by adding to K_xM_0 the rule

$$(\mathrm{AIN}\Delta) \qquad \frac{M \Longrightarrow \Delta}{M \Longrightarrow v} \ ;$$

in analogy to the case of (W), the formula v here is called the *weakening* formula to the right (*and not* a principal formula). Unless distinctions really matter, I shall write KM_0 instead of K_xM_0 and KJ_0 instead of K_xJ_0. The calculi KM_0 are called those of *minimal logic,* the calculi KJ_0 those of *intuitionistic logic.*

As KM_0 is simply KP with a distinguished variable Δ, is has the subformula property and the properties of directness and decidability, and the cut rule is admissible for KM_0.

Also KJ_0 has the subformula property and the property of directness, and the decision method of Chapter **1** remains in effect. The rule (AINΔ) may be viewed as structural and its instances will not be counted in the definition of lengths of derivations. The cut rule is also admissible for KJ_0, and the cut elimination algorithms of Chapter **2** remain in effect; the necessary additions to the cases E1 and E2 in the definition of the operator $\mathbf{E}_1$ were discussed already there. The modifications required for explicit retracing (Lemma **3**.2) and for ascending cut elimination are analogous.

There are variants KJ_{01} of the calculi KJ_0 which omit the rule (AINΔ) and, besides the axioms $\Delta \Longrightarrow \Delta$, use additional J-*axioms* $\Delta \Longrightarrow x$ (or generalized J-axioms $\Delta, M \Longrightarrow x$ in the case of KJ_{01u}). Clearly, every KJ_{01}-derivation becomes a KJ_0-derivation if the J-axioms are replaced by instances of (AINΔ).

Conversely, every KJ_0-derivation D can be transformed into a KJ_{01}-derivation D_1 by replacing instances of (AINΔ). This is obvious if D has length 0; assume that is has been established for derivation of length shorter than D. If D ends with an instance of a rule (R) which is not (AINΔ) then I transform the subderivations leading to the premisses of (R) and apply this rule again. There remains the case that (R) is (AINΔ). Observe first that a sequent $\Delta \Longrightarrow u$ has a KJ_{01}-derivation of length at most the complexity of u; this follows immediately by induction. Let D_{11} be a K_{01}-derivation of the premiss $M \Longrightarrow \Delta$ preceding the endsequent $M \Longrightarrow u$ of D, and consider the parameter tree P in D_{11} of the right occurrence of Δ. As there is no rule introducing Δ on the right, the maximal nodes of P are axioms $\Delta \Longrightarrow \Delta$. Replacing Δ by u on the right in all sequents of D_{11}, these axioms become derivable sequents $\Delta \Longrightarrow u$ and all instances of rules in P (also where instances of (E$\rightarrow$) branch leftwards out of P) remain such instances. Hence D_{11} is transformed into a derivation D_1 of $M \Longrightarrow u$.

The argument above can also be used in order to show that KJ_{01} admits (CUT). Here only the Lemma **2**.1 requires an additional consideration of generalized J-axioms. If $M \Longrightarrow v$ is such, i.e. Δ is in M, then also $M, N \Longrightarrow u$ is a generalized J-axiom. If $v, N \Longrightarrow u$ is a generalized J-axiom and $v \neq \Delta$ then again $M, N \Longrightarrow u$ is a generalized J-axiom. If $v = \Delta$ then I transform the derivation D of $M \Longrightarrow \Delta$ into a derivation D_1 of $M \Longrightarrow u$ as above. Observe that now an upper bound, for the length of the derivation obtained from a derivation D by the elimination algorithm, will also depend on the complexities of the formulas u.

3. The Intermediary Calculi KM_1 , KJ_1

Neither of the calculi KM_0, KJ_0 contains formulas $\neg v$. However, I can use the KOLMOGOROV interpretation of negation in order to introduce a *defined* operation $\sim$ by

$$\sim v = v \rightarrow \Delta \ .$$

Then

$$\frac{v, M \Longrightarrow \Delta}{M \Longrightarrow \sim v}$$

is a special case of (I$\rightarrow$) and thus an admissible rule. Also, in the special case of (E$\rightarrow$)

$$\frac{M \Longrightarrow v \qquad \Delta, M \Longrightarrow \Delta}{\sim v, M \Longrightarrow \Delta}$$

the right premiss is a generalized axiom and hence derivable; thus

$$\frac{M \Longrightarrow v}{\sim v, M \Longrightarrow \Delta}$$

is also an admissible rule.

I now consider a language $Fm(\Delta, \neg)$ which employs both the constant Δ and the connective $\neg$. On $Fm(\Delta, \neg)$ I define the auxiliary calculi K_xM_1: axioms shall be those of KP and the sequent $\Delta \Longrightarrow \Delta$, rules shall be those of K_xP together with the new logical rules

$$(I\Delta\neg)\ \frac{v, M \Longrightarrow \Delta}{M \Longrightarrow \neg v} \qquad (E\Delta\neg)\ \frac{M \Longrightarrow v}{\neg v, M \Longrightarrow \Delta} \quad \text{or} \quad \frac{\neg v, M \Longrightarrow v}{\neg v, M \Longrightarrow \Delta}$$

in the case of u-sequents; here $\neg v$ is the principal and v is the side formula. In KM_1 the the reflexivity axioms are derivable since $v \Longrightarrow v$ gives $\neg v, v \Longrightarrow \Delta$ and $\neg v \Longrightarrow \neg v$. Hence also the sequents (da): $M \Longrightarrow \neg\Delta$, mentioned in the introduction of this Chapter, are derivable with $(I\Delta\neg)$. Adding the rule $(AIN\Delta)$ to KM_1, I obtain an intuitionistic calculus KJ_1 on $Fm(\Delta, \neg)$.

Both KM_1 and KJ_1 have the property of directness; they have the subformula property if Δ is considered as subformula of every formula $\neg v$. A KM_1- or KJ_1-derivation of a positive sequent is already a KP-derivation: being direct, it cannot contain negated formulas and, therefore, no constants Δ which might have disappeared in instances of $(I\Delta\neg)$, $(E\Delta\neg)$.

I conclude this section with two Lemmas which will become important later. I first define the set of *good* $Fm(\Delta, \neg)$-formulas:

> every variable (but not Δ) is good, and if v, w are good then so are $v \wedge w$, $v \vee w$, $v \rightarrow w$, $\neg v$ and $v \rightarrow \Delta$.

In particular, there is a subset of good $Fm(\Delta)$-formulas. It follows that Δ is the only non-good formula which may occur as subformula in a good formula.

A sequent $M \Longrightarrow u$ of $Fm(\Delta, \neg)$-formulas is *semigood* if u as well as every formula in M is either good or is Δ. It is *good* if, moreover, all formulas in M are good. Observe in what follows that a KM_0- or KJ_0-derivation is also a KM_1- or KJ_1-derivation.

Given a KM_1- or KJ_1-derivation D of a semigood endsequent, it follows from the subformula principle that all its sequents are semigood again. Let g be a node with one or two upper neighbours g', g", and assume that Δ occurs in the antecedent of the sequent at g. Then one of the following situations is taking place:

1. The rule at g is (W) and introduces a copy of Δ while the antecedent at g' did not contain Δ.

2. The rule at g is (W), but not of type 1 above, or this rule is (RP), (RC), (AINΔ), a logical rule for $\wedge$, $\vee$ or $\neg$, or the rule ($I\rightarrow$). In these cases Δ in the antecedent can only be parametric.

3. The rule at g is an *uncritical* instance of ($E\rightarrow$): the principal formula is not of the form $v\rightarrow\Delta$. Again, Δ is only parametric.

4. The rule at g is a *semicritical* instance of ($E\rightarrow$): the principal formula is $v\rightarrow\Delta$, $v\neq\Delta$, and the sequent at g is $v\rightarrow\Delta, N \Longrightarrow \Delta$.

5. The rule at g is a *fully critical* instance of ($E\rightarrow$): the principal formula is $v\rightarrow\Delta$, $v\neq\Delta$, and the sequent at g is $v\rightarrow\Delta, N \Longrightarrow u$ with $u\neq\Delta$.

In both critical cases the multiplicity of Δ is risen by 1 in the antecedent of the right premiss.

LEMMA 1 There is an operator $\mathbf{D}^c$ defined for K_tM_1- and K_tJ_1-derivations D of semigood endsequents such that $\mathbf{D}^c(D)$ is a K_tM_1- or K_tJ_1-derivation of the same endsequent without fully critical instances of ($E\rightarrow$). If d is the length of D then the length of $\mathbf{D}^c(D)$ is less than 2^d.

I consider derivations starting from generalized axioms and define the operator $\mathbf{D}^c$ by recursion on the length d of D. I set $\mathbf{D}^c(D) = D$ if $d = 0$ or if $d = 1$ and if the one rule of D is not a fully critical instance of ($E\rightarrow$). If it is fully critical then the right premiss has the form

$$\Delta,N \Longrightarrow u \ , \ u\neq\Delta \quad \text{with} \quad N = \Delta,\dots,\Delta,N_0 \text{ and } \Delta \text{ not in } N_0$$

and is a generalized axiom. Thus u is in N_0 and also the endsequent $v\rightarrow\Delta,N \Longrightarrow u$ is a generalized axiom; let $\mathbf{D}^c(D)$ be the derivation consisting of this generalized axiom. If $d>1$ then let (R) be the last instance of a logical rule ending D; let D', D" be the subderivations leading to its premisses. If (R) is not a fully critical instance of ($E\rightarrow$) then I continue $\mathbf{D}^c(D')$ and $\mathbf{D}^c(D'')$ with (R) to obtain $\mathbf{D}^c(D)$. If the lengths of $\mathbf{D}^c(D')$, $\mathbf{D}^c(D'')$ are less than 2^{d-1} then that of $\mathbf{D}^c(D)$ is less than $1+2^{d-1} < 2^d$.

Assume now that (R) is a fully critical instance of ($E\rightarrow$)

$$\frac{M \Longrightarrow v \qquad \Delta,M \Longrightarrow u}{v\rightarrow\Delta, M \Longrightarrow u} \qquad u\neq\Delta$$

where the premisses shall already have the derivations $\mathbf{D}^c(D')$, $\mathbf{D}^c(D'')$. If Δ is parametric, i.e. occurs in M, then $\mathbf{D}^c(D'')$ followed by a contraction of Δ and a weakening with $v\rightarrow\Delta$ gives the derivation $\mathbf{D}^c(D)$ without changing the length. So there remains the case that Δ does not occur in M.

I abbreviate $R = D^c(D'')$ and write f for the root of R. I define the *pretty* nodes of R as follows:

f is pretty.

if g is pretty and the situation of types 2 or 3 prevails for the upper neighbours of g then these are pretty.

if g is pretty and carries a semicritical instance of (E⥽) then the left upper neighbour of g is pretty.

Thus every pretty node carries a sequent with Δ in its antecedent.

I transform R into R^+ by adding v⥽Δ and M to the antecedents of all its sequents (this would be superfluous for K_uM_0) such that f now carries $v \rightarrow \Delta, M, \Delta, M \Longrightarrow u$. R^+ is a derivation starting from generalized axioms. I transform R^+ into R^{+-} by removing, at all pretty nodes, all copies of Δ from the antecedents of their sequents; thus f now carries $v \rightarrow \Delta, M, M \Longrightarrow u$.

If g was not pretty then nothing has been changed in the sequents at g and its upper neighbours, and thus the rules from R remain in effect between them. If g is pretty in the situations of type 2 or 3 then logical rules and (AINΔ) remain in effect as remain contractions and weakenings which do not concern Δ. For all other pretty nodes g, I change R^{+-} into $R^{\S}$ by performing the following minor changes:

Instances of (RP) are assigned appropriately re-defined permutations.

Instances of (W) and (RC) which concern Δ are omitted.

Semicritical instances of (E⥽) in R have the form

$$\frac{\Delta, \ldots, \Delta, N_0 \Longrightarrow v_0 \qquad \Delta, \Delta, \ldots, \Delta, N_0 \Longrightarrow u_0}{v_0 \rightarrow \Delta, \Delta, \ldots, \Delta, N_0 \Longrightarrow u_0}$$

and in R^{+-} are transformed into

$$\frac{v \rightarrow \Delta, M, N_0 \Longrightarrow v_0 \qquad \Delta, \Delta, \ldots, \Delta, v \rightarrow \Delta, M, N_0 \longrightarrow u_0}{v_0 \rightarrow \Delta, v \rightarrow \Delta, M, N_0 \Longrightarrow u_0}$$

Inserting contractions, the right premiss becomes $\Delta, v \rightarrow \Delta, M, N_0 \Longrightarrow u_0$, and so in $R^{\S}$ another semicritical instance of (E⥽) will lead to the transformed conclusion.

It follows that $R^{\S}$ has the properties of a KM_0- or KJ_0-derivation at all non maximal nodes and at all maximal non-pretty nodes. It remains to consider the situation at the maximal pretty nodes g.

Such a node g carries in R^+ a generalized axiom $\Delta, \ldots, \Delta, A_0 \Longrightarrow a$ with Δ not in A_0. In R^{+-} and $R^{\S}$ this becomes $A_0 \Longrightarrow a$, and if $a \neq \Delta$ then this remains an axiom. If $a = \Delta$ then observe that, by construction of R^+, M is a subsequence of A_0 since Δ is not in M. The endsequent of $D^c(D')$ is $M \Longrightarrow v$; applying weakenings I obtain a derivation L_0 ending with $A_0 \Longrightarrow v$. Using

L_0 as left premiss and the axiom $\Delta, A_0 \Longrightarrow \Delta$ as right premiss, an uncritical instance of (E→) produces a derivation L_1 of $v{\to}\Delta, A_0 \Longrightarrow \Delta$. Again by construction of R^+, also $v{\to}\Delta$ will occur in A_0, and so a contraction gives a derivation L_g of $A \Longrightarrow \Delta$ which is my sequent at g in $R^§$.

I define R^* as the derivation obtained from $R^§$ by implanting the derivations L_g at the maximal pretty nodes g . R^* then ends at f with the same sequent $v{\to}\Delta, M, M \Longrightarrow u$ which ended R^{+-} and $R^§$. I define $\mathbf{D}^c(D)$ by prolonging R^* with contractions leading to the sequent $v{\to}\Delta, M \Longrightarrow u$.

It follows from the properties of $R = \mathbf{D}^c(D'')$ and $\mathbf{D}^c(D')$ that R^* and $\mathbf{D}^c(D)$ do not contain fully critical instances of (E→). The lengths of $R = \mathbf{D}^c(D'')$ and of R^+ are less than 2^{d-1}. The structural rules used to define $R^§$ from R^{+-} do not change lengths; hence also the length of $R^§$ is less than 2^{d-1}. The length of $\mathbf{D}^c(D')$ and of L_0 is less than 2^{d-1}, and the lengths of L_1 and L_g are less than or equal to 2^{d-1}. Implanting the L_g enlarges the length of $R^§$ at most by $2^{d-1}-1$, and thus the lengths of R^* and $\mathbf{D}^c(D)$ will be less than 2^d. This concludes the proof of Lemma 1.

LEMMA 2 The cut rule is admissible for KM_1 and KJ_1, and the cut elimination algorithms of Chapter **2** can be expanded so as to work with the same bounds.

I first redefine the exchange operator **E**: there shall hold $\mathbf{E}(D) = D$ also in the case that D is 1−extreme, the cut formula is Δ, and the left premiss of the cut is the conclusion of (EΔ¬). In all other situations I keep the definition of **E**, expanded for the connectives ¬ and Δ as described in the additions to the cases E1, E2. E3 in the definition of the operator $\mathbf{E}_1$; observe that in the case E1 the rule (EΔ¬) does not need to be considered.

Next I redefine the operator $\mathbf{A}_0$, preserving its property **2**.(A0), for the case that D is 1−extreme, essentially ending with a cut with the cut formula Δ and of weight (a,b). If $a \cdot b = 0$ then I argue as before in Chapter **2**. Assume now $a > 0$ and let (R) be the rule which has the left premiss as conclusion. If (R) is not (EΔ¬) then the operator **E** can be applied; **E**(D) contains one or two 1−cuts with Δ of weights (c,d) with $c \leq a-1$, $d \leq b$ such that the subderivations E', E'' leading to them are 1−extreme; hence $\mathbf{A}_0(D)$ can be defined as before. There remains the case that (R) is (EΔ¬) :

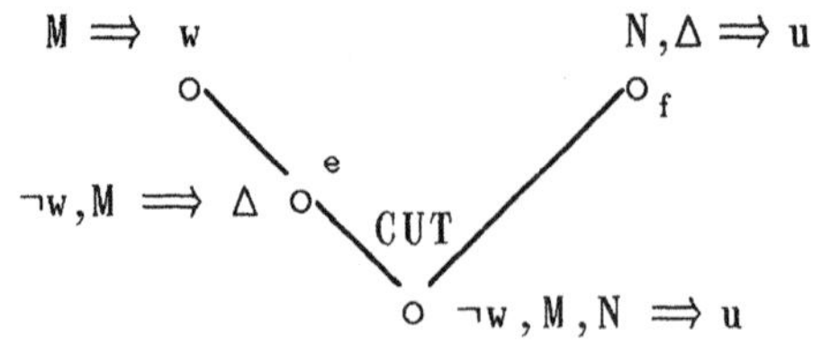

Let T be the subderivation leading to the premisses at f. I transform T into T^+ by adding ¬w and M to the antecedents of all its sequents. I call *pretty* the nodes g of T^+ with $f<g$ such that (1) Δ occurs in the antecedent of the sequent at g and (2) if $f<h\leq g$ then the sequent at h is *not* premiss of an instance of a logical rule with Δ as its left side formula. I transform T^+ into T^{+-} by removing Δ from the antecedents of the sequents at f and at pretty nodes; thus f again now carries the sequent $\neg w, M, N \Longrightarrow u$. I then expand T^{+-} into a derivation $\mathbf{A}_0(D)$ of this endsequent by the method employed in the proof of Lemma 1. Weakenings which introduced Δ at pretty nodes g I omit, as well as contractions concerning Δ, and permutations I simplify accordingly. Instances of logical rules remain preserved: they cannot have introduced Δ, and as a left side formula it had not been used. A maximal pretty g carries in T^+ a generalized axiom $\Delta, A \Longrightarrow a$ where I may assume that Δ does not occur in A. In T^{+-} this has been changed to $A \Longrightarrow a$, and if $a \neq \Delta$ then it remains an axiom. If $a = \Delta$ then, by construction of T^+, both $\neg w$ and M occur within A. Let L be the subderivation ending at the node e; applying weakenings I obtain a derivation L_g ending with $A \Longrightarrow \Delta$. Implanting the L_g, I obtain the derivation $\mathbf{A}_0(D)$.

Since D was 1-extreme, $\mathbf{A}_0(D)$ does not contain cuts anymore. The length of T and T^+ is b, namely the depth of f; the length of L and L_g is a, and so the length of $\mathbf{A}_0(D)$ here is at most $a+b$.

If it is not the case that D is 1-extreme and essentially ends with a cut with the cut formula Δ, then I define $\mathbf{A}_0(D)$ as before in Chapter **2** with help of the expanded operator **E**. Since the cut formula is different from Δ, no shift of a cut over the rule $(E\Delta\neg)$ occurs in Case 1. and the earlier definition covers all situations. This concludes the proof of Lemma 2.

As an application of the admissibility of (CUT) I observe that the deductive principle (db) is admissible in KM_1:

$$\frac{\dfrac{v, M \Longrightarrow w}{\neg w, v, M \Longrightarrow \Delta} \qquad M \Longrightarrow \neg w}{\dfrac{v, M \Longrightarrow \Delta}{M \Longrightarrow \neg v}}$$

Similarly, the translation of $(E\rightarrow)$ is admissible

$$\frac{M \Longrightarrow v \qquad M, \Delta \Longrightarrow u}{M, \neg v \Longrightarrow u}$$

4. The Calculi LM and LJ

I consider a propositional language $Fm(\neg)$ with the connectives $\wedge$, $\vee$, $\rightarrow$ and $\neg$. The sequents used so far I shall call K-sequents. I define a new kind of sequents, called L-*sequents:* every K-sequent is an L-sequent, and if M is a finite (possibly empty) sequence or set of formulas then also $M \Longrightarrow$ (without a formula on the right) shall be an L_s– or L_t–sequent. In order to improve readability, I shall write $M \Longrightarrow \blacktriangle$ instead of $M \Longrightarrow$.

I define the calculi L_xM of L–sequents with the same axioms as are used for K_xP. Their rules shall, first, be the rules of K_xP (where in the E–rules the formula u may not be present) and with the further logical rules

$$(I\neg)\quad \frac{v, M \Longrightarrow \blacktriangle}{M \Longrightarrow \neg v} \qquad (E\neg)\quad \frac{M \Longrightarrow v}{\neg v, M \Longrightarrow \blacktriangle} \text{ or } \frac{\neg v, M \Longrightarrow v}{\neg v, M \Longrightarrow \blacktriangle} \text{ for } L_uM$$

with $\neg v$ as principal formula and v as side formula. I define the calculus L_xJ by adding to L_xM the rule

$$(AIN)\quad \frac{M \Longrightarrow \blacktriangle}{M \Longrightarrow v}$$

with v as weakening formula; its instances are counted when defining the lengths of derivations. Again, the calculi LM are those of *minimal logic,* the calculi LJ those of *intuitionistic logic.*

The basic mechanics of derivations developed in Chapter 1 work as well for LM and LJ. The reflexivity axioms are derivable as in KM_1, and the Lemmata 1.1–3 remain in effect. The calculi LM and LJ have the subformula property and the property of directness, and the decision method of Chapter 1 remains in effect. In $(I\neg)$ and (AIN) the right occurrence of $\neg v$ respectively v does not correspond to an occurrence in the premiss, and for L_tM the principal occurrence of $\neg v$ does not correspond to an occurrence in the premiss; for L_uM the principal occurrence of $\neg v$ in the conclusion corresponds to that in the premiss. The notion of predecessors and that of the parameter tree then is defined as before.

In view of (AIN) the rule

$$(E\neg N)\quad \frac{M \Longrightarrow v}{\neg v, M \Longrightarrow [q]} \text{ or } \frac{\neg v, M \Longrightarrow v}{\neg v, M \Longrightarrow [q]} \text{ for } L_uJ\ , \quad q \text{ a formula or } \blacktriangle$$

is admissible in LJ; an instance of $(E\neg N)$ shall be *real* if q actually is a formula.

LEMMA 3 LJ derives the same sequents as the L–calculus with the rules from KP and the rules $(I\neg)$ and $(E\neg N)$.

Consider first an instance of (AIN) at some node f_0 of a derivation D. Then the upper neighbour f of f_0 carries a sequent $N \Longrightarrow \blacktriangle$, and in the subderivation D_f ending at f I can define the parameter tree P of this right occurrence of $\blacktriangle$. Then the maximal nodes of P are instances of $(E\neg)$. Consider now an instance of $(E\neg)$ at some node e of D carrying a sequent $N \Longrightarrow \blacktriangle$ and assume that below e there still occur nodes with sequents with formulas on their right. If f_0 is the largest of them, carrying a sequent $N \Longrightarrow u$, then an instance of (AIN) occurs at f_0; let f be the upper neigbour of f_0. In the parameter tree P of D_f I replace $\blacktriangle$ everywhere by u, replace the instances of (E$\neg$) at the maximal nodes of P by instances of (E$\neg$N) with q, and finally remove the indication of an instance of (AIN) at f. Then I obtain a derivation of the same endsequent in which the numbers of instances of (E$\neg$) and of (AIN) both have been lowered. Repeating this construction a finite number of times, I can remove all instances of (AIN) and replace all instances of (E$\neg$) by such of (E$\neg$N); all instances of (E$\neg$N) at nodes e will be real unless all nodes below e will carrying sequents with empty right sides.

Concerning the inversion rules of Chapter **1**, their proofs remain valid for the minimal calculi LM; for **JI**$\vee$ the definition of Harrop formulas is extended by by defining $\neg v$ as Harrop for any v. In the intuitionistic case, let D be a derivation of $M \Longrightarrow u$ with $u = p \rightarrow q$, $u = p \wedge q$ or $u = p \vee q$, and consider a maximal node e of the parameter tree of the right occurrence of u. If e carries $N_e \Longrightarrow u$ then it may happen that this sequent arose under (AIN) from $N_e \Longrightarrow \blacktriangle$ at an upper neighbour of e. But then (AIN) also produces $N_e \Longrightarrow p$ and $N_e \Longrightarrow q$, and so the inversion rules remain in effect for LJ with appropriately amended proofs. For the rule (I$\neg$) the argument used for (JI$\rightarrow$) gives for both LM and LJ the inversion rule

(JI$\neg$) There is an operator **JI**$\neg$ transforming any derivation D of $M \Longrightarrow \neg p$ into a derivation **JI**$\neg$(D) of $p, M \Longrightarrow \blacktriangle$ which has at most the length of D .

An inversion rule for (E$\neg$) does not hold in LM or LJ since $\neg\neg x, \neg x \Longrightarrow \blacktriangle$ can be derived from $\neg x \Longrightarrow \neg x$, but $\neg\neg x \Longrightarrow x$ cannot. There hold, however, a number of *extended* inversion rules which I collect in the following

LEMMA 4 There are operators, transforming LJ- and LJC-derivations with endsequents listed in the left column into LJ- or LJC-derivations with endsequents in the right column without enlarging their lenghts:

JS$\rightarrow$	$v \rightarrow w, M \Longrightarrow u$	$\neg v, M \Longrightarrow u$
JS$\vee$**L**	$M \Longrightarrow v \vee w$	$\neg w, M \Longrightarrow v$
JS$\vee$**R**	$M \Longrightarrow v \vee w$	$\neg v, M \Longrightarrow w$
JS$\neg\rightarrow$	$\neg(v \rightarrow w), M \Longrightarrow u$	$v, \neg w, M \Longrightarrow u$
JS$\neg\wedge$**L**	$\neg(v \wedge w), M \Longrightarrow u$	$\neg v, M \Longrightarrow u$
JS$\neg\wedge$**R**	$\neg(v \wedge w), M \Longrightarrow u$	$\neg w, M \Longrightarrow u$
JS$\neg\vee$	$\neg(v \vee w), M \Longrightarrow u$	$\neg v, \neg w, M \Longrightarrow u$
JS$\neg\neg$	$\neg\neg v, M \Longrightarrow u$	$v, M \Longrightarrow u$

I consider again the calculus L_uJ. For **JS**$\rightarrow$ I replace in a derivation D of $v \rightarrow w, M \Longrightarrow u$ the predecessors of $v \rightarrow w$ by $\neg v$, and remove at instances of (E$\rightarrow$) with a principal predecessor the entire right branch. Then (E$\neg$) leads from the transformed left premiss to the transformed conclusion.

For **JS∨L** I consider the parameter tree P of v∨w in a derivation D of M ⟹ v∨w. Then the maximal nodes of P carry instances of (I∨) with premisses either N ⟹ v or N ⟹ w . Replace v∨w in P by v and weaken all antecedents in D with ¬w.. Then in the first case those premisses coincide with their conclusion, and in the second case an instance of (E¬N) leads from ¬w, N ⟹ w to ¬w, N ⟹ v .

For **JS¬→** let D be a derivation ¬(v→w), M ⟹ u and let f_0 be a node which is maximal for carrying a sequent with a principal predecessor of this occurrence of ¬(v→w); thus f_0 carries a sequent ¬(v→w), N ⟹ ▲ . Let f be the upper neighbour of f_0 carrying ¬(v→w), N ⟹ v→w. The subderivation D_f leading to f determines the derivation D_f' = **JI→**(D_f) of v, ¬(v→w), N ⟹ w . I now replace D_f by D_f' and then weaken (1) with v all antecedents in the preserved part of D and (2) with ¬w all antecedents, both in D_f' and in the preserved part of D; let D' be the resulting object. Both the (weakened) D_f' and the (weakened) remaining part of D locally remain derivations, and f, belonging to D_f', now carries ¬w, v, ¬(v→w), N ⟹ w while f_0, belonging to the preserved part, carries ¬w, v, ¬(v→w), N ⟹ ▲ . Hence an instance of (E¬) leads from f to f_0 such that D' itself is a derivation, now of the endsequent ¬w, v, ¬(v→w), M ⟹ u . But at f now the occurrence of ¬(v→w) has became parametric, and thus the number of principal predecessors of ¬(v→w) has decreased in D'. If there are more of them in D', I take the next maximal one and produce in the same manner a derivation of ¬w, v, ¬w, v, ¬(v→w), M ⟹ u which, due to the admissibility of (RC) in L_uJ, can be simplified into one of ¬w, v, ¬(v→w), M ⟹ u again. Repeating this a finite number of steps, I arrive at a derivation of this sequent in which no predecessor of the first ¬(v→w) is principal, and this now will remain a derivation of ¬w, v, M ⟹ u if ¬(v→w) and all its predecessors are omitted.

For **JS¬∧** , **JS¬∨** and **JS¬¬** I use the same argument, considering for **JS¬∧L** a maximal principal predecessor of v∧w coming in ¬(v∧w), N ⟹ ▲ from ¬(v∧w), N ⟹ v∧w . I replace D_f by **JI∧L**(D_f) ending with ¬(v∧w), N ⟹ v and then I weaken the antecedents everywhere with ¬v obtaining D'. The step from f to f_0 becomes an instance of (E¬), D' is a derivation ending with ¬v, ¬(v∧w), M ⟹ u in which the number of principal predecessors of ¬(v∧w) has decreased.

For **JS¬∨** I cannot use inversion but have to look at the parameter tree P_f in D_f of ¬(v∨w), N ⟹ v∨w ; its maximal nodes e with ¬(v∨w), H ⟹ v∨w come either from ¬(v∨w), N ⟹ v or ¬(v∨w), N ⟹ w . Replacing in P_f the right occurrences of v∨w by ▲ and weakening all antecedents with ¬v, ¬w, the nodes f_0 and f then carry the same sequents, while the sequents at nodes e arise from those at their upper neighbours by (E¬). Thus I have again a derivation D' of ¬v, ¬w, ¬(v∨w), M ⟹ u in which the the number of principal predecessors of ¬(v∧w) has decreased.

For **JS¬¬** I replace a derivation D_f of ¬¬v, N ⟹ ¬v at f by the derivation **JI¬**(D_f) of v, ¬¬v, N ⟹ ▲ ; D' then derives v, ¬¬v, M ⟹ u .

It will be noticed that only the construction of **JS**∨ makes use of (AIN); all the other operators work already for LM.

The cut rule is admissible for LM and LJ, and the cut elimination algorithms of Chapter **2** remain in effect; the necessary additions to the cases E1, E2, E3 in the definition of the operator $\mathbf{E}_1$ were discussed already there with sequents $M \Longrightarrow \blacktriangle$ written as $M \Longrightarrow \Delta$. The modifications required for explicit retracing in L_uMC are similarly simple. If in the proof of Lemma **3**.1 the sequent J(e) arises from J(e'), J(e") under (E ¬) with the omitted occurrence of $v = \neg p$ as principal

$$\begin{array}{llll} J(e'): & \neg p, A \Longrightarrow v & \qquad J^*(e'): & A \Longrightarrow p \\ J(e): & \neg p, A \Longrightarrow \blacktriangle & \qquad J^*(e): & A \Longrightarrow \blacktriangle \end{array}$$

then I define $H_{k}'{}_e$ by taking a copy of H_k' with H(k'): p,M, A/M $\Longrightarrow \blacktriangle$, obtained by writing A/M to the right sides of all sequents in H_k', and continue with the permutation producing $p,A \Longrightarrow \blacktriangle$. I define J_e by

$$\dfrac{\dfrac{J_e'}{A \Rightarrow p} \qquad \dfrac{H_k{}'_e}{p, A \Rightarrow \blacktriangle}}{A \Longrightarrow \blacktriangle} \quad .$$

Here H_k' and $H_k{}'_e$ have at most length n−1, whence again *length*$(J_e) \leq n-1 + 2 \cdot |e|$. Thus Lemma **3**.1 remains in effect, and Lemma **3**.2 requires no change at all. The modifications required for ascending cut elimination are analogous.

For the reader's exercise I present some derivations which will become useful in later Chapters. First derivations in LM:

$$\dfrac{\begin{array}{r} x \Longrightarrow x \\ x \wedge y \Longrightarrow x \\ \neg x, x \wedge y \Longrightarrow \blacktriangle \\ \neg x \Longrightarrow \neg(x \wedge y) \\ \neg\neg(x \wedge y) \Longrightarrow \neg\neg x \end{array} \qquad \begin{array}{r} y \Longrightarrow y \\ x \wedge y \Longrightarrow y \\ \neg y, x \wedge y \Longrightarrow \blacktriangle \\ \neg y \Longrightarrow \neg(x \wedge y) \\ \neg\neg(x \wedge y) \Longrightarrow \neg\neg y \end{array}}{\neg\neg(x \wedge y) \Longrightarrow \neg\neg x \wedge \neg\neg y} \tag{1}$$

$$\begin{array}{r} \dfrac{x, y \Rightarrow x \qquad x, y \Rightarrow y}{x, y \Longrightarrow x \wedge y} \\ \neg(x \wedge y), x, y \Longrightarrow \blacktriangle \\ \neg(x \wedge y), y \Longrightarrow \neg x \\ \neg\neg x, \neg(x \wedge y), y \Longrightarrow \blacktriangle \\ \neg\neg x, \neg\neg y, \neg(x \wedge y) \Longrightarrow \blacktriangle \\ \neg\neg x, \neg\neg y \Longrightarrow \neg\neg(x \wedge y) \\ \neg\neg x \wedge \neg\neg y, \neg\neg y \Longrightarrow \neg\neg(x \wedge y) \\ \neg\neg x \wedge \neg\neg y, \neg\neg x \wedge \neg\neg y \Longrightarrow \neg\neg(x \wedge y) \\ \neg\neg x \wedge \neg\neg y \Longrightarrow \neg\neg(x \wedge y) \end{array} \tag{2}$$

$$
\begin{array}{lcr}
x \Rightarrow x & & y \Rightarrow y \\
x \Rightarrow x \vee y & & y \Rightarrow x \vee y \\
\neg(x \vee y) \Rightarrow \neg x & & \neg(x \vee y) \Rightarrow \neg y \\
& \searrow \ \swarrow & \\
& \neg(x \vee y) \Rightarrow \neg x \wedge \neg y & \\
(3) & \neg(\neg x \wedge \neg y) \Rightarrow \neg\neg(x \vee y) &
\end{array}
$$

$$
\begin{array}{lcr}
x \Rightarrow x & & y \Rightarrow y \\
x, \neg x \Rightarrow \blacktriangle & & y, \neg y \Rightarrow \blacktriangle \\
x, \neg x \wedge \neg y \Rightarrow \blacktriangle & & y, \neg x \wedge \neg y \Rightarrow \blacktriangle \\
& \searrow \ \swarrow & \\
& x \vee y, \neg x \wedge \neg y \Rightarrow \blacktriangle & \\
& \neg x \wedge \neg y \Rightarrow \neg(x \vee y) & \\
& \neg x \wedge \neg y, \neg\neg(x \vee y) \Rightarrow \blacktriangle & \\
(4) & \neg\neg(x \vee y) \Rightarrow \neg(\neg x \wedge \neg y) &
\end{array}
$$

$$
\begin{array}{lr}
& \neg w \Rightarrow \neg w \\
& \neg w, \neg\neg w \Rightarrow \blacktriangle \\
& \neg w, \neg w \wedge \neg\neg w \Rightarrow \blacktriangle \\
& \neg w \wedge \neg\neg w, \neg w \wedge \neg\neg w \Rightarrow \blacktriangle \\
& \neg w \wedge \neg\neg w \Rightarrow \blacktriangle \\
(5) & \blacktriangle \Rightarrow \neg(\neg w \wedge \neg\neg w)
\end{array}
$$

$$
\begin{array}{lcr}
& & y \Rightarrow y \\
x \Rightarrow x & & \neg y, y \Rightarrow \blacktriangle \\
& \searrow \ \swarrow & \\
& x, \neg y, x \to y \Rightarrow \blacktriangle & \\
& \neg\neg x, \neg y, x \to y \Rightarrow \blacktriangle & \\
& \neg\neg x, \neg y \Rightarrow \neg(x \to y) & \\
& \neg\neg x, \neg y, \neg\neg(x \to y) \Rightarrow \blacktriangle & \\
& \neg\neg x, \neg\neg(x \to y) \Rightarrow \neg\neg y & \\
(6) & \neg\neg(x \to y) \Rightarrow \neg\neg x \to \neg\neg y &
\end{array}
$$

and now derivations in LJ

$$
\begin{array}{lcr}
& & x, y \Rightarrow y \\
x \Rightarrow x & & x, y \Rightarrow x \to y \\
\neg x, x \Rightarrow \blacktriangle & & x, y, \neg(x \to y) \Rightarrow \blacktriangle \\
x \Rightarrow \neg\neg x & & x, \neg\neg y, \neg(x \to y) \Rightarrow \blacktriangle \\
& \searrow \ \swarrow & \\
& x, \neg(x \to y), \neg\neg x \to \neg\neg y \Rightarrow \blacktriangle & \\
& x, \neg(x \to y), \neg\neg x \to \neg\neg y \Rightarrow y & \\
& \neg(x \to y), \neg\neg x \to \neg\neg y \Rightarrow x \to y & \\
& \neg(x \to y), \neg\neg x \to \neg\neg y \Rightarrow \blacktriangle & \\
(7) & \neg\neg x \to \neg\neg y \Rightarrow \neg\neg(x \to y) \quad . &
\end{array}
$$

$$
\begin{array}{rl}
\neg\neg x, x & \Longrightarrow x \\
x & \Longrightarrow \neg\neg x \to x \\
x, \neg(\neg\neg x \to x) & \Longrightarrow \blacktriangle \\
\neg(\neg\neg x \to x) & \Longrightarrow \neg x \\
\neg\neg x, \neg(\neg\neg x \to x) & \Longrightarrow \blacktriangle \\
\neg\neg x, \neg(\neg\neg x \to x) & \Longrightarrow x \\
\neg(\neg\neg x \to x) & \Longrightarrow \neg\neg x \to x \\
\neg(\neg\neg x \to x) & \Longrightarrow \blacktriangle \\
(8) \qquad & \Longrightarrow \neg\neg(\neg\neg x \to x)
\end{array}
$$

The calculi LM and LJ are, in a way, specializations of KM_1 and KJ_1 to the case of formulas not containing Δ. Let me define a K-sequent $M \Longrightarrow v$ to be *very good* if Δ does not occur in the formulas of M and if the formula v is either Δ or Δ does not occur in v.

LEMMA 5 Every LM- or LJ-derivation becomes a KM_1- or KJ_1-derivation if sequents $M \Longrightarrow \blacktriangle$ are written as $M \Longrightarrow \Delta$.

Every KM_1- or KJ_1-derivation of a very good endsequent becomes a LM- or LJ-derivation if sequents $M \Longrightarrow \Delta$ are written as $M \Longrightarrow \blacktriangle$.

The first statement is obvious. Consider now a KM_1- or KJ_1-derivation D of a very good endsequent. I claim that every node e of D is very good, i.e. carries a very good sequent. For the root of D this holds by hypothesis. If f is an upper neighbour of e and the rule (R) leading to e is structural or is (AINΔ) then together with e also f is very good. If (R) is logical and is not ($I\Delta\neg$) then a side formula Δ would produce a principal formula containing Δ. If (R) is ($I\Delta\neg$) and $v = \neg w$ then f carries $w, M \Longrightarrow \Delta$. Thus together with e also f is always very good. - In particular, the maximal nodes of D are very good, hence cannot carry axioms $\Delta \Longrightarrow \Delta$. This proves the second statement.

For practical purposes, the calculi LM and LJ are the most convenient ones. In particular, their cut elimination algorithm is the straightforward extension of that for KP, whereas the algorithms for KM_1 and KJ_1 require the additional considerations developed in Lemma 2. This should be no surprise as $\blacktriangle$ is nothing while Δ is a formula.

The expressive power of KM_1, KJ_1 using $Fm(\Delta,\neg)$ may appear to be larger than that of LM, LJ using $Fm(\neg)$, as well as larger than that of KM_0, KJ_0 using $Fm(\Delta)$. However, these calculi can be translated into each other as follows.

I define a map ρ sending $Fm(\Delta,\neg)$-formulas v into $Fm(\Delta)$-formulas v^+: ρ shall be the identity on the set of variables, shall be homomorphic for the positive connectives $\wedge$, $\vee$, $\to$, and there shall hold

$$(\neg v)^+ = v^+ \to \Delta \quad \text{and} \quad \Delta^+ = \Delta \ .$$

In particular, ρ also maps Fm($\neg$) into Fm(Δ), and the image of Fm($\neg$) is the set of good Fm(Δ)-formulas; restricted to Fm(Δ) it is the identity. For sequences (or sets) M of formulas I write M^+ for the sequences of all m^+ with m in M. A K-sequent $M \Longrightarrow u$ or an L-sequent $M \Longrightarrow u$ with $u \neq \Delta$ shall have as *ρ-image* the K-sequent $M^+ \Longrightarrow u^+$, an L-sequent $M \Longrightarrow \blacktriangle$ shall have the ρ-image $M^+ \Longrightarrow \Delta$.

LEMMA 6 Let D be a KM_1-, LM-, KJ_1- or LJ-derivation. Replace all its sequents by their ρ-images. Then the resulting object is a KM_0- or KJ_0-derivation of the ρ-image of the endsequent of D.

Observe first that ρ maps KM_1-axioms and LM-axioms into KM_0-axioms. Also, instances of structural rules are preserved, and instances of logical rules for positive connectives are preserved because ρ, being homomorphic, maps the side formulas of a principal formulas v into the side formulas of v^+. Instances of (I$\Delta\neg$), (I$\neg$) and (E$\Delta\neg$), (E$\neg$) are replaced by those of (I$\to$) and (E$\to$) :

$$\frac{v, M \Longrightarrow \Delta}{M \Longrightarrow \neg v} \quad \text{becomes} \quad \frac{v^+, M^+ \Longrightarrow \Delta}{M^+ \Longrightarrow v^+ \to \Delta} \ ,$$

$$\frac{M \Longrightarrow v}{\neg v, M \Longrightarrow \Delta} \quad \text{becomes} \quad \frac{M^+ \Longrightarrow v^+ \qquad M^+, \Delta \Longrightarrow \Delta}{v^+ \to \Delta, M^+ \Longrightarrow \Delta} \ .$$

I next wish to define a map σ sending certain Fm($\Delta,\neg$)-formulas v into Fm($\neg$)-formulas v_+. This map is defined for good Fm($\Delta,\neg$)-formulas: it shall be the identity on the set of variables, shall be homomorphic for the connectives $\wedge$, $\vee$, $\neg$, and there shall hold

$$(v \to w)_+ = v_+ \to w_+ \ \text{ if } w \neq \Delta \ , \quad \text{and} \quad (w \to \Delta)_+ = \neg v_+ \ .$$

In particular, σ also maps Fm(Δ) into Fm($\neg$), and restricted to Fm($\neg$) it is the identity. For sequences (or sets) M of formulas I write M_+ for the sequences of all m_+ with m in M. A good K-sequent $M \Longrightarrow u$ with $u \neq \Delta$ shall have as *σ-image* the L-sequent $M_+ \Longrightarrow u_+$, and if $u = \Delta$ then it shall have the σ-image $M_+ \Longrightarrow \blacktriangle$.

For the following Lemma, observe again that a KM_0- or KJ_0-derivation is in particular a KM_1- or KJ_1-derivation.

LEMMA 7 There is an operator $\mathbf{D}_+$ defined for KM_{0t}- and KJ_{0t}-derivations D of good endsequents such that $\mathbf{D}_+(D)$ is an L_tM- or L_tJ-derivation of the σ-image of the endsequent of D. If d is the length of D then the length of $\mathbf{D}_+(D)$ is less than 2^d.

Since D has a good endsequent, I can form the derivation $D_1 = \mathbf{D}^c(D)$. Let D_2 be obtained from D_1 by replacing all good sequents of D_1 by their σ-images and omitting all non-good sequents. I claim that D_2 is an L_tM- or L_tJ-derivation. If $M \Longrightarrow u$ in D_1 is a good generalized axiom $M \Longrightarrow u$ then M is good, and since u occurs in M also u is good, hence distinct from Δ. Thus $M_+ \Longrightarrow u_+$ in D_1 is an LM-axiom. Instances of structural rules and of rules for $\wedge$, $\vee$ and $\neg$ have the property that good conclusions come from good premisses; hence they remain in effect between the σ-images in D_2. If (AINΔ) in KJ_{ot} produced a good $M \Longrightarrow u$ from $M \Longrightarrow \Delta$ then also this premiss is good, and thus (AIN) in LJ leads from $M_+ \Longrightarrow$ to $M_+ \Longrightarrow u_+$. If a sequent $M \Longrightarrow v \dashv w$ arises by (I$\dashv$) from $v, M \Longrightarrow w$ and is good, then this premiss is good; hence the σ-images form an instance of (I$\dashv$) if $w \neq \Delta$ and an instance of (I$\neg$) if $w = \Delta$. Consider now a good K-sequent $v \dashv w, M \Longrightarrow u$ obtained with (E$\dashv$) from $M \Longrightarrow v$ and $w, M \Longrightarrow u$. Then the first premiss is good also; the left will be good if $w \neq \Delta$, and in this uncritical case I can apply (E$\dashv$) to the σ-images. In the semicritical case $w = \Delta$, $u = \Delta$ the good left premiss has the σ-image $M_+ \Longrightarrow v_+$ from where (E$\neg$) leads to the image $\neg v_+, M_+ \Longrightarrow$ of the conclusion. Fully critical instances of (E$\dashv$) do not occur in D_1. [In the case of KJ_{ot} the operator $\mathbf{D}^c$ is actually not needed since fully critical instances of (E$\dashv$) can be treated directly: the good left premiss again gives $\neg v_+, M_+ \Longrightarrow$ and thus $\neg v_+, M_+ \Longrightarrow u_+$ by (AIN).]

5. K-Calculi for the Connective $\neg$

The use of L-sequents makes it possible to formulate minimal and intuitionistic calculi for the language Fm($\neg$) in a convenient manner. Still, it possible to formulate such calculi also with K-sequents. For the sake of convenience, I restrict myself to the case of t-sequents.

I define the calculus KM_2 on Fm($\neg$). It has the same axioms as KM_0, has the rules of KP for positive connectives, and has the new logical rules

$$(P_0\neg)\quad \frac{v, M \Longrightarrow w}{\neg w, M \Longrightarrow \neg v} \qquad\qquad (P_1\neg)\quad \frac{v, M \Longrightarrow \neg w}{w, M \Longrightarrow \neg v} \quad .$$

I define KM_2C as the extension of KM_2 by a cut rule. It follows from $(P_0\neg)$ alone that the reflexive axioms can be derived in KM_2.

In KM_1 the rule $(P_0\neg)$ is admissible since to its premiss I first may apply (E$\Delta\neg$) and from that then obtain the conclusion by (I$\Delta\neg$). Also $(P_1\neg)$ is admissible since

$$\dfrac{\dfrac{\dfrac{w \Longrightarrow w}{\neg w, w \Longrightarrow \Delta}}{w \Longrightarrow \neg\neg w} \qquad \dfrac{v, M \Longrightarrow \neg w}{\neg\neg w, v, M \Longrightarrow \Delta}}{\dfrac{v, w, M \Longrightarrow \Delta}{w, M \Longrightarrow \Delta}}$$

making use of the admissibility of cuts in KM_1. Thus

LEMMA 8 Every KM_2-derivation can be transformed into a KM_1- and into a LM-derivation of its endsequent and into a KM_0-derivation of the ρ-image of its endsequent.

The construction of the KM_1-derivation requires cut elimination, and so the length may grow considerably. From that I obtain an LM-derivation by Lemma 5 and a KM_0-derivation by Lemma 6.

In order to conversely transform other derivations into KM_2C-derivations, it suffices to do this for KM_1-derivations, since LM-derivations can be transformed into these by Lemma 5 (and KM_0-derivations are KM_1-derivations). So I shall need a map τ from *all* of $Fm(\Delta,\neg)$ to $Fm(\neg)$ and the problem, obviously, is to find an image for Δ. Observe that in KM_1 the generalized axiom $t,\Delta \Longrightarrow \Delta$ gives $\Delta \Longrightarrow \neg t$ for every t; hence if t happens to be KM_1-*provable,* in the sense that $\blacktriangle \Longrightarrow t$ has a KM_1-derivation (where $\blacktriangle$ now symbolizes an empty antecedent), then $(E\Delta\neg)$ derives $\neg t \Longrightarrow \Delta$ such that Δ and t are interdeducible. This suggests to use as image of Δ a fixed formula $\neg t$ such that t is KM_1-provable, e.g. $t = x \rightarrow x$ for a fixed variable x. Writing v_- for the image under τ, I so define

τ is the identity on the set of variables, is homomorphic for the connectives $\wedge, \vee, \rightarrow, \neg$, and there holds $\Delta_- = \neg t$.

The τ-image of a sequent $M \Longrightarrow u$ shall again be $M_- \Longrightarrow u_-$.

LEMMA 9 Let D be KM_1-derivation. Replace all its sequences by their τ-images. Then the resulting object can be expanded into a KM_2C-derivation of the τ-image of the endsequent of D.

Observe first that the τ-images of KM_1-axioms $u,M \Longrightarrow u$ in KM_2 are either axioms again or, if $u = \Delta$, are derivable as $\neg t, M_- \Longrightarrow \neg t$. Instances of logical rules for positive connectives are preserved. Instances of $(I\Delta\neg)$ and $(E\Delta\neg)$ are transformed to

$$\frac{v_-, M_- \Longrightarrow t}{M_- \Longrightarrow \neg v_-} \quad \text{and} \quad \frac{M_- \Longrightarrow v_-}{\neg v_-, M_- \Longrightarrow \neg t}$$

and this can be replaced by

$$\dfrac{\blacktriangle \Longrightarrow t \qquad \dfrac{v_-, M_- \Longrightarrow \neg t}{t, M_- \Longrightarrow \neg v_-}\,(P_1\neg)}{M_- \Longrightarrow \neg v_-} \qquad\qquad \dfrac{\dfrac{M_- \Longrightarrow v_-}{t, M_- \Longrightarrow v_-}}{\neg v_-, M_- \Longrightarrow \neg t}\,(P_0\neg)$$

since t was provable and (CUT) is available in KM_2C.

There is a variant $KM_{21}C$ of KM_2C in which $(P_1\neg)$ is replaced by

$$(P_2\neg) \qquad \begin{array}{l} M \Longrightarrow v \\ M \Longrightarrow \neg\neg v \end{array} \quad .$$

This rule is admissible for KM_2C since

$$\begin{array}{ccc} & & \neg v \Longrightarrow \neg v \\ M \Longrightarrow v & & v \Longrightarrow \neg\neg v \quad (P_1\neg) \\ & \diagdown \quad \diagup & \\ & M \Longrightarrow \neg\neg v & \end{array} \quad .$$

Conversely, $(P_1\neg)$ is admissible for $KM_{21}C$:

$$\begin{array}{ccc} w \Longrightarrow w & & v, M \Longrightarrow \neg w \\ w \Longrightarrow \neg\neg w \quad (P_2\neg) & & \neg\neg w, M \Longrightarrow \neg v \quad (P_0\neg) \\ & \diagdown \quad \diagup & \\ & w, M \Longrightarrow \neg v & \end{array}$$

Thus KM_2C-derivations and $KM_{21}C$-derivations can be transformed into each other.

I define the calculus KJ_2 on $Fm(\neg)$ by adding to KM_2 the rule

$$(AIN\neg) \qquad \begin{array}{cc} M \Longrightarrow w & M \Longrightarrow \neg w \\ \multicolumn{2}{c}{M \Longrightarrow v} \end{array} \quad .$$

This calculus is not at all direct, nor does it have the subformula property. Lemma 8 remains in effect for KJ_2, KJ_1, LJ and KJ_0. And also Lemma 9 remains in effect for KJ_1 and KJ_2C, because $(AIN\neg)$ is admissible in KJ_1

$$\begin{array}{ccc} & & w \Longrightarrow w \\ M \Longrightarrow w \quad M \Longrightarrow \neg w & & \neg w, w \Longrightarrow \Delta \\ M \Longrightarrow w \wedge \neg w & & w \wedge \neg w \Longrightarrow \Delta \\ & \diagdown \quad \diagup & \\ & M \Longrightarrow \Delta & \\ & M \Longrightarrow v & \end{array} \quad ,$$

and conversely in KJ_2 the τ-image of the conclusion of $(AIN\Delta)$ can be derived from the τ-image of the premiss:

$$\begin{array}{ccc} & & \Longrightarrow t \\ M_- \Longrightarrow \neg t & & M_- \Longrightarrow t \\ & \diagdown \quad \diagup & \\ & M_- \Longrightarrow v_- & \end{array} \quad .$$

In the following table I collect the calculi established so far.

name	where	how	cut	maps to
KP			y	
KM_0	Fm(Δ)		y	KM_1 trivially, LM via σ
KM_1	Fm(Δ,¬)	IΔ¬, EΔ¬	y	KM_0 via ρ, LM via $\mathbf{D}_+$ KM_2C via τ
KM_2	Fm(¬)	P_0¬, P_1¬		KM_1, KM_0, $KM_{21}C$
KM_{21}	Fm(¬)	P_0¬, P_2¬		KM_2C
LM	Fm(¬)	I¬, E¬	y	KM_1 trivially, KM_0 via ρ
KJ_0	Fm(Δ)	KM_0+AINΔ	y	
KJ_1	Fm(Δ,¬)	KM_1+AINΔ	y	
KJ_2	Fm(¬)	KM_2+AIN¬		
LJ	Fm(¬)	LM+AIN	y	

6. Tableaux

The method of tableaux, developed for positive logic in Chapter **1.5**, extends to minimal and intuitionistic logic in a straightforward manner.

Let Fm(¬) be the propositional language used for LM and LJ, and assume again that there are two copies of Fm(¬) which give rise to signed formulas, obtained by prefixing one of the letters P and Q. In addition, I request the presence of a particular P-signed formula P▲ which is considered as non-composite. I extend the classification of signed composite formulas by adding the two unramified cases with components

Q ¬v : P v
P ¬v : Q v P▲ .

I extend the definition of a (positive) tableau to that of *minimal* one by adding the clauses

(t4) If t(ε(e)) is Q¬v *and* if the P-predecessor of e carries P▲ then t(e') is the component P v .

(t5) If t(ε(e)) is P¬v then t(e') is the component Q v *and* e' has a further upper neighbour which carries P▲ .

In order to obtain intuitionistic tableaux, I extend the definition of ε permitting t(ε(e)) also to be P▲ (which is not composite) and adding the clause (t6) If $\varepsilon(e)$ is the P–predecessor of e (and $t(\varepsilon(e))$ is not already P▲) then e has one upper neighbour e' and t(e') is P▲ .

With these definitions, the operators from Lemma 1.4 can be extended such as to transform closed minimal and intuitionistic tableaux into L_uM– and L_uJ–derivations and vice versa. The proof is an obvious extension of that of Lemma 1.4 .

The following are examples, first of a minimal and then of an intuitionistic tableau:

	P x	
5	P ▲	1 , (t4)
4	Q x	
3	P ¬x	3 , (t5)
2	P x → ¬x	2 , (t1)
1	Q ¬x	0 , (t1)
0	P ¬x → (x → ¬x)	0 , (t1)

	P x	
7	P ▲	3 , (t4)
6	Q x	
5	P ¬x	5 , (t5)
4	P ▲	1 , (t4)
3	Q ¬x	2 , (t6)
2	P ¬x → x	2 , (t1)
1	Q ¬¬x	0 , (t1)
0	P ¬¬x → (¬x → x)	0 , (t1)

Here the comments "n, t(k)" indicate that the next line is obtained by dissolving line n applying clause t(k).

Different approaches to intuitionistic tableaux were developed in FITTING 69 and in FELSCHER 85 .

References

H.B.Curry: Foundations of Mathematical Logic. New York 1963

W.Felscher: Dialogues, strategies, and intuitionistic provability. Ann.Pure Appl.Logic **28** (1985) 217–254

M.C.Fitting: Intuitionistic Logic, Model Theory and Forcing. Amsterdam 1969

I.Johansson: Der Minimalkalkul, ein reduzierter intuitionistischer Formalismus. Compos.Math. **4** (1937) 119–136

A.N.Kolmogorov: On the principle of tertium non datur [Russian] . Mat.Sbornik **32** (1925) 646–667

Chapter 5. Sequent Calculi for Classical Logic

The calculus LJ was defined for L−sequents of formulas from $\mathrm{Fm}(\neg)$ with the connectives $\wedge, \vee, \rightarrow, \neg$. For the same linguistical environment I define the calculi LK, or LK_0, of *classical logic* by adding to LJ the rule

$$(K_0) \qquad \begin{array}{ccc} w, M \Rightarrow v & & \neg w, M \Rightarrow v \\ & M \Rightarrow v & \end{array} \quad .$$

CURRY 63 has proposed to read (K_0) as a rule of deductive completeness with respect to derivability simultaneously from w and from $\neg w$, but there are, obviously, neither fair nor reasonable reasons to expect such a property from deductive situations. The effect of (K_0) is the same as that of $(\mathrm{AIN}\neg)$ in KJ_2: LK is neither *direct* nor does it have the *subformula property.*

It should be noticed that the related rule

$$(M_0) \qquad \begin{array}{ccc} w, M \Rightarrow v & & \neg w, M \Rightarrow v \\ & M \Rightarrow \neg\neg v & \end{array}$$

is admissible already for LM as follows from the LM−derivation of **4**.(5) together with

$$\begin{array}{ccccc}
 & & & w, M \Rightarrow v & \neg w, M \Rightarrow v \\
 & & & \neg v, M \Rightarrow \neg w & \neg v, M \Rightarrow \neg\neg w \\
 & & & \searrow & \swarrow \\
 & & & \multicolumn{2}{c}{\neg v, M \Rightarrow \neg w \wedge \neg\neg w} \\
\blacktriangle \Rightarrow \neg(\neg w \wedge \neg\neg w) & \mathrm{CUT} & & \multicolumn{2}{c}{\neg(\neg w \wedge \neg\neg w), M \Rightarrow \neg\neg v} \\
\searrow & & \swarrow & & \\
 & M \Rightarrow \neg\neg v & & &
\end{array}$$

The cut rule is admissible for LK, and the cut elimination algorithms of Chapter **2** remain in effect. Because (K_0) does not produce a principal formula, and the cases E1 and E2 in the definition of the operator $\mathbf{E}_1$ were discussed already there.

The rule (K_0) can be replaced by the rule

$$(K_t) \qquad \begin{array}{c} w \vee \neg w \;,\; M \Rightarrow v \\ M \Rightarrow v \end{array}$$

of *tertium non datur*, giving rise to a variant LK_t. (K_t) is admissible for LK_0 since (K_0) permits to conclude from $w \Rightarrow w \vee \neg w$ and $\neg w \Rightarrow w \vee \neg w$ upon $\blacktriangle \Rightarrow w \vee \neg w$ such that a cut with the premiss of (K_t) gives its conclusion. Conversely, the premisses of (K_0) give the premiss of (K_t) with $(\mathrm{I}\vee)$, and then the conclusion of (K_t) is that of (K_0). – A special case of (K_0) is the rule

$$(K_n) \qquad \begin{array}{c} \neg v, M \Rightarrow v \\ M \Rightarrow v \end{array} \quad ;$$

conversely, (K_0) is admissible for the calculus LK_n obtained from LJ by adding (K_n) if I also use the cut rule:

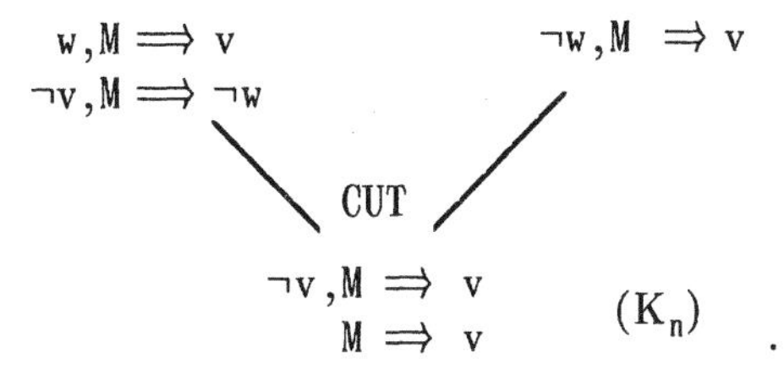

Actually, (CUT) is admissible in LK_n as will be shown later in this Chapter. The special case of K_n, in which v itself is a negation, is admissible in LM as follows from **JS**¬¬. (K_0) can also be replaced by the *double negation* rule

$$(K_d) \qquad \frac{M \Longrightarrow \neg\neg v}{M \Longrightarrow v}$$

which together with (M_0) implies the admissibility of (K_0). That (K_d) is admissible in LK_0 follows with a cut from

$$\frac{\dfrac{\dfrac{\neg v \Longrightarrow \neg v}{\neg\neg v, \neg v \Longrightarrow \blacktriangle}}{\neg\neg v, \neg v \Longrightarrow v} \qquad \neg\neg v, v \Longrightarrow v}{\neg\neg v \Longrightarrow v}\ (K_0)$$

and that (K_d) is admissible in LK_n follows by applying **JS**¬¬ to a derivation of $M \Longrightarrow \neg\neg v$, obtaining a derivation of $\neg v, M \Longrightarrow \blacktriangle$, hence of $\neg v, M \Longrightarrow v$ from where (K_n) leads to $M \Longrightarrow v$; here it is easily verified that the addition of (K_n) to LJ does not affect the proof of **JS**¬¬.

Classical calculi KK_1 and KK_2 with K–sequents arise by adding (K_0) to the intuitionistic calculi KJ_1 over $Fm(\Delta,\neg)$ and KJ_2 over $Fm(\neg)$. The cut elimination algorithm established for KJ_1 remains in effect for KK_1. The translations between KJ_1, KJ_2, L_tJ established in Lemma **4**.5 and **4**.7 remain in effect between KK_1, KK_2, L_tK .

Each of these classical calculi uses an intuitionistic rule in the form (AIN), (AINΔ) or (AIN¬). That this is necessarily so follows from the observation that the calculi LD, KD_1, KD_2, obtained by adding (K_0) to LM, KM_1, KM_2, are weaker than LK, KK_1, KK_2. Because in each of the latter the sequent $x, \neg x \Longrightarrow y$ can be derived; but if x and y are distinct variables then it cannot be derived in the former. Because assume that D is an L_tD–derivation, say, of this sequent, and let B be the branch of D which is the farthest to the right. It then suffices to show that for every node e of B there holds the following statement A(e):

> e carries a sequent $M \Longrightarrow y$ and the only non–negated formula possibly occurring in M is x ;

once this has been established, it follows from $x \neq y$ that the maximal node of B cannot carry an axiom. A(e) is trivial for the smallest node of B; assume A(e) and consider the (right) upper neighbour f of e. If (W), (RP) or (RC) leads from f to e then also A(f) will hold. Since the composite formulas in M are negated, the only logical rule leading from f to e could be $(E\neg)$, and this is impossibly since it would produce an empty right side at e. Thus there remains (K_0), and as this uses a sequent $\neg w, M \Longrightarrow y$ at f, the statement A(f) holds.

For a second example, let x be distinct from y and consider the LK-derivation

$$
\begin{array}{ccc}
 & x \Longrightarrow x & \\
 & x,\neg x \Longrightarrow \blacktriangle & \\
 & x,\neg x \Longrightarrow y & x \Longrightarrow x \\
 & \neg x \Longrightarrow x \to y & \neg x, x \Longrightarrow x \\
 & \diagdown \qquad \diagup & (E\to) \\
x, (x\to y)\to x \Longrightarrow x & \neg x, (x\to y)\to x \Longrightarrow x & \\
\diagdown \qquad \diagup & (K_0) & \\
(x\to y)\to x \Longrightarrow x & &
\end{array}
$$

The assumption, that there is a L_tD-derivation D of its endsequent, again leads to a contradiction. Because given D, let B be its branch which, ascending from the root, contains for each of its nodes e the right upper neighbour of e if e carries an instance of (K_0), and the left or single upper neighbour of e in every other case. It then suffices to show that for every node e of B there holds the following statement A(e):

e carries a sequent $M \Longrightarrow v$ such that

the only non-negated formulas possibly occurring in M are x, $(x\to y)\to x$;
v is one of x, y, $x\to y$;
if $v = x$ then x is not in M.

Because once this has been established, it follows that the left side of an axiom in B can only be x while its right side must be y. A(e) is trivial for the smallest node of B; assume A(e) and consider the upper neighbour f of e in B. If (W), (RP) or (RC) leads from f to e then also A(f) will hold. Of the logical rules, $(E\neg)$ again is impossible, and so there remains $(I\to)$ with the principal formula $v = x\to y$ or $(E\to)$ with the principal formula $(x\to y)\to x$. In the first case, f carries $x, M \Longrightarrow y$ and A(f) holds. In the second case, f carries $M' \Longrightarrow x\to y$ with a subsequence M' of M, and A(f) holds again. Finally, if (K_0) leads to e then f carries a sequent $\neg w, M \Longrightarrow v$ and A(e) once more implies A(f).

In this second example, the sequent $(x\to y)\to x \Longrightarrow x$ is *positive,* i.e. it belongs to the language of KP. The calculi LM, LJ, KM_1 and KJ_1 were *conservative* over KP: a derivation of positive sequent by these calculi was already a KP-derivation. In contrast to this agreeable behaviour, the calculi

LK, KK_1 and KK_2 are *progressive*: there are positive sequents derivable by them which are not derivable by KP (and thus their LK-derivations necessarily employ negated formulas).

1. The Multiple Calculus MK

I continue to consider the language Fm(¬) with the connectives ∧, ∨, →, ¬. I define a new kind of sequents, called *M-sequents:* they shall be ordered pairs <M, N>, written as $M \Longrightarrow N$, and consisting of two (possibly empty) sequences or sets M, N of formulas, the latter of which also is called the *succedent* of the M-sequent. Again, ▲ shall be used to symbolize an empty antecedent or empty succedent.

I define the calculi M_xK of M-sequents with the same axioms as are used for K_xP. The rules of M_xK arise from those of L_xJ by admitting finite sets N of parameters in the succedent. Thus M_tK will have the structural rules

(W) $\dfrac{M \Longrightarrow N}{v, M \Longrightarrow N}$ (AI) $\dfrac{M \Longrightarrow N}{M \Longrightarrow u, N}$

(RCL) $\dfrac{v, v, M \Longrightarrow N}{v, M \Longrightarrow N}$ (RCR) $\dfrac{M \Longrightarrow u, u, N}{N \Longrightarrow u, N}$

(RPL) $\dfrac{M \Longrightarrow N}{\pi(M) \Longrightarrow N}$ (RPR) $\dfrac{M \Longrightarrow N}{M \Longrightarrow \pi(N)}$

for every permutation π, and the logical rules

(I∧) $\dfrac{M \Longrightarrow v, N \quad M \Longrightarrow w, N}{M \Longrightarrow v \wedge w, N}$ (E∧l) $\dfrac{v, M \Longrightarrow N}{v \wedge w, M \Longrightarrow N}$ (E∧r) $\dfrac{w, M \Longrightarrow N}{v \wedge w, M \Longrightarrow N}$

(I∨l) $\dfrac{M \Longrightarrow v, N}{M \Longrightarrow v \vee w, N}$ (I∨r) $\dfrac{M \Longrightarrow w, N}{M \Longrightarrow v \vee w, N}$ (E∨) $\dfrac{v, M \Longrightarrow N \quad w, M \Longrightarrow N}{v \vee w, M \Rightarrow N}$

(I→) $\dfrac{v, M \Longrightarrow w, N}{M \Longrightarrow v \rightarrow w, N}$ (E→) $\dfrac{M \Longrightarrow v, N \quad w, M \Longrightarrow N}{v \rightarrow w, M \Longrightarrow N}$

(I¬) $\dfrac{v, M \Longrightarrow N}{M \Longrightarrow \neg v, N}$ (E¬) $\dfrac{M \Longrightarrow v, N}{\neg v, M \Longrightarrow N}$.

The changes necessary in (I∧), (I∨l), (I∨r), (I→), (I¬), (E¬) for M_uK are obvious. Observe that the intuitionistic rule (AI) now has the innocuous form of a weakening rule for the succedent. Also, every LJ-derivation is a MK-derivation. In analogy to the case of KP, also for MK a variant can be formulated which has the 2-premiss rules in mixed parameter form.

The basic mechanics of derivations developed in Chapter 1 work as well for MK. The reflexivity axioms are derivable, and the Lemmata 1.1-2 remain in effect. The technical advantage of MK is that does have the *subformula property* as well as the property of *directness.*

The notion of predecessors and that of the parameter tree is defined as before. But now the parameter tree of u in a derivation of M $\Longrightarrow$ u,N may contain instances of I-rules. This does not affect the proofs of the inversion rules (JI$\rightarrow$), (JI$\wedge$), (JI$\neg$), but that of (JI$\vee$) must fail since MK derives $\Longrightarrow$ $v\vee\neg v$: while the parameter tree of $p\vee q$ in the first derivation

$$
\begin{array}{ccc}
v \Longrightarrow v & w \Longrightarrow w & v \Longrightarrow v \\
\blacktriangle \Longrightarrow v,\neg v & \blacktriangle \Longrightarrow w,\neg w & v \Longrightarrow v\vee\neg v \\
\blacktriangle \Longrightarrow v,\neg v\vee\neg w & \blacktriangle \Longrightarrow w,\neg v\vee\neg w & \blacktriangle \Longrightarrow \neg v\ ,\ v\vee\neg v \\
\multicolumn{2}{c}{\blacktriangle \Longrightarrow v\wedge w\ ,\ \neg v\vee\neg w} & \blacktriangle \Longrightarrow v\vee\neg v\ ,\ v\vee\neg v \\
\multicolumn{2}{c}{\neg(v\wedge w) \Longrightarrow \neg v\vee\neg w} & \blacktriangle \Longrightarrow v\vee\neg v
\end{array}
$$

does branch, in the second derivation it branches, so to speak, *internally*, producing two immediate predecessors of $v\vee\neg v$ at the instance of (RCR). However, making use of (AI) at the maximal nodes of P, I find the inversion rule

(JI$\vee$) There is an operator **JI**$\vee$ transforming any derivation D of M $\Longrightarrow$ $p\vee q$,N into a derivation **JI**$\vee$(D) of M $\Longrightarrow$ p,q,N which has at most the length of D.

sending the above derivations into

$$
\begin{array}{ccc}
v \Longrightarrow v & w \Longrightarrow w & v \Longrightarrow v \\
\blacktriangle \Longrightarrow v,\neg v & \blacktriangle \Longrightarrow w,\neg w & v \Longrightarrow v,\neg v \\
\blacktriangle \Longrightarrow v,\neg v,\neg w & \blacktriangle \Longrightarrow w,\neg v,\neg w & \blacktriangle \Longrightarrow \neg v,v,\neg v \\
\multicolumn{2}{c}{\blacktriangle \Longrightarrow v\wedge w,\ \neg v,\neg w} & \blacktriangle \Longrightarrow v,\neg v,v,\neg v \\
\multicolumn{2}{c}{\neg(v\wedge w) \Longrightarrow \neg v,\neg w} & \blacktriangle \Longrightarrow v,\neg v
\end{array}
$$

.

The inversion rule (JE$\rightarrow$L) remains in effect, and I now also have an inversion rule

(JE$\rightarrow$R) There is an operator **JE$\rightarrow$R** transforming any derivation D of $p\rightarrow q$,M $\Longrightarrow$ N into a derivation **JE$\rightarrow$R**(D) of M $\Longrightarrow$ p,N which has at most the length of D .

Consider again M_uK and set $b = p\rightarrow q$. Replace, for every node h of D, the sequent $M_h \Longrightarrow N_h$ at h by $M_h{}^{\dagger}{}_b \Longrightarrow p,N_h$ where $M_h{}^{\dagger}{}_b$ arises from M_h by removing the (predecessor of the) occurrence of b. Also, if $M_h \Longrightarrow N_h$ is the conclusion of an E$\rightarrow$-instance with (this occurrence of) b as principal formula, then remove the entire subderivation leading to the right upper neighbour of h. In the resulting object instances of rules, not using a predecessor as principal occurrence, remain such instances. Instances of E→ with a predecessor as principal occurrence

$$\frac{p{\to}q, M \Longrightarrow p, N \qquad q, M \Longrightarrow N}{p{\to}q, M \Longrightarrow N} \quad \text{become} \quad \frac{M \Longrightarrow p, p, N}{M \Longrightarrow p, N}$$

and thus are instances of (RCR).

The inversion rules (JE∧) and (JE∨) remain in effect, and modifying the above proof of (JE→R), I find the inversion rule

(JE¬) There is an operator **JE¬** transforming any derivation D of ¬p,M ⟹ N into a derivation **JE¬**(D) of M ⟹ p,N which has at most the length of D .

As for the cut rule, it now has to be formulated for M–sequents as one of

$$(\mathrm{CUT}_0) \qquad \frac{M \Longrightarrow v, P \qquad v, M \Longrightarrow P}{M \Longrightarrow P}$$

$$(\mathrm{CUT}_1) \qquad \frac{M \Longrightarrow v, P \qquad v, N \Longrightarrow Q}{M, N \longrightarrow P, Q}$$

$$(\mathrm{CUT}_2) \qquad \frac{M \Longrightarrow v, P \qquad v, N \Longrightarrow Q}{M, N^{\dagger} \Longrightarrow P^{\dagger}, Q} \quad .$$

If one of the premisses of (CUT_0) is a generalized axiom or a reflexivity axiom then so is the conclusion. For (CUT_1) and (CUT_2), however, the argument of Lemma 2.1 does not work for K–sequents. Still there holds the

LEMMA 1 Let D' and D" be derivations of M ⟹ x,P and x,N ⟹ Q . Then I can construct a derivation D of $M,N^{\dagger} \Longrightarrow P^{\dagger},Q$ such that *length*(D) ≤ *length*(D') + *length*(D").

Consider first the case that D" has length 0 whence x,N ⟹ Q is a generalized axiom, hence x is in Q. In the derivation D' I remove all predecessors of occurrences of x in the succedent x,P of its endsequent and, further, add to all sequents of D' the sequence $N^{\dagger}$ to the right of their antecedent and the sequence Q to the right of their succedent; let $D^{\S}$ be the resulting object. Thus $D^{\S}$ ends with $M,N^{\dagger} \Longrightarrow P^{\dagger},Q$; axioms of D' not affected by the removal of x become generalized axioms, and axioms x ⟹ x affected become $x,N^{\dagger} \Longrightarrow Q$; also they are generalized axioms since x is in Q. From $D^{\S}$ I form D in two steps:

1. remove instances of (AI) and (RC) which introduced or contracted a removed occurrence of x, and remove instances of transpositions (RP) which affected a removed occurrence of x,

2. add instances of (AI) which restore x to all premisses of instances of logical rules using x as a side formula.

Then D is a derivation of $M,N^\dagger \Longrightarrow P^\dagger,Q$ of the same length as D'.

Consider now the case that D" has positive length. In D" I remove all predecessors of occurrences of x in the antecedent x,N of its endsequent and, further, add to all sequents of D" the sequence M to the left of their antecedent and the sequence $P^\dagger$ to the left of their succedent; let $D^\S$ be the resulting object. Thus $D^\S$ ends with $M,N^\dagger \Longrightarrow P^\dagger,Q$; axioms of D" not affected by the removal of x become generalized axioms, and axioms $x \Longrightarrow x$ affected become $M \Longrightarrow P^\dagger,x$. Applying the case already treated to D' and the derivation consisting only of an axiom $A: x \Longrightarrow x$, I see that $M \Longrightarrow P^\dagger,x$ has a derivation D_A of the same length as D'. From $D^\S$ I form D in three steps:

0. Implant the derivations D_A for the affected axioms A,

1. remove instances of (W) and (RC) which introduced or contracted a removed occurrence of x, and remove instances of transpositions (RP) which affected a removed occurrence of x,

2. add instances of (W) which restore x to all premisses of instances of logical rules using x as a side formula.

Then D is a derivation of $M,N^\dagger \Longrightarrow P^\dagger,Q$ whose length is at most the sum of the lengths of D' and D".

Let MKC_i be again the calculus obtained from MK by adding the rule (CUT_i) and, in case of M_uK and $1 \leq i \leq 2$, also the rule (RC).

LEMMA 2 The cut elimination algorithms of Chapter **2** can be expanded for M_sKC_i with $i \leq 2$, M_tKC_2 and MP_uC_0, M_uKC_2 such that the length of **A**(D) in Theorem **2**.1 is at most $e_4(k,n)$.

There are only three changes necessary in the definitions of the operators connected with **A** in the proof of Theorem **2**.1. First, when defining $\mathbf{A}_0(D)$ for $a \cdot b = 0$, Lemma 1 will be used in place of Lemma **2**.1. Next, in the definition of $\mathbf{E}_1$ the rules $(I\neg)$ and $(E\neg)$ must be considered also in the Case E1 and are treated there in complete analogy to the case E2. Finally, in the case E3 the cut formula in the succedent may be parametric, and so the changed derivations become more complicated. For $v = p \wedge q$

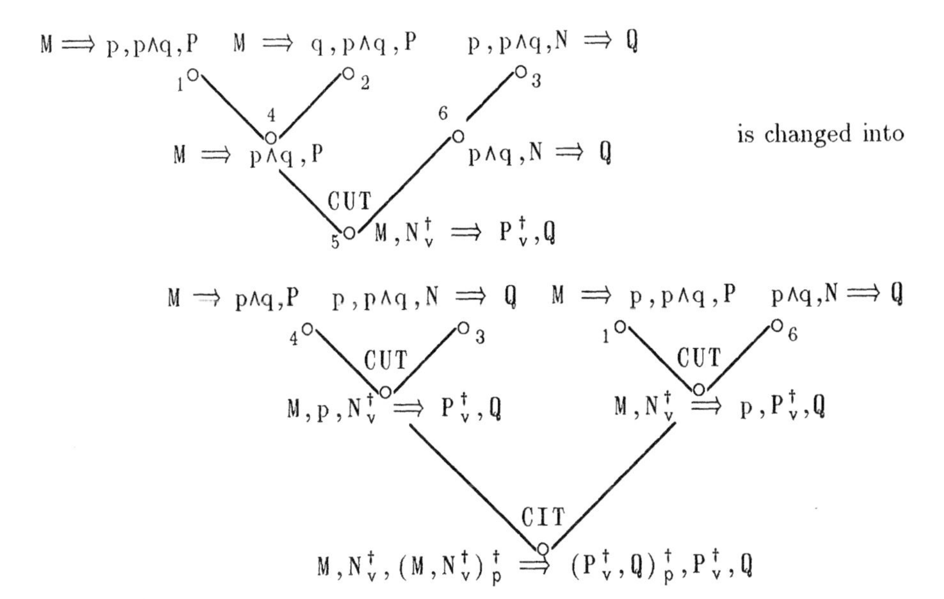

from where contractions lead to the earlier endsequent. Thus an (a,b)–cut is replaced by an (a,d)–cut with d<b and a (c,b)–cut with c<a, and below them the length may grow by 1. For v = p∨q

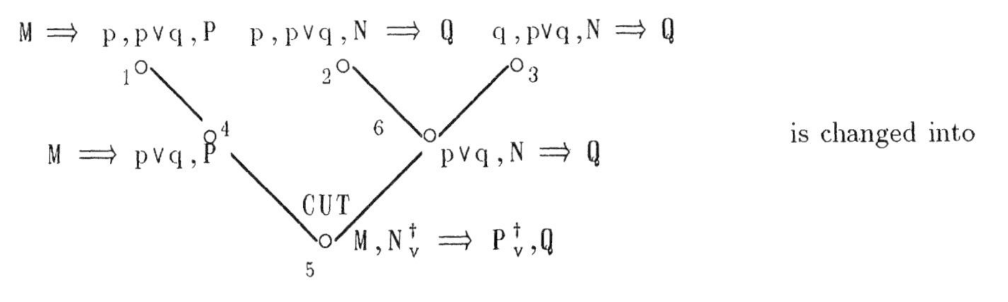

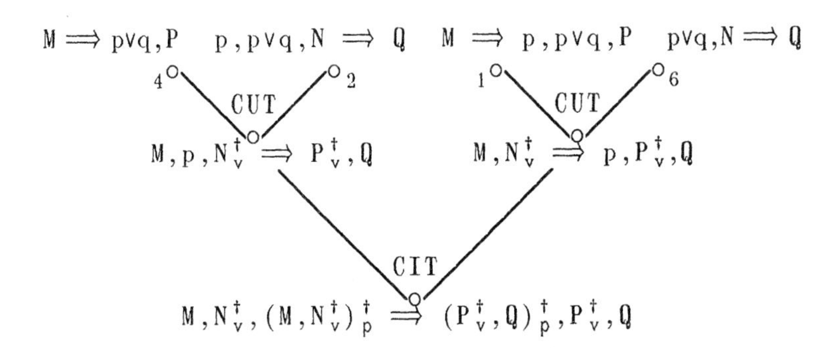

from where contractions lead to the earlier endsequent. Thus an (a,b)–cut is replaced by a (a,d)–cut with d<b and a (c,b)–cut with c<a, and below them the length may grow by 1. For v = p→q with p≠q

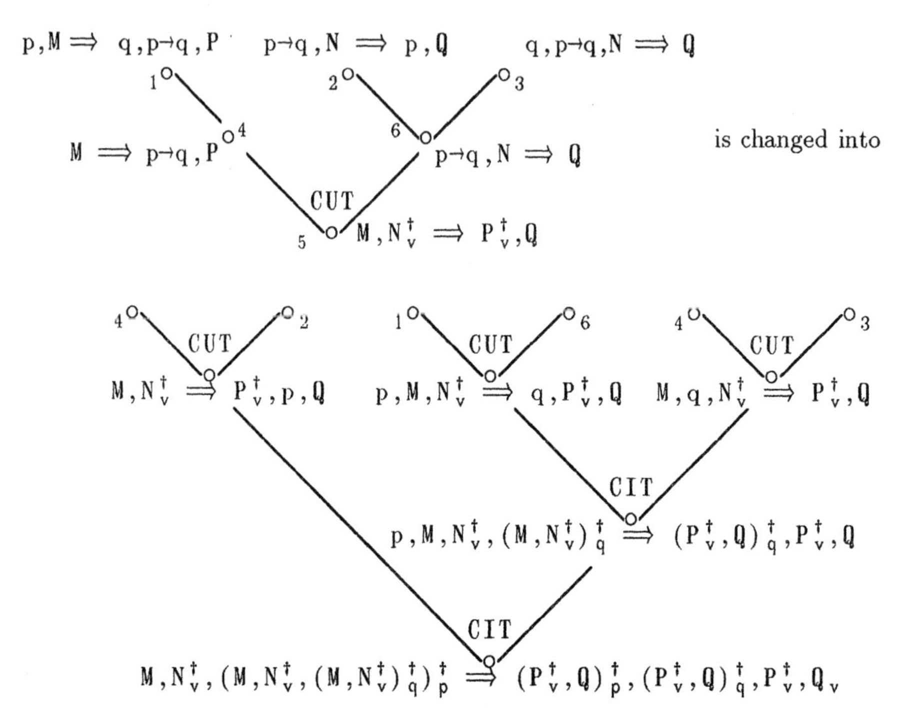

from where contractions lead to the earlier endsequent. Thus an (a,b)-cut is replaced by a (c,b)-cut with c<a and two (a,d)-cuts with d<b, and below them the length may grow by 2. For v = ¬p

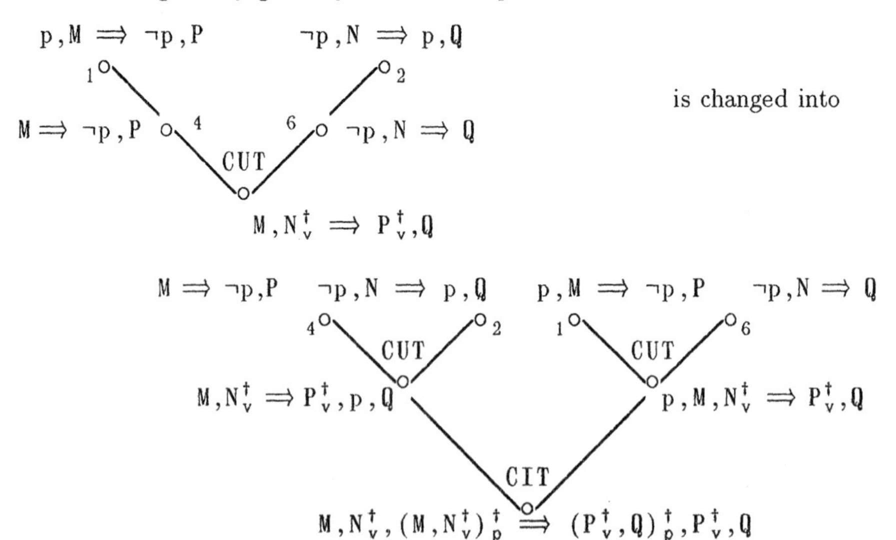

from where contractions lead to the earlier endsequent. Thus an (a,b)-cut is replaced by an (a,d)-cut with d<b and a (c,b)-cut with c<a, and below them the length may grow by 1.

It follows that the properties (e_0), (e_1) of **2**.(AE) still hold, but the situation for $v = p \rightarrow q$ shows that now there may hold $i = 2$ in (e_0) when both cases of (e_1) occur. This has the effect that in **2**.(A0) the upper bound a+2b for the length of $\mathbf{A}_0(D)$ must be replaced by 2(a+b). Because repeating the argument establishing **2**.(A0), I then know by induction that the lengths of $\mathbf{A}_0(E')$, $\mathbf{A}_0(E'')$ are at most 2(c+d) and thus are bounded by

$$2(a+b) - 2$$

in the two cases of (e_1). Thus

$$\mathit{length}(\mathbf{A}_0(D)) \leq 2 + \mathit{max}(\mathit{length}\mathbf{A}_0(D'), \mathit{length}\mathbf{A}_0(D'')) \leq 2(a+b) .$$

Consequently, in **2**.(A1) the upper bound 3^n for the length of $A_1(\mathbf{D})$ can be replaced by 4^n since $2(4^{n-1} + 4^{n-1}) = 4^n$.

2. Cut Elimination with Inversion Rules for MK

The inversion operators acting on MK-derivations are

JI→	$M \Longrightarrow p \rightarrow q, N$	$p, M \Longrightarrow q, N$	
JE→L	$p \rightarrow q, M \Longrightarrow N$	$q, M \Longrightarrow N$	
JE→R	$p \rightarrow q, M \Longrightarrow N$	$M \Longrightarrow p, N$	only for MK
JI∧L	$M \Longrightarrow p \wedge q, N$	$M \Longrightarrow p, N$	
JI∧R	$M \Longrightarrow p \wedge q, N$	$M \Longrightarrow q, N$	
JE∧	$p \wedge q, M \Longrightarrow N$	$p, q, M \Longrightarrow N$	
JI∨	$M \Longrightarrow p \vee q, N$	$M \Longrightarrow p, q, N$	only for MK
JE∨L	$p \vee q, M \Longrightarrow N$	$p, M \Longrightarrow N$	
JE∨R	$p \vee q, M \Longrightarrow N$	$q, M \Longrightarrow N$	
JI¬	$M \Longrightarrow \neg p, N$	$p, M \Longrightarrow N$	
JE¬	$\neg p, M \Longrightarrow N$	$M \Longrightarrow p, N$	only for MK .

Making use of them, I shall show that there is an operator **B** acting on M_xKC_i-derivations D, $i \leq 2$, such that with the cut degree function *cdg* from Chapter **2** there holds the

THEOREM 1 **B**(D) is an MK-derivation with the same endsequent as D and there holds

$$\mathit{length}(\mathbf{B}(D)) \leq 2^{2^{\mathit{cdg}(D)} \cdot \mathit{length}(D)} \cdot \mathit{length}(D) .$$

This result was mentioned in GORDEEV 87, § 2.4; the following proof is due to HUDELMAIER 89, 92. The use of inversion operators does not require to exchange an instance of (CUT) with one of (RC) and, therefore, permits to work right away with the simple cut rule (CUT_1).

The operator **B** will be defined from an operator $\mathbf{B}_1$ and a numerical function h, both acting on MKC–derivations D, such that

(B0) for every D : $h(\mathrm{D}) \leq 2^{cdg(\mathrm{D})} \cdot length(\mathrm{D})$;

(B1) $\mathbf{B}_1(\mathrm{D})$ is a derivation with the same endsequent as D and with

(B10) $length(\mathbf{B}_1(\mathrm{D})) \leq 2 \cdot length(\mathrm{D})$,

(B11) if D is not cut free then $h(\mathbf{B}_1(\mathrm{D})) < h(\mathrm{D})$.

Once this has been established, I define $\mathbf{B}(\mathrm{D})$ as the $h(\mathrm{D})$–fold iteration of $\mathbf{B}_1$ applied to D: it is cut free because of (B11), and the bound for its length follows from (B10) and (B0).

I define $\mathbf{B}_1$ by recursion on the length d of D. If d=0 then I set $\mathbf{B}_1(\mathrm{D}) = \mathrm{D}$. If d>0 then let (R) be the non–structural rule essentially ending D. Let D', and in case of a two premiss rule also D'', be the subderivation(s) of D leading to the premiss(es) of that (R)–instance. Since they have a length shorter than D, both $\mathbf{B}_1(\mathrm{D'})$ and $\mathbf{B}_1(\mathrm{D''})$ are defined.

Let $\mathrm{D}^{\ddagger}$ be the derivation obtained from D if D' is replaced by $\mathbf{B}_1(\mathrm{D'})$ and D'' is replaced by $\mathbf{B}_1(\mathrm{D''})$.

I set $\mathbf{B}_1(\mathrm{D}) = \mathrm{D}^{\ddagger}$ in all situations in which it is *not* the case that

(A) (R) is a cut and D', D'' are cut free.

Otherwise, D' and D'' end with sequents $\mathrm{M} \Longrightarrow \mathrm{v,P}$ and $\mathrm{v,N} \Longrightarrow \mathrm{Q}$. I distinguish five cases and consider $\mathrm{M_tKC_1}$:

0. v is a variable x.

   Let $\mathbf{B}_1(\mathrm{D})$ be the derivation constructed in Lemma 1.

1. $\mathrm{v} = \mathrm{p} \wedge \mathrm{q}$. Let $\mathbf{B}_1(\mathrm{D})$ be the derivation

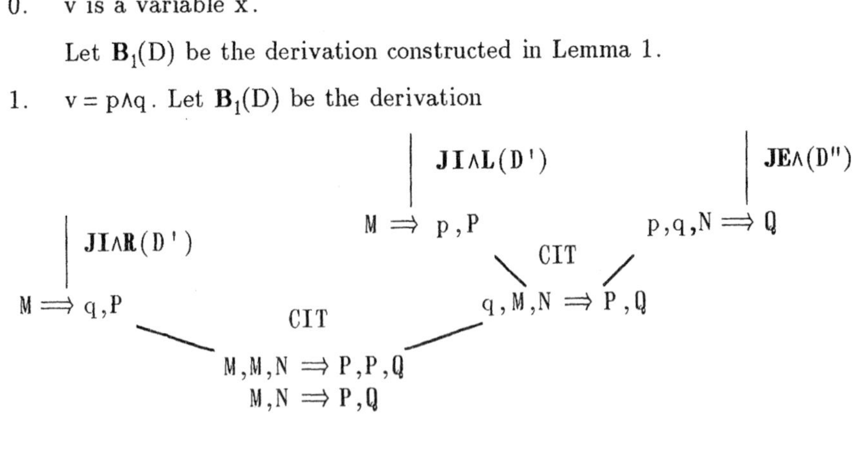

2. v = p∨q. Let $\mathbf{B}_1$(D) be the derivation

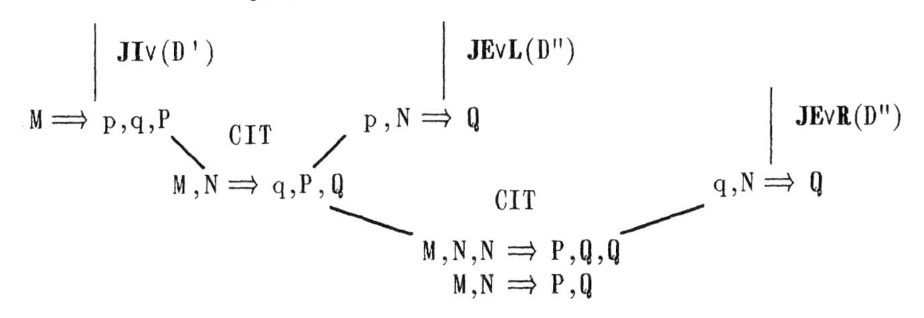

3. v = p→q. Let $\mathbf{B}_1$(D) be the derivation

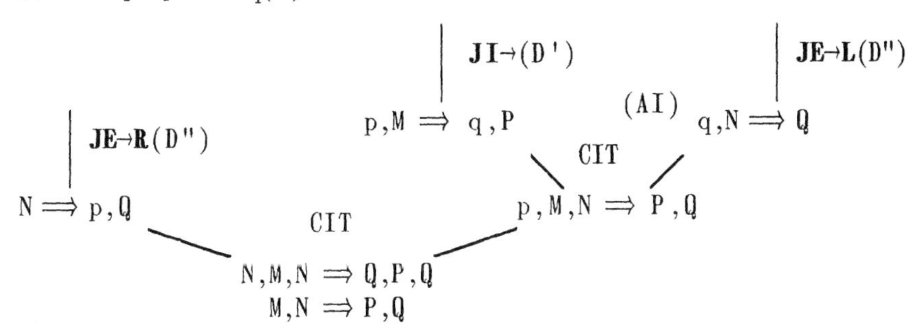

4. v = ¬p. Let $\mathbf{B}_1$(D) be the derivation

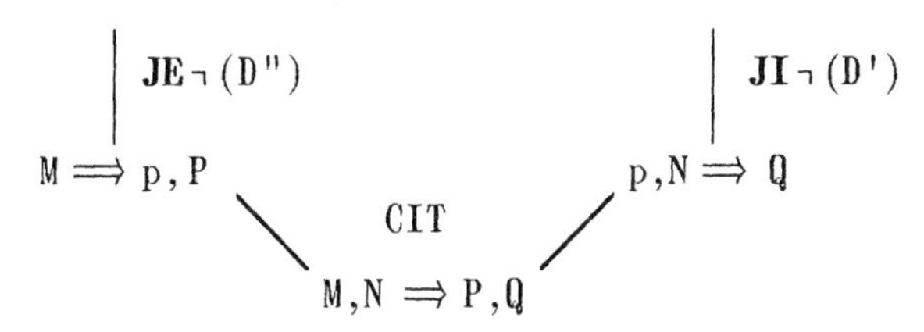

This concludes the definition of the operator $\mathbf{B}_1$. Now Lemma 1 implies:

(1) If D ends with a cut with a variable as cut formula and if the subderivations of the premisses are cut free then $\mathbf{B}_1$(D) is cut free and $length(\mathbf{B}_1(D)) \leq 2 \cdot (length(D) - 1)$.

I next prove (B10) by induction on $d = length(D)$, and the cases $d = 0$, $d = 1$ are trivial; assume that (3) holds for derivations shorter than D. If $\mathbf{B}_1$(D) is defined in case (A) then (3) holds by (1) and (2). Otherwise $\mathbf{B}_1$(D) is defined as $D^{\ddagger}$ and $length(D^{\ddagger})$ is $1 + max(length(\mathbf{B}_1(D')), length(\mathbf{B}_1(D'')))$ and so is bounded by $1 + max(2 \cdot length(D'), 2 \cdot length(D'')) = 1 + 2(d-1) < 2d$.

In order to approach the function h I begin by defining the *length* $|v|_1$ of a formula v as $|x|_1 = 1$, $|\neg p|_1 = |p|_1 + 1$, $|p\%q|_1 = |p|_1 + |q|_1 + 1$ for binary connectives %. Length is related to the usual complexity $|v|$ of v by

$$|v|_1 < 2^{|v|+1}$$

since $1<2^{0+1}$ for variables, and for $v = p\%q$ induction gives

$$|p|_1+|q|_1+1 \leq (2^{|p|+1}-1)+(2^{|q|+1}-1)+1 \leq 2\cdot 2^{\max(|p|,|q|)+1}-1$$
$$= 2\cdot 2^{|v|}-1 < 2^{|v|+1} .$$

For every branch B of a derivation D I define

$$w(B) = \Sigma < |v|_1 \mid v \text{ is cut formula of a cut occurring on } B >$$

and with $BR(D)$ as the set of all branches of D I define

$$h(D) = max < w(B) \mid B \in BR(D) > .$$

Thus D is cut free if, and only if, $h(D)$ is 0.

Each member $|v|_1$ of the sum $w(B)$ is bounded by $2^{|v|+1}$, and the maximum k of all $|v|+1$ is the cut degree $cdg(D)$. If d_B is the length of B then $w(B)$ is bounded by $2^k\cdot d_B$, and since $d_B \leq length(D)$ there follows (B0).

(2) If D ends with a cut with a composite cut formula v and if the subderivations of the premisses are cut free then $h(\mathbf{B}_1(D)) < h(D)$ and $length(\mathbf{B}_1(D)) \leq length(D) + 1$.

Because it follows from the above definitions that here the branches B of $\mathbf{B}_1(D)$ arise from the cut free branches of D' and D" by prolonging them with at most two cuts with direct subformulas of the cut formula v. Every branch B_0 of D contains this cut and thus $w(B)<w(B_0)$ for every B and every B.

Finally, I prove (B11) by induction on the length d of D. As D is not cut free, d is at least 1, and if d = 1 then the situation of (1) occurs whence $h(\mathbf{B}_1(D)) = 0 < 1 = h(D)$. Assume that (B11) holds for derivations shorter than D.

If $\mathbf{B}_1(D)$ is defined by case (A) then (B11) holds by (2). Otherwise $\mathbf{B}_1(D)$ is defined as $D^{\ddagger}$, and since D is not cut free, it follows that at least one of D', D" is not cut free, irrespective whether the rule (R) is or is not a cut. Thus by inductive hypothesis $h(\mathbf{B}_1(D'))\leq h(D')$ and $h(\mathbf{B}_1(D''))\leq h(D'')$ where equality occurs only if D' or D" is cut free. Let k be the weight of the cut formula at the root of D if (R) is a cut, and let k be zero otherwise. Then

the branches B of $\mathbf{B}_1(D) = D^{\ddagger}$ are in bijective correspondence to the branches B' of $\mathbf{B}_1(D')$ and the branches B" of $\mathbf{B}_1(D'')$ from which they arise by prolonging them with the root of D carrying (R), and then $w(B) = k+w(B')$ or $w(B) = k+w(B'')$ respectively,

the branches B_0 of D are in bijective correspondence to the branches B_0' of D' and the branches B_0'' of D" from which they arise by prolonging them with the root of D carrying (R), and then $w(B_0) = k+w(B_0')$ or $w(B_0) = k+w(B_0'')$ respectively.

Thus

$$
\begin{aligned}
h(D^{\ddagger}) &= max\langle k+w(B'),\ k+w(B'')\mid B'\in BR(\mathbf{B}_1(D')),\ B''\in BR(\mathbf{B}_1(D''))\rangle\\
&= k + max\langle w(B'),\ w(B'')\mid B'\in BR(\mathbf{B}_1(D')),\ B''\in BR(\mathbf{B}_1(D''))\rangle\\
&= k + max(\, max\langle w(B')\mid B'\in BR(\mathbf{B}_1(D'))\rangle,\\
&\qquad\qquad max\langle w(B'')\mid B''\in BR(\mathbf{B}_1(D''))\rangle)\\
&= k + max(h(\mathbf{B}_1(D')),\ h(\mathbf{B}_1(D''))),\\
h(D) &= max\langle k+w(B_0'),\ k+w(B_0'')\mid B_0'\in BR(D'),\ B_0''\in BR(D'')\rangle\\
&= k + max(\, max\langle w(B_0')\mid B_0'\in BR(D')\rangle,\\
&\qquad\qquad max\langle w(B_0'')\mid B_0''\in BR(D'')\rangle)\\
&= k + max(h(D'),\ h(D'')),
\end{aligned}
$$

Now the numbers $a = h(\mathbf{B}_1(D'))$, $b = h(D')$, $c = h(\mathbf{B}_1(D''))$, $d = h(D'')$ are such that ($a = b = 0$ and $c<d$) or ($a<b$ and $c = d = 0$) or ($a<b$ and $c<d$). Hence always $max(a,c) < max(b,d)$, and thus $h(\mathbf{B}_1(D)) = h(D^{\ddagger}) < h(D)$. This concludes the proof of (B11) and, therefore, of the Theorem.

3. MK as a Calculus for Classical Logic

Calculi for classical logic were introduced as LK and as KK_0, KK_2. L-sequents are special cases of M-sequents, and the structural and logical rules of LK are special cases of the corresponding rules of MK; in particular, (AIN) is a special case of (AI). There remains the rule (K_0) which, even in the generalized form

$$
(K_0')\qquad \frac{w,M \Longrightarrow v,N \qquad \neg w,M \Longrightarrow v,N}{M \Longrightarrow v,N}
$$

is admissible in MK since

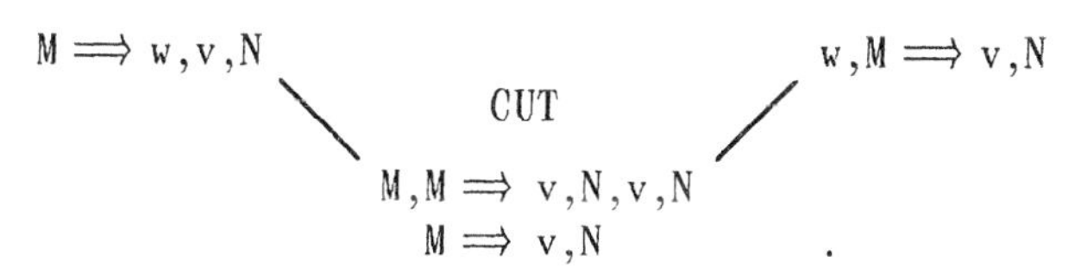

and since **JE**¬, applied to a derivation D'' of (K_0')'s right premiss, produces a derivation of $M \Longrightarrow w,v,N$ not longer than D''. Defining the degree of classicality *ccdg*(D) of an LK-derivation D to be 2 plus the maximum of all $|w|$ for formulas w in instances of (K_0), I so obtain the

LEMMA 3 Every LK-derivation D can be transformed into an MKC-derivation C of the same endsequent with *length*(C) ≤ *length*(D) and *cdg*(C) = *ccdg*(D), hence also into a MK-derivation **M**(D) whose length is bounded exponentially in *length*(D) and *ccdg*(D) by Theorem 1.

Recall that in Chapter **1.10** I defined generalized disjunctions and conjunctions $\bigvee\lambda$ and $\bigwedge\lambda$ of finite sequences λ of formulas: $\bigvee\lambda = \lambda(0)$ if $\mathrm{def}(\lambda) = 1$ and $\bigvee\lambda = \bigvee\lambda_{n-1} \vee \lambda(n-1)$ if $\mathrm{def}(\lambda) = n$, $n>1$, where λ_{n-1} consists of the first $n-1$ members of λ . Already in KP I can derive the *commutative* laws

(3) $\quad \bigvee\lambda \Rightarrow \bigvee\lambda\cdot\pi \quad$ and $\quad \bigwedge\lambda \Rightarrow \bigwedge\lambda\cdot\pi$

for every sequence λ and every permutation π of $n = \mathrm{def}(\lambda)$. To do so, observe that

$$\dfrac{\dfrac{\dfrac{x \Rightarrow x}{x \Rightarrow (y\vee z)\vee x} \qquad \dfrac{\dfrac{y \Rightarrow y}{y \Rightarrow y\vee z}}{y \Rightarrow (y\vee z)\vee x}}{x\vee y \Rightarrow (y\vee z)\vee x} \qquad \dfrac{\dfrac{z \Rightarrow z}{z \Rightarrow y\vee z}}{z \Rightarrow (y\vee z)\vee x}}{(x\vee y)\vee z \Rightarrow (y\vee z)\vee x}$$

derives the first of

(4) $\quad (x\vee y)\vee z \Rightarrow (y\vee z)\vee x \quad$ and $\quad$ (5) $\quad (x\vee y)\vee z \Rightarrow (x\vee z)\vee y$

and the second one follows analogously. For $i<n$ I define the subsequences λ_k, λ^k of λ by $\lambda = \lambda_k + \lambda(k) + \lambda^k$ (where $+$ denotes concatenation). I claim that for $\xi = \lambda_k + \lambda^k + \lambda(k)$ the sequent

(6) $\quad \bigvee\xi \Rightarrow \bigvee\lambda$

is derivable. Because for $n = 3$ and $k = 1$ this is (5), and for $k = 0$ I derive

$$\dfrac{(y\vee z)\vee x \Rightarrow (z\vee x)\vee y \qquad (z\vee x)\vee y \Rightarrow (x\vee y)\vee z}{(y\vee z)\vee x \Rightarrow (x\vee y)\vee z} \qquad \begin{array}{l}\text{both by (4)}\\ \text{with (CUT)}.\end{array}$$

Assume now $n>3$ and that (6) holds for $n-1$. If $k = n-1$ then (6) holds by inductive hypothesis; so assume $k<n-1$ and abbreviate λ_{n-1} as μ . Thus

$$\mu_k = \lambda_k \quad , \quad \mu(k) = \lambda(k) \quad , \quad \lambda^k = \mu^k + \lambda(n-1)$$

and the inductive hypothesis for λ_{n-1} and $\eta = \mu_k + \mu^k + \mu(k)$ shows that

$$\bigvee\eta \Rightarrow \bigvee\lambda_{n-1}$$

is derivable, hence with ($I\vee$) and ($E\vee$) also

$$\bigvee\eta \vee \lambda(n-1) \Rightarrow \bigvee\lambda_{n-1} \vee \lambda(n-1) , \quad \text{i.e.}$$

$$\bigvee(\mu_k + \mu^k + \lambda(k) + \lambda(n-1)) \Rightarrow \bigvee\lambda .$$

But $\xi = \lambda_k + \lambda^k + \lambda(k) = \lambda_k + \mu^k + \lambda(n-1) + \lambda(k)$, and (5) gives a derivation of

$$\bigvee(\lambda_k + \mu^k + \lambda(n-1) + \lambda(k)) \Rightarrow \bigvee(\lambda_k + \mu^k + \lambda(k) + \lambda(n-1)) , \quad \text{i.e.}$$

$$\bigvee\xi \Rightarrow \bigvee(\lambda_k + \mu^k + \lambda(k) + \lambda(n-1)) .$$

Applying a cut, I obtain a derivation of (6).

I now turn to the proof of (3). The case $n = 2$ follows immediately with ($I\vee$), ($E\vee$); assume now $n>2$ and that (3) holds for $n-1$. If $\pi(n-1) = n-1$

then π induces a permutation ρ of n−1, and $\lambda_{n-1}\cdot\rho + \lambda(n-1) = \lambda\cdot\pi$. By inductive hypothesis

$$\mathbb{W}\lambda_{n-1} \Longrightarrow \mathbb{W}\lambda_{n-1}\cdot\rho$$

is derivable from where (I∨), (E∨) derive (3). If $\pi(n-1)\neq n-1$ then let k be such that $\pi(k) = n-1$ and set $S = \{i \mid i<n \text{ and } i\neq k\}$; let ρ be the restriction of π to S. Observe that

ρ is a bijection from S onto n−1,

the map ϑ with $\vartheta(i) = i$ for $i<k$, $\vartheta(i) = i+1$ for $i\geq k$, is a bijection from n−1 onto S

and so $\rho\cdot\vartheta$ is a permutation of n−1. Thus by inductive hypothesis

$$\mathbb{W}\lambda_{n-1} \Longrightarrow \mathbb{W}\lambda_{n-1}\cdot\rho\cdot\vartheta$$

is derivable, and with (I∨), (E∨) then also

(7) $\quad \mathbb{W}\lambda_n \Longrightarrow \mathbb{W}\lambda_{n-1}\cdot\rho\cdot\vartheta \vee \lambda(n-1)$

is derivable. Now

$$\lambda_{n-1}\cdot\rho\cdot\vartheta(i) = \lambda(\pi(i)) \text{ for } i<k \quad\text{and}\quad \lambda_{n-1}\cdot\rho\cdot\vartheta(i) = \lambda(\pi(i+1)) \text{ for } i\geq k$$

by definition of ρ and ϑ; thus $\lambda_{n-1}\cdot\rho\cdot\vartheta$ is $(\lambda\cdot\pi)_k + (\lambda\cdot\pi)^k$ while $\lambda(n-1)$ is $(\lambda\cdot\pi)(k)$. Hence (7) becomes

(8) $\quad \mathbb{W}\lambda_n \Longrightarrow \mathbb{W}((\lambda\cdot\pi)_k + (\lambda\cdot\pi)^k + (\lambda\cdot\pi)(k))$.

By (6) there is a derivation of

$$\mathbb{W}((\lambda\cdot\pi)_k + (\lambda\cdot\pi)^k + (\lambda\cdot\pi)(k)) \Longrightarrow \mathbb{W}\ \lambda\cdot\pi$$

and so a cut will derive (3).

For every M-sequent $M \Longrightarrow N$ which is not already an L-sequent itself, I define L-sequents, called its L-*transforms*, as follows. In the case of t-sequents the only L-transform shall be $M \Longrightarrow \mathbb{W}N$. In the case of s-sequents, the L-transforms shall be the sequents $M \Longrightarrow \mathbb{W}\nu$ for every bijection ν from the unique number $n = \mathrm{card}(N)$ onto N. It follows from (3) that for any two such ν_0, ν_1 there is a derivation of $\mathbb{W}\nu_0 \Longrightarrow \mathbb{W}\nu_1$ such that every derivation of $M \Longrightarrow \mathbb{W}\nu_0$ can be continued to a derivation of $M \Longrightarrow \mathbb{W}\nu_1$. For this reason, I shall also in the case of s-sequents denote by $M \Longrightarrow \mathbb{W}N$ an arbitrary one of the L-transforms of $M \Longrightarrow N$.

The L-transform of an MK-derivation shall be the object obtained by replacing all M-sequents by (some of) their L-transforms.

LEMMA 4 The L-transform of an MK-derivation can be expanded to an LK-derivation.

It suffices to show that, for each of the rules of MK, an LK-derivation of the L-transforms of the premisses can be extended to an LK-derivation of the L-transform of the conclusion. This is clear for the rules (W), (RCL), (RPL), (E$\wedge$), (E$\vee$) in which the succedent is not changed. For (RPR) it follows from (3). For (AI) I make use of (3) for $k = 0$ in

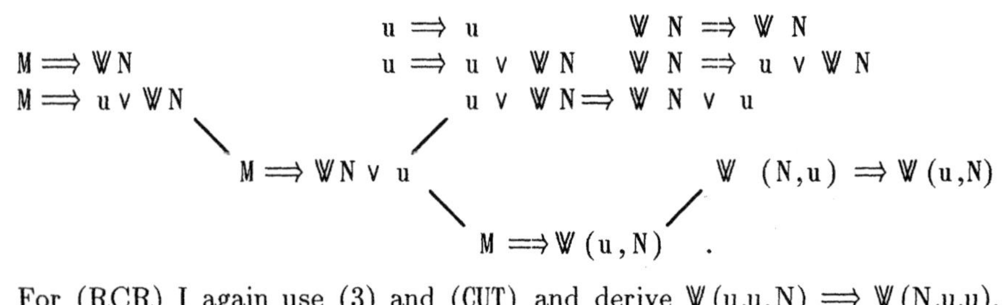

For (RCR) I again use (3) and (CUT) and derive $\mathbb{W}(u,u,N) \Longrightarrow \mathbb{W}(N,u,u)$, $\mathbb{W}(u,u,N) \Longrightarrow (\mathbb{W}N \vee u) \vee u$, $\mathbb{W}(u,u,N) \Longrightarrow \mathbb{W}N \vee (u \vee u)$, employ a derivation of $u \vee u \Longrightarrow u$ for $\mathbb{W}(u,u,N) \Longrightarrow \mathbb{W}N \vee u$, use again (3) for $\mathbb{W}(N,u) \Longrightarrow \mathbb{W}(u,N)$ and finally obtain derivations of $\mathbb{W}(u,u,N) \Longrightarrow \mathbb{W}(u,N)$ and of $M \Longrightarrow \mathbb{W}(u,N)$.

Observe that so far I have used only arguments which hold already for KP. I now turn to the logical rules. The rules (I$\vee$) I handled in analogy to (RCR). As for (I$\wedge$), I make use of (3) and (CUT) and prolong derivations of the L-transforms of the premisses by

$$\dfrac{\begin{array}{l} M \Longrightarrow \mathbb{W}(v,N) \\ M \Longrightarrow \mathbb{W}(N,v) \\ M \Longrightarrow \mathbb{W}N \vee v \\ M \Longrightarrow v \vee \mathbb{W}N \end{array} \qquad \begin{array}{l} M \Longrightarrow \mathbb{W}(w,N) \\ M \Longrightarrow \mathbb{W}(N,w) \\ M \Longrightarrow \mathbb{W}N \vee w \\ M \Longrightarrow w \vee \mathbb{W}N \end{array}}{M \Longrightarrow (v \vee \mathbb{W}N) \wedge (w \vee \mathbb{W}N)}\,(\mathrm{I}\wedge) \quad .$$

Replacing d in the KP-derivation

$$\dfrac{\dfrac{\begin{array}{l} v,w \Longrightarrow v \qquad v,w \Longrightarrow w \\ v,w \Longrightarrow v \wedge w \\ v,w \Longrightarrow (v \wedge w) \vee d \end{array} \qquad \begin{array}{l} d \Longrightarrow d \\ d \Longrightarrow (v \wedge w) \vee d \end{array}}{v \vee d, w \Longrightarrow (v \wedge w) \vee d}\,(\mathrm{E}\vee) \qquad \begin{array}{l} d \Longrightarrow d \\ d \Longrightarrow (v \wedge w) \vee d \end{array}}{\begin{array}{l} v \vee d,\ w \vee d \Longrightarrow (v \wedge w) \vee d \\ (v \vee d) \wedge (v \vee d) \Longrightarrow (v \wedge w) \vee d \end{array}}\,(\mathrm{E}\vee)$$

by $\mathbb{W}N$, a cut with the endsequent of the preceding derivation gives a derivation of $M \Longrightarrow (v \wedge w) \vee \mathbb{W}N$, hence with (3) one of $M \Longrightarrow \mathbb{W}(v \wedge w, N)$. Here neither intuitionistic nor classical rules have been used.

For (I$\rightarrow$) I prolong a derivation of the L-transform of the premiss by

$$\begin{array}{l} v,M \Longrightarrow \mathbb{W}(w,N) \\ v,M \Longrightarrow w \vee \mathbb{W}N \end{array}$$

and replace d in the KP-derivation

$$
\begin{array}{lcl}
v, w \Longrightarrow w & & \\
w \Longrightarrow v \rightarrow w & & d \Longrightarrow d \\
w \Longrightarrow (v \rightarrow w) \vee d & (E\vee) & d \Longrightarrow (v \rightarrow w) \vee d \\
\hline
 & w \vee d \Longrightarrow (v \rightarrow w) \vee d &
\end{array}
$$

by $\mathbb{W}N$. A cut with the preceding derivation gives the left derivation in

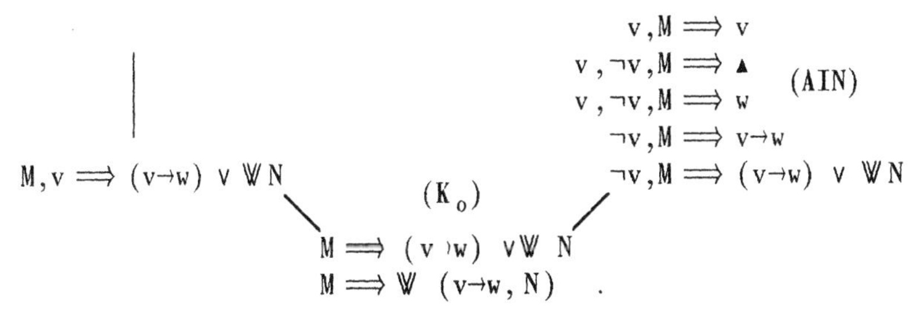

$$
\begin{array}{lcl}
 & & v, M \Longrightarrow v \\
 & & v, \neg v, M \Longrightarrow \blacktriangle \quad (AIN) \\
 & & v, \neg v, M \Longrightarrow w \\
 & & \neg v, M \Longrightarrow v \rightarrow w \\
M, v \Longrightarrow (v \rightarrow w) \vee \mathbb{W} N & (K_0) & \neg v, M \Longrightarrow (v \rightarrow w) \vee \mathbb{W} N \\
\hline
 & M \Longrightarrow (v \rightarrow w) \vee \mathbb{W} N & \\
 & M \Longrightarrow \mathbb{W} (v \rightarrow w, N) & .
\end{array}
$$

So here the full strength of LK has been used.

For ($E\rightarrow$) I replace d in the KP-derivation

$$
\begin{array}{lcl}
v \Longrightarrow v \qquad v, w \Longrightarrow w & & \\
v, v \rightarrow w \Longrightarrow w & & d, v \rightarrow w \Longrightarrow d \\
v, v \rightarrow w \Longrightarrow w \vee d & (E\vee) & d, v \rightarrow w \Longrightarrow w \vee d \\
\hline
 & v \vee d, v \rightarrow w \Longrightarrow w \vee d &
\end{array}
$$

by $\mathbb{W} N$. Then I prolong a derivation of the L-transform of the left premiss by

$$
\begin{array}{lcl}
M \Longrightarrow \mathbb{W} (v, N) & & \\
M \Longrightarrow v \vee \mathbb{W} N & (CUT) & v \vee \mathbb{W} N, v \rightarrow w \Longrightarrow w \vee \mathbb{W} N \\
\hline
 & v \rightarrow w, M \Longrightarrow w \vee \mathbb{W} N &
\end{array}
$$

and prolong a derivation of the L-transform $w, M \Longrightarrow \mathbb{W} N$ of the right premiss by $w \vee \mathbb{W} N, M \Longrightarrow \mathbb{W} N$ such that another cut gives a derivation of

$$v \rightarrow w, M \Longrightarrow \mathbb{W} N$$

which is the L-transform of the conclusion. Again, neither intuitionistic nor classical rules have been used.

For ($I\neg$) I use the full strength of LK in

$$
\begin{array}{lcl}
v, M \Longrightarrow \mathbb{W} N & & \neg v, M \Longrightarrow \neg v \\
v, M \Longrightarrow \neg v \vee \mathbb{W} N & (K_0) & \neg v, M \Longrightarrow \neg v \vee \mathbb{W} N \\
\hline
 & M \Longrightarrow \neg v \vee \mathbb{W} N & \\
 & M \Longrightarrow \mathbb{W} (\neg v, N) & .
\end{array}
$$

For (E$\neg$) I replace d in the LJ-derivation

$$
\begin{array}{lcr}
v, M \Longrightarrow v & & \\
\neg v, v, M \Longrightarrow \blacktriangle & & \\
\neg v, v, M \Longrightarrow d & (E\vee) & \neg v, d, M \Longrightarrow d \\
& \neg v, v \vee d, M \Longrightarrow d &
\end{array}
$$

by $\bigvee N$ and then prolong a derivation of the L-transform of the premiss by

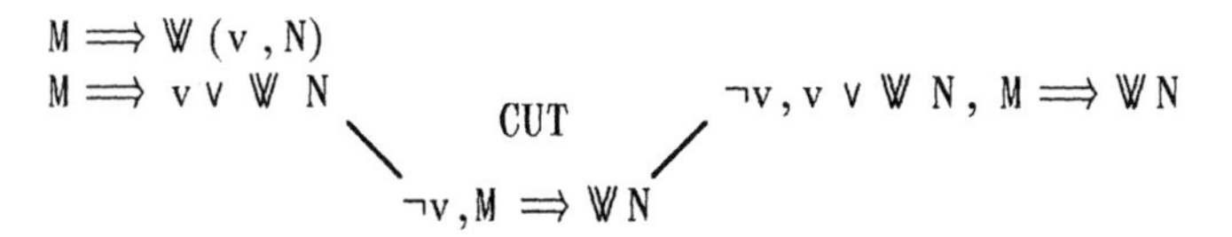

$$
\begin{array}{lcr}
M \Longrightarrow \bigvee(v, N) & & \\
M \Longrightarrow v \vee \bigvee N & \text{CUT} & \neg v, v \vee \bigvee N, M \Longrightarrow \bigvee N \\
& \neg v, M \Longrightarrow \bigvee N &
\end{array}
$$

In this case, therefore, only the intuitionistic rule is required.

This completes the proof of Lemma 4. In view of the abundance of cuts having to be eliminated in order to transform the LKC-derivations into LK-derivations, no reasonable bound for the length of the LK-derivation obtained from an MK-derivation seems to be available.

The Lemmata 3 and 4 provide algorithms transforming LK- and MK-derivations into each other under preservation of their endsequents. This having been established, I now may call MK a calculus of classical logic as well.

From a purely technical point of view, MK has the obvious advantage over LK to have both the subformula property and that of directness. But while the step from K-sequents to L-sequents was evident and, from the technical point of view, trivial when rewriting the Fm($\neg$)-part of KJ_1 as LJ, here the step from L-sequents to M-sequents produces sequents of a new, surprising efficiency. When introducing MK, I did not give a deductive interpretation of its sequents, but now I know that the MK-deducibility of

$$M \Longrightarrow N$$

is equivalent to the LK-deducibility of

$$M \Longrightarrow \bigvee N \quad ,$$

i.e. to the possibility to establish this deductive situation in LK (albeit at a larger technical price). But while the rules of LK, with the exception of (K_0), were fair and reasonable for deductive situations $M \Longrightarrow v$, in MK now the formally same rules are applied to quite different deductive situations $M \Longrightarrow \bigvee N$, and it is not at all clear that to do so permits a reasonable deductive interpretation: the premiss $v, M \Longrightarrow v$ under the rule (I$\neg$) leads to the intuitionistically untenable conclusion $M \Longrightarrow v \vee \neg v$.

The proof of Lemma 4 above did show that only three of the rules of MK required more than KP in order to remain preserved for the L-transforms of its sequents: $(E\neg)$ required (AIN), and only $(I\rightarrow)$ and $(I\neg)$ used (K_0) and so the full strength of LK. Consequently, the algorithm of Lemma 4 transforms an MK-derivation D into an LJ-derivation if in D the rules $(I\rightarrow)$ and $(I\neg)$ are applied only to sequents which are already L-sequents. It thus is the application of the rules $(I\rightarrow)$ and $(I\neg)$ to M-sequents proper which effects the transgression from intuitionistic, deductively interpretable situations to dubious classical ones.

Conversely, the fact that the classical misbehaviour can be isolated in $(I\rightarrow)$ and $(I\neg)$ will permit to set up calculi using M-sequents also for positive, minimal and intuitionistic logic, as will be dicussed in the next section.

It should further be noticed that the calculus MK *cannot* be subjected to restrictions on the size of the succedent N in its sequents $M \Longrightarrow N$ - say to contain 27 members only. This is illustrated b the following example, due to ULRICH MAYER. The endsequent of the MK-derivation

$$x \Longrightarrow x,y \qquad\qquad y \Longrightarrow x,y$$
$$\blacktriangle \Longrightarrow \neg x, x,y \qquad\qquad \blacktriangle \Longrightarrow \neg y, x,y$$
$$x \Longrightarrow \neg x\wedge\neg y,\ x$$
$$\blacktriangle \Longrightarrow \neg x\wedge\neg y,\ x,\neg x \qquad\qquad \blacktriangle \Longrightarrow \neg x\wedge\neg y,\ x,y$$
$$\blacktriangle \Longrightarrow \neg x\wedge\neg y,\ x,\ \neg x\wedge y$$
$$\blacktriangle \Longrightarrow (\neg x\wedge\neg y)\vee(\neg x\wedge y),\ x$$

cannot be the conclusion from MK-derivable premisses with succedents of only 2 members. Because the only possible rules from which to obtain it would be $(I\wedge)$ or $(I\vee)$, and in the first case the premisses would have to be the sequents

$$\blacktriangle \Longrightarrow x \quad \text{or} \quad \blacktriangle \Longrightarrow (\neg x\wedge\neg y)\vee(\neg x\wedge y)$$

and in the second case

$$\blacktriangle \Longrightarrow x, \neg x\wedge\neg y \quad \text{or} \quad \blacktriangle \Longrightarrow x,\ \neg x\wedge y$$

all of which are underivable. In the same way, the MK-derivable sequent

$$\blacktriangle \Longrightarrow (\neg x_0\wedge\neg x_1\wedge \ldots \wedge\neg x_{25}\wedge\neg x_{26}) \vee (\neg x_0\wedge\neg x_1\wedge \ldots \wedge\neg x_{25}\wedge x_{26}),\ x_0,x_1,\ldots,x_{25}$$

cannot be the conclusion from MK-derivable premisses with succedents of only 27 members.

4. The Calculi MP, MM and MJ

An M-sequent $M \Longrightarrow N$ shall be called *sharp* if N is not empty. I define the calculus M_xP by

admitting only sharp M-sequents,

the structural rules of MK,

the rules (E∧), (E∨), (E→) of M_xK,

the rules (I∧), (I∨) of M_sK, M_tK, and also for M_uK only those in the M_tK-formulation (the occurrence of the principal formula not being parametric),

the rule (I→) only for L-sequents (i.e. $v,M \Longrightarrow w$ and $M \Longrightarrow v{\to}w$).

Then every KP-derivation is an MP-derivation, and the L-transform of an MP-derivation can be expanded to a MP-derivation. MP has the subformula property and that of directness. The inversion rule (JE→R) fails since, due to the restriction on (I→), $\blacktriangle \Longrightarrow p,p{\to}q$ has no MP-derivation whence $p{\to}q \Longrightarrow p{\to}q$ cannot be inverted. Thus cut elimination by Theorem 1 will not work. The inversion rule (JI∨), however, remains in effect. Because consider the parameter tree of $p\vee q$ in a derivation of $M \Longrightarrow p\vee q,N$. Then P cannot contain an instance of (I→), and thus the replacement of $p\vee q$ by p,q in the sequents of P will not affect instances of (I→).

Cut elimination by Theorem **2**.1 for cut rules formulated with (sharp) M-sequents fails in view of the restriction on (I→). Still, these cut rules remain admissible. Because MK-derivations of the premisses of (CUT_1) can be prolonged by

$$\begin{array}{ll} M \Longrightarrow v,P & v,N \Longrightarrow Q \\ M \Longrightarrow v,P,Q & v,N \Longrightarrow P,Q \end{array} .$$

I then proceed with KP-derivations of the L-transforms and obtain a KPC-derivation ending with

$$\dfrac{M \Longrightarrow v \vee \mathbb{W}(P,Q) \qquad \dfrac{v,N \Longrightarrow \mathbb{W}(P,Q)}{v \vee \mathbb{W}(P,Q),N \Longrightarrow \mathbb{W}(P,Q)}}{M,N \Longrightarrow \mathbb{W}(P,Q)}\ \text{CUT} .$$

Cut elimination in KPC then gives a KP-derivation of that endsequent. As this is also an MP-derivation, **JI∨** then produces an MP-derivation of $M,N \Longrightarrow P,Q$.

I define the calculus M_xM for M-sequents as follows:

the structural rules of MK, but (AI) restricted to sharp sequents,

the rules (E∧), (E∨), (E→) of M_xK,

the rules (I∧), (I∨) of M_sK, M_tK, and also for M_uK only those in the M_tK-formulation (the occurrence of the principal formula not being parametric),

the rules (I→), (I¬), (E¬) only for L-sequents.

Then every LM-derivation is an MM-derivation, and the L-transform of an MM-derivation can be expanded to an LM-derivation. The cut rules for M-sequents remain admissible.

I define the calculus M_xJ for M-sequents

the structural rules of MK , but (AI) restricted to sharp sequents,

the rules (E∧), (E∨), (E→), (E¬) of M_xK,

the rules (I∧), (I∨) of M_sK, M_tK , and also for M_uK only those in the M_tK formulation (the occurrence of the principal formula not being parametric),

the rules (I→), (I¬) only for L-sequents.

Then (AI) is admissible for arbitrary sequents. Because let D be a derivation of M ⟹ ▲ from which I wish to obtain a derivation of M ⟹ N; consider the parameter tree P of ▲ , i.e. the subtree ascending from the root with sequents M_h ⟹ ▲ at its nodes h. Maximal nodes of P cannot be axioms and so must be conclusions ¬v, M_h ⟹ ▲ of premisses M_h ⟹ v . These sequents are sharp whence the restricted form (AI) will give a derivation of M_h ⟹ v, N from where (E¬) gives ¬v, M_h ⟹ N . Writing P as the succedent in all sequents of P, I so obtain a derivation of M ⟹ N.

It now follows again that every LJ-derivation is an MJ-derivation, and that the L-transform of an MJ-derivation can be expanded to a LJ-derivation. The cut rules for M-sequents remain admissible.

5. The Peirce Rule

The discussion of classical logic took place in a language Fm(¬). Let KP be the calculus of positive logic defined on the sublanguage Fm of formulas not containing ¬. Already in the first section of this Chapter I mentioned that KK_0, LK and MK have the unpleasant property of being progressive over KP, illustrated by the *Peirce sequent* ▲ ⟹ ((x→y)→x)→x being derivable in LK but not in LJ or even LD. Adding to KP the the *Peirce rule*

$$(K_p)\qquad \frac{v{\to}w,\ M \Longrightarrow v}{M \Longrightarrow v}\ ,$$

I obtain a calculus KPP in which this sequent does have a derivation, viz.

$$\begin{array}{c}
x{\to}y \Longrightarrow x{\to}y \qquad\qquad x \Longrightarrow x\\
(x{\to}y){\to}x,\ x{\to}y \Longrightarrow x\\
x{\to}y,\ (x{\to}y){\to}x \Longrightarrow x\\
(x{\to}y){\to}x \Longrightarrow x \quad (K_p)\\
\blacktriangle \Longrightarrow ((x{\to}y){\to}x){\to}x\ .
\end{array}$$

Also, KPP will not derive more than LK since (K_p) is admissible in KK_0 as well as in LK. Because a derivation H of its premiss is also an MK-derivation, and so **JE→R** produces an MK-derivation H' of $M \Longrightarrow v,v$, hence of $M \Longrightarrow v$, and that can be transformed into an LK- or a KK_0-derivation by Lemma 4. Thus there is an operator $\mathbf{P}_q$ which maps KPP-derivations into LK-derivations of the same endsequent.

Clearly, KPP does not have the subformula property. I now shall show that every LK-derivable positive sequent is KPP-derivable:

THEOREM 2 The is an operator $\mathbf{P}_p$ which transforms an LK-derivation D of a positive endsequent into a KPP-derivation of the same endsequent, and $\mathbf{P}_p(D)$ contains, besides $\rightarrow$, only connectives occurring in the endsequent of D.

For a first step, I let KPPC arise from KPP by adding the cut rule and prove the

LEMMA 5 There is an operator $\mathbf{P}_{po}$ such that, for an LK-derivation D with a positive endsequent, $\mathbf{P}_{po}(D)$ is a KPPC-derivation of the endsequent of D which contains, besides $\rightarrow$, only connectives occurring in that endsequent. The length of $\mathbf{P}_{po}(D)$ depends linearly on that of D and on the number and the complexities of formulas introduced by (AIN) in D.

Given D, let z be a propositional variable which does not occur in any of the formulas in sequents of D. Let φ be the map which is the identity on variables, is homomorphic for the positive connectives, and which maps $\neg v$ to $v^\varphi \rightarrow z$. Thus φ is the identity for positive formulas. Let D_1 arise from D by replacing every K-sequent $N \Longrightarrow u$ by its φ-image $N^\varphi \Longrightarrow u^\varphi$ and every sequent $N \Longrightarrow \blacktriangle$ by $N^\varphi \Longrightarrow z$. Thus the object D_1 ends with the same endsequent as D, and axioms and instances of positive rules are preserved. Instances of $(I\neg)$ and $(E\neg)$ become instances of $(I\rightarrow)$ and $(E\rightarrow)$ by

$$\frac{N^\varphi, v^\varphi \Longrightarrow z}{N^\varphi \Longrightarrow v^\varphi \rightarrow z} \quad \text{and} \quad \frac{N^\varphi \Longrightarrow v^\varphi \qquad z \Longrightarrow z}{N^\varphi, v^\varphi \rightarrow z \Longrightarrow z} \quad .$$

I shall expand D_1 into an object D_2 taking care of the transformed instances of (AIN) and of (K_0). Let $e_0,\dots,e_{k-1}$ be the nodes of D (and D_1) which carry instances of (AIN) introducing formulas $u_0,\dots,u_{k-1}$. I obtain D_2 by adding

$$h_i = z \rightarrow u_i^\varphi$$

for every $i<k$ to the antecedent at e_i and at every node g below e_i. Thus

$$\frac{N_i \Longrightarrow \blacktriangle}{N_i \Longrightarrow u_i} \quad \text{becomes} \quad \frac{N_i^{\varphi} \Longrightarrow z}{N_i^{\varphi} \Longrightarrow u_i^{\varphi}} \quad \text{and then} \quad \frac{h_p, \dots h_r, N_i^{\varphi} \Longrightarrow z}{h_0, \dots h_r, z \to u_i^{\varphi}, N_i^{\varphi} \Longrightarrow u_i^{\varphi}}$$

where the $h_p, \dots h_r$ come from the nodes $e_p, \dots e_r$ above e_i. Now the transformed conclusions of (AIN) can be obtained as conclusions of ($E\to$) in

$$\frac{h_p, \dots h_r, N_i^{\varphi} \Longrightarrow z \qquad h_p, \dots h_r, N_i^{\varphi}, u_i^{\varphi} \Longrightarrow u_i^{\varphi}}{h_0, \dots h_r, z \to u_i^{\varphi}, N_i^{\varphi} \Longrightarrow u_i^{\varphi}} \quad .$$

if I enlarge D_2 to D_3 by adding new upper right neighbours of the e_i with the appropriate reflexivity axioms (and their derivations). Observe that the (images of) instances of all other rules are not affected by the addition of the h_i. – As for (K_0), its instances

$$\frac{p, N_i \Longrightarrow q \qquad \neg p, N_i \Longrightarrow q}{N_i \Longrightarrow q}$$

in D become

$$\frac{h_p, \dots h_r, p^{\varphi}, N_i^{\varphi} \Longrightarrow q^{\varphi} \qquad h_p, \dots h_r, p^{\varphi} \to z, N_i^{\varphi} \Longrightarrow q^{\varphi}}{h_p, \dots h_r, N_i^{\varphi} \Longrightarrow q^{\varphi}}$$

in D_3. I now expand D_3 to D_4 as follows. A subderivation leading to the image of the left premiss I continue with ($E\to$) amd ($I\to$) as

$$\frac{\dfrac{h_p, \dots h_r, p^{\varphi}, N_i^{\varphi} \Longrightarrow q^{\varphi} \qquad h_p, \dots h_r, N_i^{\varphi}, z \Longrightarrow z}{h_p, \dots h_r, q^{\varphi} \to z, p^{\varphi}, N_i^{\varphi} \Longrightarrow z}}{h_p, \dots h_r, q^{\varphi} \to z, N_i^{\varphi} \Longrightarrow p^{\varphi} \to z}$$

introducing the three new nodes of which the rightmost one carries a generalized axiom. A cut of the lowest sequent with the image of the right premiss of my instance of (K_0) gives

$$h_p, \dots h_r, q^{\varphi} \to z, N_i^{\varphi} \Longrightarrow q^{\varphi}$$

from where (K_p) leads to

$$h_p, \dots h_r, N_i^{\varphi} \Longrightarrow q^{\varphi} \quad ,$$

i.e. the image of the conclusion. Performing this for all images of (K_0)-instances, I arrive at D_4. This now is a derivation containing only instances of positive rules and of (K_p). Also, none of its formulas contains $\neg$, and the change of v to v^{φ} and the addition of the h_i does not add new connectives besides $\to$. Thus D_4 is a KPPC derivation. If $M \Longrightarrow r$ was the endsequent of D then the endsequent of D_4 is

$$z \to u_0^{\varphi}, \; z \to u_1^{\varphi}, \; \dots \; , \; z \to u_{k-1}^{\varphi}, \; M \Longrightarrow r \; .$$

I transform D_4 into D_5 by replacing the variable z everywhere by the formula r from this endsequent and, where necessary, implant derivations of reflexivity axioms $r, P' \Longrightarrow r$ arising from generalized axioms $z, P \Longrightarrow z$. Then also D_5 remains a KPCC-derivation and has the endsequent

$$r \to u_0^{\varphi}{}_{-}, \; r \to u_1^{\varphi}{}_{-}, \; \dots \; , \; r \to u_{k-1}^{\varphi}{}_{-}, \; M \Longrightarrow r$$

where $u_i^{\varphi}{}_-$ is the image of u_i^{φ} under this replacement. Finally, I prolong D_5 by k instances of (K_p) and obtain $\mathbf{P}_{po}(D)$ deriving $M \Longrightarrow r$.

The growth of lengths from D_2 to D_3 and from D_5 to $\mathbf{P}_{po}(D)$ depends linearly on the number and the complexity of the formulas u_i. The step from D_3 to D_4, however, uses cuts

It so remains to be shown that cuts can be eliminated in KPPC, and here I shall consider the extremal calculus K_uPPC. I shall present a proof by the method used in Theorem **2**.1, but with $\mathbf{A}_0$ defined by explicit retracing as in the Lemmata **3**.1 and **3**.2. Observe first that Lemma **3**.1 and its proof remain in effect for K_uPPC since the omitted predecessors of the occurrence of v in the endsequent of J cannot occur as p or q in an instance

$$(I_p) \qquad \frac{q \rightarrow p, M \Longrightarrow q}{M \Longrightarrow q}$$

of (K_p) in J which so remains an instance of (K_p) in a derivation J_e. But matters are different for Lemma **3**.2. In its place, I shall define an operator $\mathbf{F}_p$ acting on pairs H, J of K_uPPC-derivations such that there holds the

LEMMA 6 If H ends with $M \Longrightarrow v$, and if J ends with $v, M \Longrightarrow u$, and if all cuts in H and J have cut degrees below the degree of v,

then $\mathbf{F}_p(H,J)$ is a K_uPPC-derivation of $M \Longrightarrow u$, all cuts in $\mathbf{F}_u(H,J)$ have degrees below the degree of v , and $0 < \mathit{length}(H)$ implies

$$\mathit{length}(\mathbf{F}_p(H,J)) \leq \mathit{length}(H) + 2 \cdot \mathit{length}(H) \cdot \mathit{length}(J)$$

while $\mathit{length}(\mathbf{F}_p(H,J)) \leq \mathit{length}(J)$ for $0 = \mathit{length}(H)$.

If v is a variable or is principal formula of a rule producing the endsequent of H then the operator $\mathbf{F}_u$ of Lemma **3**.1 can be taken as $\mathbf{F}_p$. For the general case, I first have to define a new predecessor tree of v .

To this end, I extend the notion of corresponding occurrences to the case of (K_p): the right occurrence in the premiss corresponds to the right occurrence in the conclusion and an occurrence i+1 in the antecedent of the premiss corresponds to the occurrence i in the conclusion.

I now consider the derivation H of $M \Longrightarrow v$ and denote by i the right occurrence of v here. I define the *right* and the *left predecessors* of i in the sequents of H as follows.

In the endsequent i is its own right predecessor, and there are no left predecessors. Let s be a sequent in D which does not lay on a maximal node and thus is conclusion of an instance (R) of a rule. If s does not contain right or left predecessors of i then neither shall its premisses. Assume now that s contains at least one of (1) an occurrence j of a right predecessor and (2) occurrences k of left predecessors of i.

If (R) is not ($E\rightarrow$), or is ($E\rightarrow$) such that the principal occurrence is not a left predecessor of i, then
any occurrence in a premiss which corresponds to j or some k shall also be a right or left predecessor of i.

If (R) is (I_p) and if the right occurrence in the conclusion is a predecessor of i then the occurrence 0 in the antecedent of the premiss is a left predecessor of i,

If (R) is ($E\rightarrow$) such that the principal occurrence, carrying $p\rightarrow q$, is a left predecessor of i, then the right occurrence of p in the left premiss is a right predecessor of i, and any occurrence in a premiss which corresponds to j or some k shall also be a right or left predecessor of i.

I define the *predecessor tree* P of i to be the subtree of H, formed from sequents containing left or right predecessors of i. Observe that if e is in P then it is either the root of H or its lower neighbour is also in P. Hence if k is not in P then no node above k is in P.

It follows by induction on the height of nodes in P that every predecessor of i carries the same formula v as i itself and that every left predecessor carries a formula $v\rightarrow q$. Also, every left predecessor occurs both in premisses and conclusions of rules different from (R_p) and, therefore, is not side occurrence of a logical rule.

As all cuts in H have cut formulas of a complexity below that of v, it follows that all right, and even more so all left, predecessors of i occurring in some cut occur there parametrically.

For every node e of P, the sequent H(e) is of the form $B \Longrightarrow b$. Let B^u arise from B by replacing the occurrences of left predecessors $v\rightarrow q$ by $u\rightarrow q$; let b^u be u if b = v and the right occurrence is a predecessor, and otherwise $b^u = b$. Let the object H^* arise from H by replacing, for every e in P, H(e) by $B^u \Longrightarrow b^u$.

I shall construct, by recursion on the depth |k| of nodes in H, a derivation $H_k{}^*$ of the sequent $H^*(k)$.

If k is maximal in H and not in P then nothing has been changed and I set $H_k{}^* = H_k$. If k is maximal in H and in P then H(k) is a generalized axiom, hence b is a variable. Also $B^u \Longrightarrow b$ is a generalized axiom since the change from B to B^u does not affect left occurrences of variables. So if b is not a predecessor then $b^u = b$, and I set $H_k{}^* = H_k$. If b is a predecessor, hence b = v, then M still will be a subsequence of B^u since it is a subsequence of B and does not contain left predecessors. Hence the derivation $H_k{}^\dagger$ consisting of the generalized axiom $B^u \Longrightarrow v$ alone, together with the derivation J_k, obtained from J by weakening with the sequence B^u/M of formulas in B^u not in M, are of the form such that $\mathbf{F}_u$ can be applied, and $H_k{}^* = \mathbf{F}_u(H_k{}^\dagger, J_k)$ then ends with $B^u \Longrightarrow u$, i.e. with $H^*(k)$.

Assume now that k is not maximal in H and that H_f^* has been constructed for the upper neighbours f of k. If k is not in P then the f are neither; hence $H^*(f) = H(f)$ and $H^*(k) = H(k)$, and I obtain H_k^* from the H_f^* by the same rule which lead from the H_f to H_k. Moreover, induction shows that $H_k^* = H_k$. It remains to consider the case that k is in P.

If k is maximal in P (but not maximal in H) then v is introduced in H(k): $M_k \Longrightarrow v$ as principal formula of a logical rule. M_k cannot contain left predecessors since these would force upper neighbours of k to belong to P. Thus M_k is M_k^u, and H_k and the derivation J_k, obtained from J by weakening with the sequence M_k/M of formulas in M_k not in M, are of the form such that $\mathbf{F}_u$ can be applied, and $H_k^* = \mathbf{F}_u(H_k, J_k)$ then ends with $M_k^u \Longrightarrow u$ i.e. with $H^*(k)$.

If k is not maximal in P, and if all left and right predecessors occurring in H(k) are parametric with respect to the upper neighbours f of k, then I define H_k^* from the H_f^* by the same rule which lead from the H_f to H_k. In particular, for (K_p) this includes the case that v is right predecessor

$$\begin{array}{llcll} H(f): & v \to q,\ B \Longrightarrow v & \qquad & H^*(f): & u \to q,\ B^u \Longrightarrow u \\ H(e): & B \Longrightarrow v & & H^*(e) & B^u \Longrightarrow u \end{array}$$

and the case that v is not right predecessor

$$\begin{array}{llcll} H(f): & v \to q,\ B \Longrightarrow v & \qquad & H^*(f): & v \to q,\ B^u \Longrightarrow v \\ H(e): & B \Longrightarrow v & & H^*(e) & B^u \Longrightarrow v\ . \end{array}$$

If a left predecessor in H(k) is not parametric, hence principal, then I proceed in the same manner with

$$\frac{H(f)\ :\ v \to q,\ B \Longrightarrow v \qquad H(g):\ v \to q, q, B \Longrightarrow b}{H(e):\ v \to q,\ B \Longrightarrow b}$$

$$\frac{H^*(f):\ u \to q,\ B^u \Longrightarrow u \qquad H^*(g):\ u \to q, q, B^u \Longrightarrow b^u}{H(e):\ u \to q,\ B^u \Longrightarrow b^u}$$

There remains the *critical* case of k that a right predecessor is not parametric, hence principal, and that, k not being maximal in P, there are parametric left predecessors. If $v = p \to r$ then $H(k): B \Longrightarrow v$ and $H(f): p, B \Longrightarrow r$ for the upper neighbour f of k. Thus $H^*(f): p, B^u \Longrightarrow r$, and I prolong H_f^* to a derivation $H_k^\dagger$ of $B^u \Longrightarrow v$. Since M is still a subsequence of B^u, the derivation $H_k^\dagger$ together with the derivation J_k, obtained from J by weakening with the sequence B^u/M of formulas in B^u not in M, are of the form such that $\mathbf{F}_u$ can be applied, and $H_k^* = \mathbf{F}_u(H_k^\dagger, J_k)$ then ends with $B^u \Longrightarrow u$, i.e. with $H^*(k)$

$$\begin{array}{ll} H(f): & p, B \Longrightarrow r \\ H(k): & B \Longrightarrow v \end{array} \qquad \frac{\begin{array}{ll} H^*(f): & p,\ B^u \Longrightarrow r \\ H_k^\dagger\ : & B^u \Longrightarrow v \end{array} \qquad J_k\ :\ v, M, B^u/M \Longrightarrow u}{H^*(k)\ :\ B^u \Longrightarrow u}\ .$$

The cases $v = p \vee q$ or $v = \neg p$ are treated analogously, the case $v = p \wedge q$ by

$$\dfrac{\dfrac{H^*(f) : B^u \Rightarrow p \qquad H^*(g) : B^u \Rightarrow q}{H_k^{\dagger} : B^u \Rightarrow v} \qquad J_k : v, M, B^u/M \Rightarrow u}{H^*(k) : B^u \Rightarrow u} \quad .$$

This concludes the recursive definition of the derivations H_k^*. I set $\mathbf{F}_p(H,J) = H_h^*$ for the root h of H. I show by induction on the depth $|k|$ in H that

(1) $\quad length(H_k^*) \leq |k| + 2\cdot(max(1, |k|)\cdot length(J)$.

If k is maximal in H, i.e. $length(H_k) = 0$, then either $H_k^* = H_k$ or H_k^* is $\mathbf{F}_u(H_k^{\dagger}, J_k)$. Here $H_k^{\dagger}$ has length 0, and so

$$length(\mathbf{F}_u(H_k^{\dagger}, J_k)) \leq |k| + length(J)$$

implies (1). If k is maximal in P, but not maximal in H, then H_k^* is $\mathbf{F}_u(H_k, J_k)$ and so (1) holds again. If k is in the critical case, then the derivations H_f^*, H_g^* of the upper neighbours satisfy (1) by inductive hypothesis, hence, say,

$$\begin{aligned} length(H_f^*) &\leq |f| + 2\cdot(max(1, |f|)\cdot length(J) \\ &< |k| - 1 + 2\cdot((max(1, |k| - 1))\cdot length(J) . \end{aligned}$$

Since $H_k^{\dagger}$ prolongs H_f^* and H_g^* there follows

$$\begin{aligned} length(H_k^{\dagger}) &= 1 + max(length(H_f^*), length(H_g^*)) \\ &\leq |k| + 2\cdot((max(1, |k| - 1))\cdot length(J) \end{aligned}$$

whence

$$\begin{aligned} length(H_k^*) &= length(\mathbf{F}_u(H_k^{\dagger}, J_k)) \\ &\leq |k| + 2\cdot((max(1, |k| - 1))\cdot length(J) + 2\cdot length(J) \\ &= |k| + 2\cdot(|k|)\cdot length(J) . \end{aligned}$$

In all other cases, H_k^* prolongs the derivations H_f^*, H_g^* in the same way in which H_k prolongs the derivations H_f, H_g, and so (1) trivially holds for H_k^* if it holds for H_f^*, H_g^*. This concludes the proof of (1), and if h is the root of H then $|h| = length(H_h)$ gives the bound for $length(H_h^*)$.

In the recursion defining the derivations H_k^* the operator $\mathbf{F}_u$ has been iterated whenever a right predecessor was principal. If the construction of $\mathbf{F}_p$ is to be arithmetized, a uniform recursive definition of $\mathbf{F}_p$ would be desirable. This can be achieved in the following, admittedly artificial, manner.

I first define operators $\mathbf{F}_{v0}$, $\mathbf{F}_{v1}$ acting on derivations D, G and on *names* R of 1-premiss respectively 2-premiss rules such that

$\mathbf{F}_{v0}(D, R)$ is the prolongation of D by the 1-premiss rule (R) provided the endsequent of D is of the appropriate form, and is D otherwise,

$\mathbf{F}_{v1}(D, G, R)$ is the prolongation of D and G by the 2–premiss rule (R) provided the endsequents of D and G are of the appropriate form, and is D otherwise.

In particular, $\mathbf{F}_{vo}$ will also accept the permutation rule together with a fixed permutation.

Given now H and J as stated in the hypotheses of Lemma 6, I define a function φ on the nodes of H such that $\varphi(k)$ is

0 if k is maximal in H and is not in P ,

1 if k is maximal in P and and maximal in H ,

2 if k is maximal in P and not maximal in H ,

3 if k is not in P and not maximal in H , or if k is in P , not maximal, and if a right predecessor occurring in H(k) is parametric with respect to the upper neighbours of k ,

4 is k is in P , not maximal, and if a right predecessor occurring in H(k) is not parametric with respect to the upper neighbours of k.

Next I define operators $\mathbf{F}_{wo}$, $\mathbf{F}_{w1}$ acting on numbers z, on derivations D, G, and on names of rules R:

$\mathbf{F}_{wo}(0,D,R) = D$,

$\mathbf{F}_{wo}(1,D,R) = \mathbf{F}_u(D^u, J_D)$ if D has length 0 , D^u is obtained by replacing v by u in the antecedent B of the endsequent of D , and J_D is obtained from J by weakening with B^u/M ,

$\mathbf{F}_{wo}(2,D,R) = \mathbf{F}_u(D, J_D)$ where J_D is obtained from J by weakening with B/M if B is the antecedent of the endsequent of D ,

$\mathbf{F}_{wo}(3,D,R) = \mathbf{F}_{vo}(D,R)$,

$\mathbf{F}_{w1}(3,D,G,R) = \mathbf{F}_{vo}(D,G,R)$,

$\mathbf{F}_{wo}(4,D,R) = \mathbf{F}_u(\mathbf{F}_{vo}(D, R), J_D)$ where J_D is obtained from J by weakening with C/M if C is the antecedent of the endsequent of $\mathbf{F}_{vo}(D,R)$,

$\mathbf{F}_{w1}(4,D,G,R) = \mathbf{F}_u(\mathbf{F}_{v1}(D,G,R), J_D)$ where J_D is obtained from J by weakening with C/M if C is the antecedent of the endsequent of $\mathbf{F}_{v1}(D,G,R)$.

In particular, for z = 0,1,2 the value $\mathbf{F}_{wo}(z,D,R)$ does not depend on R. For all arguments not listed above, the operators may be left undefined.

Define now $\mathbf{F}_p(H_k, J)$ recursively for the nodes k of H as

$\mathbf{F}_{wo}(\varphi(k), H_k, R)$ if k is maximal in H ,

$\mathbf{F}_{wo}(\varphi(k), \mathbf{F}_p(H_f, J), R)$ or $\mathbf{F}_{w1}(\varphi(k), \mathbf{F}_p(H_f, J), \mathbf{F}_p(H_g, J), R)$ if k has one upper neighbour f or two upper neighbours f, g and (R) is the rule acting at k in H.

In this way, $\mathbf{F}_p$ is defined by recursion on the nodes k, and the previous proof shows that, for the root h of H, the derivation $H_h{}^*$ constructed there is $\mathbf{F}_p(H_h, J)$.

Let d be the function defined by $d(0,n) = n$ and $d(k+1,n) = 2^{2^{d(k,n)}-2}$. I now turn to the elimination algorithm and shall show that there is an operator $\mathbf{A}$ acting on K_uPPC-derivations such that there holds the

THEOREM 3 $\mathbf{A}(D)$ is a K_uPP-derivation with the same endsequent as D. If n is the length and k is the cut degree of D then the length of $\mathbf{A}(D)$ is at most $d(k,n)$.

The operator $\mathbf{A}$ will be defined from an operator $\mathbf{A}_1$ acting on K_uPPC-derivations D such that

(A1) $\mathbf{A}_1(D)$ is a derivation with the same endsequent as D; if n is the length and k is the cut degree of D then $\mathbf{A}_1(D)$ is at most of cut degree k−1 and there holds

$$\mathit{length}(\mathbf{A}_1(D)) \leq 2^{2^n-2}.$$

Because given $\mathbf{A}_1$, $\mathbf{A}$ can be defined by recursion on its cut degree k as the k-fold iteration of $\mathbf{A}_1$.

I define $\mathbf{A}_1$ by recursion on the length d of D. If d=0 then I set $\mathbf{A}_1(D) = D$. If d>0 then let (R) be the non-structural rule essentially ending D. Let D', and in case of a two premiss rule also D'', be the subderivations of D leading to the premiss(es) of that (R)-instance. Since they have a length shorter than D, both $\mathbf{A}_1(D')$ and $\mathbf{A}_1(D'')$ are defined.

Let $D^\ddagger$ be the derivation obtained from D if D' is replaced by $\mathbf{A}_1(D')$ and D'' is replaced by $\mathbf{A}_1(D'')$. If (R) is not a k-cut then I set $\mathbf{A}_1(D) = D^\ddagger$. If (R) is a k-cut then I set $\mathbf{A}_1(D) = \mathbf{F}_p(\mathbf{A}_1(D'), \mathbf{A}_1(D''))$. It remains to verify the bound for $\mathit{length}(\mathbf{A}_1(D))$.

This length is 0 for n = 0 and is at most 1 for n = 1. For n>1 I apply induction, and that is trivial if $\mathbf{A}_1(D) = D^\ddagger$ because then $\mathit{length}(\mathbf{A}_1(D))$ is 1 plus the maximum of $\mathit{length}(\mathbf{A}_1(D'), \mathit{length}(\mathbf{A}_1(D'')$. If $\mathbf{A}_1(D)$ is defined with $\mathbf{F}_p$ then Lemma 6 implies

$$\begin{aligned}\mathit{length}(\mathbf{A}_1(D)) &\leq \mathit{length}(\mathbf{A}_1(D') + 2\cdot \mathit{length}(\mathbf{A}_1(D')\cdot \mathit{length}(\mathbf{A}_1(D'')) \\ &\leq 2^{2^{n-1}-2} + 2\cdot 2^{2^{n-1}-2}\cdot 2^{2^{n-1}-2} \\ &= 2^{2^{n-1}-2} + 2^{1+2\cdot 2^{n-1}-4} = 2^{2^{n-1}-2} + 2^{2^n-3} \\ &< 2\cdot 2^{2^n-3} = 2^{2^n-2}\end{aligned}$$

since $2^{2^{n-1}-2} < 2^{2^n-3}$ for n>1.

The method used for the proof of Lemma 6 and Theorem 3 can be applied almost verbally to set up a cut elimination operator for the calculus LK_nC obtained from LJ by adding

$$(K_n) \qquad \frac{\neg v, M \Longrightarrow v}{M \Longrightarrow v}$$

and a cut rule; obviously, I here shall define $\neg v$ in the premiss to be a left predecessor of the right occurrence in the conclusion. Thus there holds the

COROLLARY 1 There is a cut elimination operator for LK_nC which has the same bounds as that for KPPC.

As observed already in the first section of this Chapter, (K_n) is admissible in LK and (K_0) is admissible in LK_nC. Hence there holds the

COROLLARY 2 LK and LK_n derive the same sequents.

While KPP arose from KP, let now LK_p be the calculus by adding (K_p) to LJ. Then every LK_p-derivation D can be transformed into an LK_nC-derivation

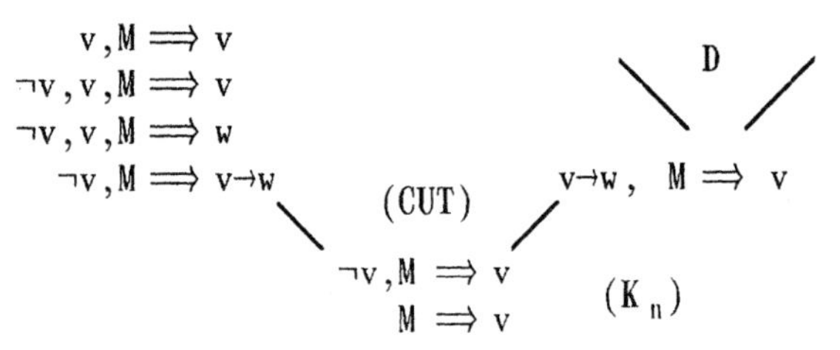

and so into an LKC- and an LK-derivation. Also, Theorem 3 remains in effect for L_uK_p. Thus there also holds the

COROLLARY 3 LK and LK_p derive the same sequents.

Of course, now LK_n and LK_p derive the same sequents, but there also is a direct transformation of LK_n-derivations into LK_p-derivations. Given an LK_n-derivation D, let e_0 be a node which is maximal for carrying an instance of (K_n); I shall show, proceeding along the lines of the proof of Lemma 4.4 for **JS**$\neg\rightarrow$, that the subderivation D_0 leading to $M \Longrightarrow v$ at e_0 can be replaced by an LK_p-derivation; in this way then the instances of (K_n) can be replaced stepwise. The upper neighbour e of e_0 carries $\neg v, M \Longrightarrow v$, and if $\neg v$ here does not have principal predecessors then I may omit it everywhere in the subderivation D_e leading to e and replace D_0 by this transform of D_e. If the number of principal predecessors is positive then I shall show that D_e can be replaced by a derivation $D_e^\dagger$ in which this number has decreased. Also since the succedent at e is not empty, I may assume that the instances of $(E\neg)$ considered are actually instances of $(E\neg N)$ as was shown in Lemma 4.3. So let g with $e \leq g$ be maximal for carrying a sequent $\neg v, N \Longrightarrow q_0$ with a principal predecessor of this occurrence of $\neg v$. Let f be the upper neighbour of g carrying $\neg v, N \Longrightarrow v$. I then weaken the antecedents of all sequents of D_e with $v \to q_0$, obtaining an object D_e^w. This changes D_e into a derivation D_e' of $v \to q_0, \neg v, M \Longrightarrow v$ which is not better than D_e, and f and g now carry $v \to q_0, \neg v, N \Longrightarrow v$ and $v \to q_0, \neg v, N, \Longrightarrow q_0$. But while in D_e (and, after permutations, still in D_e') I use $(E\neg)$ to go from f to g, I now insert a new node f' above f carrying the generalized axiom $q_0, v \to q_0, \neg v, N \Longrightarrow q_0$. Thus I obtain a new object D_e'', and here $(E\to)$ will lead from the premisses at f and f' to the sequent at g. Consequently, D_e'' becomes a derivation of $v \to q_0, \neg v, M \Longrightarrow v$ in which the number of principal predecessors of $\neg v$ has been decreased, and prolonging it by (K_p) to a derivation $D_e^\dagger$ of $\neg v, M \Longrightarrow v$, I can replace D_e by $D_e^\dagger$.

GORDEEV 87, making ingenious use of inversion rules, has established a cut elimination algorithm for K_uPPC which is much more efficient than that of Theorem 3. If (K_p) in LK_p is strengthened to the multiple Peirce rule

$$(K_p^m) \qquad \frac{v \to w_0, \ldots, v \to w_{s-1}, M \Longrightarrow v}{M \Longrightarrow v}$$

then GORDEEVs algorithm gives the upper bound $2^{1 + 3 \cdot 2^{k \cdot n}} - 2$

6. Tableaux

The method of tableaux for classical logic is much simpler and much more efficient than in the non-classical cases. As it here also has an obvious semantical meaning, it actually may serve for an introduction to (classical) logic without any prerequisites.

As the rules of MK were completely symmetrical with respect to formulas on the the left and on the right, also the rôle of P- and Q-signed formulas in classical tableaux will be symmetrical. Composite signed formulas are classified as

unramified			ramified		
$Q\, v \wedge w$:	$Q v$	$Q w$	$P\, v \wedge w$:	$P v$	$P w$
$P\, v \vee w$:	$P v$	$P w$	$Q\, v \vee w$:	$Q v$	$Q w$
$P\, v \to w$:	$Q v$	$P w$	$Q\, v \to w$:	$P v$	$Q w$
$Q\, \neg v$:	$P v$				
$P\, \neg v$:	$Q v$	.			

A *classical tableau* again consists of a finite 2-ary tree T, a root piece R of T, and two functions t and ε such that

t assigns signed formulas to the nodes of T ,

ε assigns nodes to those nodes e of T which simultaneously (a) are non-maximal in T and (b) should they belong to R are (the one) maximal (node in) R.

$\varepsilon(e)$ is a node such that $\varepsilon(e) \leq e$, $t(\varepsilon(e))$ is composite, and there holds:

(t1) If $t(\varepsilon(e))$ is unramified, or is ramified with identical components, then e has one upper neighbour e' only, and t(e') is one of the components of $t(\varepsilon(e))$.

(t2) If $t(\varepsilon(e))$ is ramified with distinct components then e has two upper neighbours e', e", and t(e'), t(e") are the two components of $t(\varepsilon(e))$.

A branch B of a tableau is called *closed* if it contains oppositely signed variables: there is a variable x such that $t(f) = Px$, $t(e) = Qx$ for nodes f, e of B. To every node e of T, above or equal to the maximal node r_0 of the root piece, I assign an M-sequent $d(e)$: $M \Longrightarrow N$ where M consists of all Q-signed and N consists of all P-signed t(f) with $f \leq e$; the tableau then is a *tableau for* the sequent $d(r_0)$.

Again, there are operators, effecting, for any M-sequent s, transformations of the closed tableaux for s into $M_u K$-derivations of s and vice versa; the proof of Lemma 1.4 clearly remains in effect with obvious simplifications.

In classical tableaux no P-predessors need to be observed; e.g. the tableau for $(((x \to y) \to x) \to x) \to y \Longrightarrow y$, studied on p. 23, closes already at its node e_4. This circumstance permits to search efficiently whether a sequent has a closed tableau.

A tableau is said *to test a* node e if t(e) either is a signed variable or if, for every branch B through e, both components of t(e) are on B in case t(e) is unramified, and at least one component of t(e) is on B in case t(e) is ramified. A tableau is called a *test tableau* if it tests each of its nodes.

THEOREM 4 For every M-sequent s, a test tableau for s can be constructed. If there exists a closed tableau for s then every test tableau for s is closed as well.

To prove the first part of the Theorem, I define for every signed formula c the number $\mu(c)$ as 0 if it is a signed variable, as $1+\mu(c')+\mu(c'')$ if c is composed with distinct components c', c", and as $1+\mu(c')$ if c has identical components c' or only has one component. If w is the complexity of the unsigned part of c then $\mu(c) \leq 2^w - 1$.

In the following, I shall start from a choice function, chosing an element (e.g. the smallest one) in non-empty and linearly ordered sets of nodes of a tableau. For my construction, I shall consider tableaux T with a family of *employment* functions $<\delta_e \mid e \epsilon \mathrm{MAX}(T)>$ where MAX(T) is the set of maximal nodes e of T and δ_e is defined on the branch of all f with $f \leq e$; δ_e assigns to these f two distinct values ∞ and *0* such that (at least the) nodes carrying signed variables obtain the value ∞ ; depending on δ_e the nodes with value ∞ are *used* and the others are *open*. Let G(T) be the set of maximal nodes e of T for which there are open nodes f with $f \leq e$; assume that G(T) is not empty. The function φ is defined on G(T), and $\varphi(e)$ shall be δ_e-open with $\varphi(e) \leq e$. For e in G(T), I define the *solvabily* $\sigma(e)$ as the sum of all $\mu(c)$ for formulas c on δ_e-open nodes below e :

$$\sigma(e) \;=\; \Sigma<\mu t(f) \mid f \leq e \text{ and } \delta_e(f) = 0> \;.$$

Given the tableau T together with $<\delta_e \mid e \epsilon \mathrm{MAX}(T)>$; I define its *dirvarct extension* T' with $<\delta_k \mid k \epsilon \mathrm{MAX}(T')>$ by adding at most two nodes on top of every node e of G(T). The node $\varphi(e)$ being open, the formula $t\varphi(e)$ is composite. If it is unramified with distinct components then I add a new upper neighbour n on top of e and a new upper neighbour m on top of n; I set $\varepsilon'(e) = \varphi(e)$, $\varepsilon'(n) = \varphi(e)$. If $t\varphi(e)$ has identical components then I add only n and define $\varepsilon'(e) = \varphi(e)$. If $t\varphi(e)$ is ramified with distinct components then I add two new upper neighbours n, m on top of e and set $\varepsilon'(e) = \varphi(e)$. This ends the definition of the tree underlying T'. All nodes f of T' with not $e \leq f$ for e in G(T) are in T, and I define $\varepsilon'(f) = \varepsilon(f)$; also, t' shall be the extension of t which assigns the respective components of $\varphi(e)$ to n and m. This concludes the definition of T'. The function δ_k coincides with δ_e if k was a maximal node e of T already; for the new maximal nodes n and m above e, I set $\delta_n(\varphi(e)) = \infty$ and $\delta_m(\varphi(e)) = \infty$; for all other nodes f already in T, I set $\delta_n(f) = \delta_e(f)$, $\delta_m(f) = \delta_e(f)$, and $\delta_n(n)$, $\delta_m(m)$ obtain the value ∞ if, and only if, n respectively m carry a composite formula under t'. - It now will be important to notice that, for new maximal nodes n, m and the solvability σ' defined for T', $<\delta_k \mid k \epsilon \mathrm{MAX}(T')>$, there holds

$$(\sigma) \qquad \sigma'(n) < \sigma(e) \quad \text{and} \quad \sigma'(m) < \sigma(e)$$

because even in the worst case $e < n < m$ the sum $\sigma'(e'')$ lacks the summand $\mu t\varphi(e)$ which is larger by 1 than the new summand $\mu t'(n) + \mu t'(n)$.

Observe further that, if T with $<\delta_e \mid e \epsilon \mathrm{MAX}(T)>$ tests each of its used nodes, then so does T' with $<\delta_k \mid k \epsilon \mathrm{MAX}(T')>$.

Given now a sequent s, or equivalently a set of signed formulas; I recursively define a sequence $<T_n \mid n \epsilon \omega>$ of tableaux for s, together with a sequence $<\delta_k{}^n \mid k \epsilon \mathrm{MAX}(T_n)>$ of employment functions. T_0 shall be its own root piece: a linearly ordered finite set with a maximal node o, the nodes of which carry the signed formulas for s, and $\delta_r{}^0$ declares as used only the nodes carrying signed variables. Let T_{n+1} with $<\delta_h{}^{n+1} \mid k \epsilon \mathrm{MAX}(T_{n+1})>$ be the direct extension of T_n with $<\delta_k{}^n \mid k\epsilon MAX(T_n)>$ if for T_n the set $G(T_n)$ is not empty; otherwise let it be T_n with $<\delta_k{}^n \mid k\epsilon MAX(T_n)>$ again. Thus every T_n will have T_o as its root piece.

Let next r be the number of formulas in s (i.e. on T_0) and let q be the maximum of their complexities; I set $p = r\cdot(2^q-1)$. For the maximal node r of T_0 the solvability $\sigma_0(r)$ is bound by p, because each of its at most r summands is bound by 2^q-1. Thus (σ) implies that in T_p the maximal nodes h have a solvability $\sigma_p(h) = 0$, meaning that there are no open nodes below them and, therefore, no open nodes at all in T_p. Consequently, if $m \geq p$ then $T_m = T_{m+1}$. As T_0 by definition tests its used nodes, so does each T_n; hence T_p tests each of its nodes and, therefore, is a test tableau. Thus T_p is a test tableau for s.

For the actual determination of a test tableau T_p, the choice function φ must be explicitly specified, for instance as one of the functions

α assigning to e the smallest open node f below e,
β assigning to e the smallest open *and* unramified node f below e, provided there is such f, and else the smallest open node f below e,
γ assigning to e the largest open node f below e.

Each of these functions leads to a test tableau T_p whose size (respectively the speed with which it is obtained) depends for different functions on the particular form of the formulas in the sequent s. For example, if s is of the particular form $\blacktriangle \Longrightarrow v$ and if v is a conjunction of alternatives

$$(x_0 \vee \neg x_0) \wedge (x_1 \vee \neg x_1) \wedge (x_2 \vee \neg x_2) \wedge (x_3 \vee \neg x_3)$$

then the construction with β produces a much smaller T_p than that with α, while for Hauber's tautology

$$\begin{aligned}((x_0 \rightarrow y_0) \wedge (x_1 \rightarrow y_1) \wedge (x_2 \rightarrow y_2) \wedge (x_0 \vee x_1 \vee x_2) \\ \wedge \neg(y_0 \wedge y_1) \neg(y_1 \wedge y_2) \neg(y_2 \wedge y_0)) \\ \rightarrow ((y_0 \rightarrow x_0) \wedge (y_1 \rightarrow x_1) \wedge (y_2 \rightarrow x_2))\end{aligned}$$

the construction with α produces a much smaller T_p than that with β.

It remains to prove the second part of Theorem 4. As a given closed tableau for s may be quite different from one of the test tableaux constructed with one of the choice functions above, a new method is required for this proof. Consider the table, attributing components to signed formulas and read Q for *true* and P for *false*; then this table expresses the connection between the classical truth values of a composite formula and the classical truth values of its components. More formally, let h be a homomorphism from the algebra of (unsigned) formulas into the Boolean algebra *2*, and extend h to the signed formulas by $h(Pv) = h(v)$, $h(Qv) = -h(v)$; hence always $h(Qv) = -h(Pv)$. Abbreviate the situation that $h(u) = 0$ by saying that h *falsifies* the signed formula u. Then the table attributing components, together with the fact that h is a Boolean homomorphism, leads to the *basic observation* on signed formulas u:

> if u is unramified then h falsifies u if, and only if, h falsifies both of its components,
>
> if u is ramified then h falsifies u if, and only if, h falsifies one of its components.

Given a tableau T and a homomorphism h from formulas into $\mathit{2}$; let e be a non-maximal node of T. Then the basic observation leads to:

> if $t\varepsilon(e)$ is unramified and h falsifies $t\varepsilon(e)$ then h falsifies $t\varepsilon(e')$ for the upper neighbour e' of e,
>
> if $t\varepsilon(e)$ is ramified and h falsifies $t\varepsilon(e)$ then h falsifies one of $t\varepsilon(e')$ and $t\varepsilon(e')$ for the upper neighbours e', e'' of e (where e' = e'' if the components are identical).

Consequently, if h falsifies all t(f) with $f \leq e$ then h falsifies the signed formula on at least one upper neighbour of e. In the case of a tableau, let me say that h falsifies a set of its nodes if it falsifies each of the signed formulas on these nodes. It follows that, if h falsifies the root piece of T then I can define by recursion a branch of T whose nodes all are falsified by h. Of course, h cannot falsify two formulas Px, Qx simultaneously as they occur on a closed branch. Hence, if the root piece of a tableau can be falsified then I can find a non-closed branch. Thus I have proven (a) in

LEMMA 7 (a) If the root piece of a tableau can be falsified then it is not closed

(b) If a test tableau is not closed then I can define a homomorphism h which falsifies its root piece.

As for (b), let T be a test tableau, a branch B of which is not closed. B being not closed, the set of variables x with Px on B is disjoint to the set of variables y with Qx on B; define a valuation by sending these x to 0 and all other variables to 1, and let h be the homomorphism to $\mathit{2}$ extending this valuation. Thus h falsifies the signed variables on B, and I now show by induction on signed formulas u that h either falsifies u or u is not on B. Because if u is composite and on B then both components of u are on B if u is unramified, hence they both are falsified by h, and thus h falsifies u; if u is ramified then one component of u is on B, hence falsified by h, and so again h falsifies u. Thus h falsifies all of B and, in particular, falsifies the root piece at the bottom of B.

It is clear that the second part of Theorem 4 is an immediate consequence of Lemma 7. Moreover, the concepts developed now provide a classical semantical interpretation of the calculus MK. To this end, let s be a sequent

$M \Longrightarrow N$ and let T be a tableau for s, meaning that its root piece consists of the Pn for $n \epsilon N$ and the Qm for $m \epsilon N$. If follows from the definitions that a homomorphism h falsifies this root piece if, and only if, $h(m) = 1$ for every m and $h(n) = 0$ for every n, i.e. if h satisfies M and does not satisfy $\bigvee N$. Thus h falsifies the root piece if, and only if, it provides a counterexample to the semantical implication $M \Vdash \bigvee N$ (in the sense of Chapter **1.10**), and Lemma 7 can be restated as

LEMMA 7a (a) If there is a counterexample for $M \Vdash \bigvee N$ then every tableau for s is not closed.

(b) If a test tableau for s is not closed then there is a counterexample for $M \Vdash \bigvee N$.

But it is equivalent that no counterexample for $M \Vdash \bigvee N$ exists and that there holds $M \Vdash \bigvee N$. Hence also

LEMMA 7b (a) If T is a closed tableau for s then $M \Vdash \bigvee N$.

(b) If $M \Vdash \bigvee N$ then any test tableau for s is closed.

Observe though that the deduction of Lemma 7b from Lemma 7a is heavily indirect. – I now summarize these insights in :

COROLLARY 4 The MK–derivations of M–sequents s correspond, by the extended Lemma 1.4, to the closed tableaux for s.

If an M–sequent s: $M \Longrightarrow N$ has an MK–derivation then $M \Vdash \bigvee N$ (by Lemma 7b (a)).

If $M \Vdash \bigvee N$ then I find a test tableau for s: $M \Longrightarrow N$ (by Theorem 4) which, being closed (by Lemma7b (b)), corresponds to an MK–derivation of s.

Thus s: $M \Longrightarrow N$ has an MK–derivation if, and only if, there holds $M \Vdash \bigvee N$.

The last observation will be taken up by another approach at the end of section 6 in the next Chapter, but also there the one direction of its proof will be indirect.

Classical tableaux were, under the name of *semantical tableaux*, investigated by SMULLYAN 68 , cf. also SMULLYAN 63 .

References

H.B.Curry: Foundations of Mathematical Logic. New York 1963

L.Gordeev: On Cut Elimination in the Presence of the Peirce Rule. Archiv Math.Logic **26** (1987) 147–164

J.Hudelmaier: Dissertation Univ. Tübingen 1989

J.Hudelmaier: Bounds for Cut Elimination in Intuitionistic Propositional Logic. Archiv Math. Logic **31** (1992) 331–353

R.M.Smullyan: A unifying principle in quantification theory. Proc.Nat.Acad.Sci.U.S.A. **49** (1963) 828–832

R.M.Smullyan: First Order Logic. Berlin 1968

Chapter 6. Classes of Algebras Associated to a Calculus

In the preceding Chapters I have studied several (propositional) sequential calculi. Their algebras of formulas were term algebras T with respect to the connectives considered.

Given one of these calculi, two formulas v,w from T are called *interdeducible* if there are derivations of $v \Longrightarrow w$ as well as of $w \Longrightarrow v$. The class **A** of algebras *associated* to a calculus shall be the equational class defined by the equations [v,w] such that v and w are interdeducible.

In this Chapter I shall characterize the classes $\mathbf{A}_f$, $\mathbf{A}_m$, $\mathbf{A}_i$, $\mathbf{A}_c$ associated to the calculi KP, LM, LJ, LK ; $\mathbf{A}_c$ is also the class associated to MK. The algebras of all these classes are lattices with an additional operation $\supset$, and for each class I shall present finite sets of defining equations. Not surprisingly, at the one extreme $\mathbf{A}_c$ will turn out to be the class of Boolean algebras with the additional operation $a \supset b = -a \cup b$. On the other side, $\mathbf{A}_g$ is preceded by the classes $\mathbf{A}_d$ and $\mathbf{A}_e$ belonging to the fragments KP_d and KP_e of KP which arise if KP is restricted to formulas containing only the connective $\rightarrow$ or only the connectives $\rightarrow$ and $\wedge$. The algebras in the classes **A** all carry an order for which there is largest element e and, in complete analogy to the case of Boolean algebras, each **A** determines a semantic consequence operator $cn^{\mathbf{A}}$ such that, for finite sets of sequences M of formulas, the derivability of $M \Longrightarrow v$ in a calculus is equivalent to $v \in cn^{\mathbf{A}}$ for the associated class A .

1. d-Algebras and d-Frames

Let T_d be the term algebra generated by the variables in X under the operation $\rightarrow$ alone. Let KP_d be the restriction of KP to T_d, let R_d be the relation of interdeducibility with respect to KP_d, and let $\mathbf{A}_d$ be the class of all models of R_d; the algebras in $\mathbf{A}_d$ I call d-*algebras.* My first aim are two characterizations of $\mathbf{A}_d$ of which the second is equational:

THEOREM 1_d $\mathbf{A}_d$ is the class of all algebras $A = \langle u(A), \supset \rangle$ with a distinguished element e in A which satisfy

(a_{mp}) if $e \supset a = e$ then $a = e$
(a_{do}) $a \supset (b \supset a) = e$
(a_{d1}) $(a \supset (b \supset c)) \supset ((a \supset b) \supset (a \supset c)) = e$
(a_d) $(a \supset b) \supset ((b \supset a) \supset a) = (b \supset a) \supset ((a \supset b) \supset b)$,

and $\mathbf{A}_d$ is also the class of all algebras $A = \langle u(A), \supset \rangle$ defined by the following set D_d of equations

(a_0) $\quad a \supset a = b \supset b$
(a_1) $\quad (a \supset a) \supset a = a$
(a_2) $\quad a \supset (b \supset c) = (a \supset b) \supset (a \supset c)$
(a_d) $\quad (a \supset b) \supset ((b \supset a) \supset a) = (b \supset a) \supset ((a \supset b) \supset b)$.

It is a matter of straightforward verification that the equations from D_d consist of interdeducible formulas and so belong to R_d, e.g. (a_2):

$$\dfrac{v, v\to(v\to u) \Rightarrow v \qquad \dfrac{v,w,w\to u \Rightarrow v \qquad \dfrac{v,w \Rightarrow w \qquad v,w,u \Rightarrow u}{v,w,w\to u \Rightarrow u}}{v,w,v\to(w\to u) \Rightarrow u}}{v, v\to w, v\to(w\to u) \Rightarrow u}$$

$$v\to w, v\to(w\to u) \Rightarrow v\to u$$
$$v\to(w\to u) \Rightarrow (v\to w)\to(v\to u)$$

$$\dfrac{\dfrac{v,v,w \Rightarrow w}{v,w \Rightarrow v\to w} \qquad \dfrac{w,v \Rightarrow v \qquad w,v,u \Rightarrow u}{w,v,v\to u \Rightarrow u}}{w,v,(v\to w)\to(v\to u) \Rightarrow u}$$

$$v,(v\to w)\to(v\to u) \Rightarrow w\to u$$
$$(v\to w)\to(v\to u) \Rightarrow v\to(w\to u) \quad .$$

Hence $\mathbf{A}_d$ is contained in $\mathbf{Mod}(D_d)$.

The algebras described in the first part of the Theorem I call *special* d-*frames.* Their distinguished element e is uniquely determined by the equations (a_{do}) (or (a_{d1})). Every algebra A in $\mathbf{Mod}(D_d)$ is a special d–frame with $e = a \supset a$. Because $a \supset a$ is independent of a by (a_o). Then (a_1) implies $e \supset a = a$, and this implies (a_{mp}). Further, (a_{do}) will follow from

$$a \supset (b \supset a) = e \ .$$

This holds since $a \supset (b \supset a) = (a \supset b) \supset (a \supset a) = ((a \supset b) \supset a) \supset ((a \supset b) \supset a) = e$ with two applications of (a_2) and one of (a_o). Finally, (a_{d1}) is an immediate consequence of the definition of e and (a_2).

The remaining proof now proceeds in two steps. I first show that the special d–frames belong to $\mathbf{Mod}(D_d)$, and I show then that they belong to $\mathbf{A}_d$. This requires some computational efforts, and in order to avoid duplications, I shall perform them in the slightly more general frame of d–*frames.*

Let A be an algebra with a 2–ary operation $\supset$; let Q be a subset of A. The pair A,Q shall be called a d–*frame* if there holds for all a,b,c:

(f_{mp}) $\quad$ if both a and $a \supset b$ are in Q then so is b,
(f_{do}) $\quad$ $a \supset (b \supset a)$ is in Q,
(f_{d1}) $\quad$ $(a \supset (b \supset c)) \supset ((a \supset b) \supset (a \supset c))$ is in Q.

I then call A *the algebra* and Q *the filter* of the frame. I define two relations:

$a \lesssim b$ if $a \supset b$ is in Q,

$a \approx b$ if $a \lesssim b$ and $b \lesssim a$.

LEMMA 1 For a d-frame A, Q the relation $\lesssim$ is a quasiorder; the relation $\approx$ is a congruence relation with respect to $\supset$ for which Q is a full congruence class. The quotient algebra $A/\approx$ of [the reduct for $\supset$ of] A is a special d-frame and is a model of D_d . The special d-frames are precisely the d-frames for which Q consists of one element e only.

That $\lesssim$ is reflexive follows from

(f0) $a \supset a$ is in Q

which holds since

1. $a \supset ((a \supset a) \supset a)$ is in Q by (f_{do})
2. $(a \supset ((a \supset a) \supset a)) \supset ((a \supset (a \supset a)) \supset (a \supset a))$ is in Q by (f_{d1})
3. $(a \supset (a \supset a)) \supset (a \supset a)$ is in Q by 1, 2 and (f_{mp})
4. $a \supset (a \supset a)$ is in Q by (f_{do})
5. $a \supset a$ is in Q by 4, 3 and (f_{mp}) .

It follows from (f_{do}) by an application of (f_{mp}) that for all b

(f1) if $q \in Q$ then $b \lesssim q$.

In particular, if both a,b are in Q then $a \lesssim b$ and $b \lesssim a$; hence the relation $\lesssim$ will not be antisymmetric if Q contains more than one element. From (f_{mp}) there follows immediately

(f2) if $q \in Q$ and $q \lesssim b$ then $b \in Q$.

Observe next that $b \lesssim c$, i.e. $b \supset c \in Q$, implies $a \supset (b \supset c) \in Q$ by (f1). But then (f_{d1}) and (f_{mp}) imply $(a \supset b) \supset (a \supset c) \in Q$. This proves

(f3) if $b \lesssim c$ then $a \supset b \lesssim a \supset c$.

More generally, since $a \supset (b \supset c) \lesssim (a \supset b) \supset (a \supset c)$ by (f_{d1}), it follows from (f3) that $(b \supset c) \supset (a \supset (b \supset c)) \lesssim (b \supset c) \supset ((a \supset b) \supset (a \supset c))$. But $(b \supset c) \supset (a \supset (b \supset c))$ is in Q by (f_{do}), hence

(f4) $b \supset c \lesssim (a \supset b) \supset (a \supset c)$.

It follows from (f3) that $\lesssim$ is transitive: if $b \lesssim c$ and $a \lesssim b$ then $a \lesssim c$; thus $\lesssim$ is a quasiorder. It follows from (f2) that Q is a full equivalence class with respect to the equivalence relation $\approx$.

(f5) $a \lesssim (a \supset b) \supset b$.

Since $(a \supset b) \supset (a \supset b)$ is in Q, it follows from (f_{d1}) that $(a \supset b) \supset a \lesssim (a \supset b) \supset b$. Thus $a \supset ((a \supset b) \supset a) \lesssim a \supset ((a \supset b) \supset b$ by (f3), whence $a \supset ((a \supset b) \supset a) \in Q$ implies that $a \supset ((a \supset b) \supset b)$ is in Q.

(f6) if $q \in Q$ then $q \supset a \approx a$.

Here $a \leq q \supset a$ holds by (f_{do}); $q \leq (q \supset a) \supset a$ holds by (f5) and implies $q \supset a \leq a$.

(f7) $a \supset (b \supset c) \approx b \supset (a \supset c)$.

I have $a \supset (b \supset c) \leq (a \supset b) \supset (a \supset c) \leq (b \supset (a \supset b)) \supset (b \supset (a \supset c))$ by (f_{d1}) and (f4). By (f_{do}), the last element is of the form $q \supset (b \supset (a \supset c))$ with $q \in Q$, hence $q \supset (b \supset (a \supset c)) \leq b \supset (a \supset c)$ by (f6). Thus $a \supset (b \supset c) \leq b \supset (a \supset c)$ from where (f7) follows by symmetry.

(f8) $b \supset c \leq (c \supset a) \supset (b \supset a)$.

Since $(c \supset a) \supset ((b \supset c) \supset (b \supset a))$ is in Q by (f4), also $(b \supset c) \supset ((c \supset a) \supset (b \supset a))$ will be in Q by (f7).

(f9) $a \supset (b \supset c) \approx (a \supset b) \supset (a \supset c)$.

One half of this equivalence is (f_{d1}); in order to prove the other, abbreviate $(a \supset b) \supset (a \supset c)$ as d. It follows from (f_{do}) that $d \leq b \supset d$ and, once more from (f_{do}), that $b \supset d \leq a \supset (b \supset d)$; thus $d \leq a \supset (b \supset d)$. By definition of d, ($f_{d1}$) gives

$$b \supset d \leq (b \supset (a \supset b)) \supset (b \supset (a \supset c))$$

hence $b \supset d \leq b \supset (a \supset c)$ by (f6) since $b \supset (a \supset b)$ is in Q by (f_{do}). Thus

$$d \leq a \supset (b \supset d) \leq a \supset (b \supset (a \supset c))$$

by (f3). But $b \supset (a \supset c) \leq a \supset (b \supset c)$ by (f7) whence

$$d \leq a \supset (b \supset (a \supset c)) \leq a \supset (a \supset (b \supset c))$$

by (f3). But $a \supset (a \supset (b \supset c)) \leq (a \supset a) \supset (a \supset (b \supset c)) \leq a \supset (b \supset c)$ by (f_{d1}) and (f6), hence $d \leq a \supset (b \supset c)$.

(f10) $(a \supset b) \supset ((b \supset a) \supset a) \approx (b \supset a) \supset ((a \supset b) \supset b)$.

It suffices to show this with $\leq$ instead of $\approx$ because then the reverse inequality follows by symmetry. It follows from (f_{d1}) that $(a \supset b) \supset (a \supset b) \leq ((a \supset b) \supset a) \supset ((a \supset b) \supset b)$, and so together with $(a \supset b) \supset (a \supset b)$ also $((a \supset b) \supset a) \supset ((a \supset b) \supset b)$ is in Q. Consequently, also $(b \supset a) \supset (((a \supset b) \supset a) \supset ((a \supset b) \supset b))$ is in Q. It follows again from (f_{d1}) that

$$(b \supset a) \supset (((a \supset b) \supset a) \supset ((a \supset b) \supset b)) \leq ((b \supset a) \supset ((a \supset b) \supset a)) \supset ((b \supset a) \supset ((a \supset b) \supset b)) ,$$

thus also the latter element is in Q whence

$$(b \supset a) \supset ((a \supset b) \supset a) \leq (b \supset a) \supset ((a \supset b) \supset b) .$$

But $(a \supset b) \supset ((b \supset a) \supset a) \leq (b \supset a) \supset ((a \supset b) \supset a)$ by (f7), and so the desired inequality follows by transitivity.

It follows from (f4), (f8) that $\approx$ is a congruence relation with respect to $\supset$: if $a \approx a'$ and $b \approx b'$ then $a \supset b \approx a' \supset b'$. Because $a \approx a'$ implies $a \supset b \approx a' \supset b$ by (f8) and $b \approx b'$ implies $a' \supset b \approx a' \supset b'$ by (f4). Thus I may form the quotient algebra $A/\approx$ for the operation $\supset$, and if e denotes Q as an element of this algebra, then it follows from (f10) that $A/\approx,\{e\}$ is a special d–frame.

While every special d–frame is, in particular, a d–frame with $Q=\{e\}$, it now follows that a d–frame A,Q is special if Q contains one element only. For $a\approx b$ then implies $a=b$ because $a\supset b=e$, $b\supset a=e$ imply with (f6) that the left side of (f10) is equivalent to a and the right side is equivalent to b. This useful property of (f10) and (a_d) was discovered by DIEGO 61 . This completes the proof of Lemma 1.

It now also follows from (f0), (f6) and (f9) that the equations D_d hold in special d–frames; hence $\mathbf{Mod}(D_d)$ coincides with the class of special d-frames. Before continuing with the proof of Theorem 1, I first need some more auxiliary considerations.

Let T be a term algebra for the signature of A, with an operation $\rightarrow$ corresponding to $\supset$, and let t,s be terms containing *only* this operation. I shall say that an equation $t\equiv s$ *holds* in A,Q if $h(t)\approx h(s)$ for every homomorphism h from T into A; for special d–frames this coincides with the usual notion. More generally I may consider formal expressions $t\precsim s$ as *inequalities,* and then I shall say that such an inequality *holds* in a d–frame if $h(t)\precsim h(s)$ for every homomorphism h from T_d into A. Clearly, $t\equiv s$ holds if, and only if, both $t\precsim s$ and $s\precsim t$ hold.

An inequality, respectively an equation, holds in a d–frame A,Q if, and only if, it holds in the contracted d–frame $A/\approx,\{e\}$. By Lemma 1 these are precisely the special d–frames. Consequently, inequalities and equations hold

in all d–frames if, and only if, they hold
in all special d–frames if, and only if, they hold
in all algebras from $\mathbf{A}_d$.

I proceed with further computations. Consider a d–frame which I now may assume to be special. Let μ be a finite sequence of length m of elements of A; if $m>0$ let by μ_0 be the sequence obtained from μ by dropping $\mu(0)$. For any b in A, I define $\mu\supset^* b$ as

$$b \text{ if } m=0 \quad , \quad \mu(0)\supset(\mu_0\supset^* b) \text{ if } m>0 .$$

I now claim that these elements are independent of the order of μ, i.e.

(f11) if π is a permutation of m then $\mu\supset^* b=\mu\pi\supset^* b$.

This follows for $m=2$ from (f7); assume it to hold for m–1 and $m>2$. If π is a permutation of m such that $\pi(0)=0$ then the statement follows from the inductive assumption applied to $\mu_0\supset^* b$. If $\pi(0)\neq 0$ then it will suffice to treat the case that π is a transposition since every permutation it a product of such. Consider first the transposition ρ which exchanges 0 and 1. It follows from (f7) that $\beta\supset^* b=\beta\rho\supset^* b$ for every β. Hence if $\pi=\rho$ then I am done. Otherwise, taking σ to be the transposition which exchanges 1 and $\pi(0)$, I find

$\pi = \rho\sigma\rho$.

Now $\beta\supset^* b = \beta\sigma\supset^* b$ since σ leaves 0 unchanged; thus $\mu\rho\sigma\rho\supset^* b = \mu\rho\sigma\supset^* b = \mu\rho\supset^* b = \mu\supset^* b$.

(f12) If $m>0$ then $\mu_o\supset^* b \leq \mu\supset^* b$.

This follows from (f_{do}) since $(\mu_o\supset^* b) \supset (\mu(0) \supset (\mu_o\supset^* b)) = e$.

(f13) If $m>1$ and $\mu(0) = \mu(1)$ then $\mu\supset^* b = \mu_o\supset^* b$.

If $m - 2$ then this follows from (f7). If $m>2$ then let μ_1 the sequence obtained from μ_o by dropping $\mu(1)$. Thus

$$\begin{aligned} \mu\supset^* b &= \mu(0)\supset(\mu_o\supset^* b) \\ &= \mu(0)\supset(\mu(1)\supset(\mu_1\supset^* b)) \\ &= \mu(1)\supset(\mu_1\supset^* b) && \text{by (f7)} \\ &= \mu_o\supset^* b \end{aligned}$$

(f7*) $a\supset(\mu\supset^* b) = \mu\supset^*(a\supset b)$.

I proceed by induction on the length m of μ ; if $m = 1$ then (f7*) is (f7). The induction step is

$$\begin{aligned} a\supset(\mu\supset^* b) &= a\supset(\mu(0)\supset(\mu_o\supset^* b)) \\ &= \mu(0)\supset(a\supset(\mu_o\supset^* b)) \\ &= \mu(0)\supset(\mu_o\supset^*(a\supset b)) \\ &= \mu\supset^*(a\supset b) \ . \end{aligned}$$

(f14) If $\mu\supset^* a = e$ then $b\supset(\mu\supset^* c) = (a\supset b)\supset(\mu\supset^* c)$.

Because

$$\begin{aligned} (a\supset b)\supset(\mu\supset^* c) &= \mu\supset^*((a\supset b)\supset c) && \text{by (f7*)} \\ &= ((\mu\supset^* a)\supset(\mu\supset^* b))\supset\mu\supset^* c) && \text{by (f9*)} \\ &= (e\supset(\mu\supset^* b))\supset\mu\supset^* c) \\ &= (\mu\supset^* b)\supset(\mu\supset^* c) && \text{by (f6)} \\ &= \mu\supset^*(b\supset c) && \text{by (f9*)} \\ &= b\supset(\mu\supset^* c) && \text{by (f7*)} \ . \end{aligned}$$

(f4*) $b\supset c \leq (\mu\supset^* b)\supset(\mu\supset^* c)$.

For $m = 1$ this is (f4), and the induction follows from $(\mu_o\supset^* b)\supset(\mu_o\supset^* c) \leq (\mu(0)\supset(\mu_o\supset^* b))\supset(\mu(0)\supset (\mu_o\supset^* c))$ by (f4).

(f9*) $\mu\supset^*(b\supset c) = (\mu\supset^* b) \supset (\mu\supset^* c)$.

For $m = 1$ this is (f9), and the induction step is

$$\begin{aligned} \mu\supset^*(b\supset c) &= \mu(0)\supset(\mu_o\supset^*(b\supset c)) \\ &= \mu(0)\supset((\mu_o\supset^* b)\supset(\mu_o\supset^* c)) \\ &= (\mu(0)\supset(\mu_o\supset^* b)) \supset (\mu(0)\supset(\mu_o\supset^* c)) \\ &= (\mu\supset^* b) \supset (\mu\supset^* c) \ . \end{aligned}$$

At this point I can return to complete the proof of Theorem 1: the special d–frames belong to the class $\mathbf{A}_d$ associated to KP_d. This amounts to show that the equations $[v,w]$ from R_d hold in all special d–frames (or in all algebras from $\mathbf{Mod}(D_d)$): $h(v) = h(w)$ for every homomorphism h from T_d into a special frame A. To this end it will suffice to show that, if a sequent $v \Longrightarrow w$ has a derivation in KP_d, then $h(v) \lesssim h(w)$ in every A. Here I shall use the form K_tP_d of KP_d and use induction on the derivation of $v \Longrightarrow w$; therefore it becomes necessary to consider *general* sequents $M \Longrightarrow u$ and to look for a description of *their* images under the homomorphism h. It then will suffice to prove that, for every sequent $M \Longrightarrow u$ of a K_tP_d–derivation D there holds

(*) $\quad h \cdot M \supset^* h(u) = e$.

For a sequent with $M = v,N$ the left side in (*) is $h(v) \supset (h \cdot N \supset^* h(u))$ *by definition* of $\mu \supset^* b$. So consider D; for reflexive axioms $v \Longrightarrow v$ the statement (*) follows from (f0). If (*) holds for the premiss(es) of an instance of the permutation rule then it follows from (f11) that it holds for the conclusion; if (*) holds for the premiss of an instance of the weakening or the contraction rule then it follows from (f12) and (f13) that it also holds for the conclusion. For instances

$$\frac{v, M \Longrightarrow w}{M \Longrightarrow v \dashrightarrow w}$$

of (I⇢) the premiss goes into $h(v) \supset (h \cdot M \supset^* h(w))$ and the conclusion into $h \cdot M \supset^* h(v \dashrightarrow w) = h \cdot M \supset^* (h(v) \supset h(w))$, and by (f7*) these elements are the same. For instances

$$\frac{M \Longrightarrow v \qquad w, M \Longrightarrow u}{v \dashrightarrow w, M \Longrightarrow u}$$

of (E⇢) the premisses go into $h \cdot M \supset^* h(v)$, $h(w) \supset (h \cdot M \supset^* h(u))$, and if the first of these two is e then it follows from (f14) that the second equals $(h(v) \supset h(w)) \supset (h \cdot M \supset^* h(u))$ which is the transform of the conclusion. Thus together with (the first and) the second premiss also the conclusion will satisfy (*).

This completes the proof of Theorem 1. The last part actually contains an algorithm which translates a KP_d–derivation D into an equational derivation of the identities (*) in special d–frames; the instances of rules are translated into instances of the auxiliary equational deductions of the various formulas (f–) employed – e.g. for (I⇢) the formula (f7*).

COROLLARY 1_d The algebra T_d, together with the set Q of all v such that $\blacktriangle \Longrightarrow v$ is LK_d–derivable, is a d–frame. The relation R_d coincides with the congruence relation $\approx$ of that frame and is an invariant congruence relation on T_d. The special d–frame $F_d = T_d/\!\approx\, = T_d/R_d$ is the free algebra in $\mathbf{A}_d$ genera–

ted by the image of the set X of variables under the canonical homomorphism p_d from T_d onto F_d. It is equivalent that

$w \Longrightarrow v$ is derivable in KP_d
$v \leq w$ in the d-frame T_d, Q
$p_d(v) \leq p_d(w)$ in the special d-frame F_d

and that

$M \Longrightarrow v$ is derivable in KP_d
$M \rightarrow^* v$ is in Q
$p_d \cdot M \supset^* p_d(v) = e$ in the special d-frame F_d.

Consider the set Q. If also $\blacktriangle \Longrightarrow v \rightarrow w$ is derivable then $v \Longrightarrow w$ is derivable by **JI→**, and by (CUT) then $\blacktriangle \Longrightarrow w$ is derivable. Hence T_d, Q has the property (f_{mp}); (f_{do}) holds since from $x,y \Longrightarrow x$ I may conclude upon $x \Longrightarrow y \rightarrow x$, and (f_{d1}) follows from the first derivation after the statement of Theorem 2. Thus T_d, Q is a d-frame. Since the derivability of $v \Longrightarrow w$ is equivalent to that of $\blacktriangle \Longrightarrow v \rightarrow w$ by (I→) and **JI→** (and so to $v \leq w$), the relations R_d and $\approx$ are the same. Thus the special d-frame F_d is the contraction $T_d/\approx$, and so $v \rightarrow w \in Q$ is also equivalent to $p_d(v) \leq p_d(w)$. If $M \Longrightarrow v$ is derivable then $p_d \cdot M \supset^* p_d(u) = e$ by (*), and by definition of F_d this means that $M \rightarrow^* v$ is in Q. But then $\blacktriangle \Longrightarrow M \rightarrow^* v$ is derivable as just proven, and from that **JI→** gives a derivation of $M \Longrightarrow v$.

For every endomorphism g of T_d, any derivation of $M \Longrightarrow v$ can be extended to a derivation of $g(M) \Longrightarrow g(v)$, making use of reflexive axioms. Thus the congruence relation R_d is invariant and, therefore, already the set of *all* T_d-equations holding in $\mathbf{Mod}(R_d) = \mathbf{A}_d$. Also, as R_d is invariant, the equations from R_d hold in $F_d = T_d/R_d$, and it follows from Chapter **1.7** that F_d then is the free algebra in $\mathbf{A}_d$, generated by image of X under the canonical homomorphism p_d from T_d onto F_d. Since for distinct variables x, y there simply is no derivation of $x \Longrightarrow y$, the restriction of p_d to X is injective.

THEOREM 2_d A sequent $M \Longrightarrow v$ is KP_d-derivable if, and only if, for every algebra A in $\mathbf{A}_d$ with the distinguished element e_A and for every homomorphism h from T_d in A, there holds: if $h(m) = e_A$ for every m in M then $h(v) = e_A$.

Thus the KP_d-derivability of $M \Longrightarrow v$ is equivalent to v being a semantical consequence of M under the consequence operator cn defined by the class $\mathbf{A}_d$ of algebras with their distinguished elements. For the proof, consider first the case that $M \Longrightarrow v$ is derivable, hence $p_d \cdot M \supset^* p_d(v) = e_F$ in F_d. Since F_d is the free algebra in $\mathbf{A}_d$, every homomorphism h from T_d into an A from $\mathbf{A}_d$ factors as $g \cdot p_x$. Thus also $h \cdot M^* \supset h(u) = e_A$ in A. If M is empty then $h \cdot M \supset^* h(u) = h(u)$ by definition, and if $M = <m_0, \ldots, m_{m-1}>$ and $M_i = <m_i, \ldots, m_{m-1}>$ for $i<m$ then $h \cdot M \supset^* h(u)$ is

$$h(m_0) \supset (h \cdot M_1 \supset^* h(u)) .$$

It follows from the equation (a_1) that

$$\text{if } h(m_0) = e_A \text{ then } h \cdot M \supset^* h(u) = h \cdot M_1 \supset^* h(u)$$

whence by induction for $i<m$

$$\text{if } h(m_0) = h(m_1) = \ \ldots\ h(m_{i-1}) = e_A \text{ then } h \cdot M \supset^* h(u) = h \cdot M_i \supset^* h(u)$$

and

$$\text{if } h(m_0) = h(m_1) = \ \ldots\ h(m_{m-1}) = e_A \text{ then } h \cdot M \supset^* h(u) = h(u) .$$

Thus

$$\text{if } h(m_0) = h(m_1) = \ \ldots\ h(m_{m-1}) = e_A \text{ then}$$
$$h \cdot M \supset^* h(u) = e_A \text{ if, and only if, } h(u) = e_A .$$

The proof of the reverse implication is indirect (as are all such *completeness* proofs): I shall show that if $M \Longrightarrow v$ is *not* derivable, i.e. if $p_d \cdot M \supset^* p_d(v)$ is *not* e_F in F_d, then there is an B in $\mathbf{A}_d$ and a homomorphism h into B with $h(m) = e_B$ for m in M and $h(u) \neq e_B$. The algebra B will be a quotient algebra of F_d and h the canonical homomorphism onto B.

If A is an algebra in $\mathbf{A}_d$ I call *filter* in A any subset F such that $e_A \in F$ and such that $a \in F$ and $a \supset b \in F$ implies $b \in F$. Thus A,F then is a d-frame which determines the congruence relation $\approx$, and F becomes the element e_B of the quotient algebra $B = A/\approx$. Consider now a non empty finite set or sequence $N = <n_0, \ldots, n_{n-1}>$ of elements of A and let F be the set of all a in A such that $N \supset^* a = e_A$; by (f11) this is independent on the order of N. Then e_A is in F: this is clear for $n = 1$ and follows by induction from the definition of $N \supset^* a$. If a and $a \supset b$ are in F then $N \supset^*(a \supset b) = (N \supset^* a) \supset (N \supset^* b)$ by (f9*) implies $e_A = e_A \supset (N \supset^* b)$ whence $N \supset^* b = e_A$. Hence F is a filter. Finally, every n_i is in F : by (f11) again it suffices to show this for $i = n-1$, and if $n = 1$ then it follows from $n_0 \supset n_0 = e_A$, and for $n>1$ it follows by induction from the definition of $N \supset^* n_{n-1}$. It is easily seen that F is the smallest filter of A containing the elements of N.

I now apply this construction to $A = F_d$ and $N = p_d \cdot M$. If g is the homomorphism from A onto $B = A/\approx$ then $h = g \cdot p_d$ maps all of N and, therefore, all of M onto e_B. But $p_d \cdot M \supset^* p_d(v) \neq e_F$ implies that $p_d(v)$ is not in F, hence $h(v) = g(p_d(v))$ is not e_B.

2. e-Algebras, e-Frames and RPC-Semilattices

Let T_e be the term algebra generated by the variables in X under the operations $\rightarrow$ and $\wedge$. Let KP_e be the restriction of KP to T_e, let R_e be the relation of interdeducibility with respect to KP_e, and let $\mathbf{A}_e$ be the class of all models of R_e; the algebras in $\mathbf{A}_e$ I call e–*algebras*. There holds the

THEOREM 1_e $\mathbf{A}_e$ is the class of all algebras $A = \langle u(A), \supset, \cap \rangle$ with a distinguished element e in A which, in addition to (a_{mp}), (a_{do}), (a_{d1}), (a_d) satisfy

(a_{eo}) $(a \cap b) \supset a = e$
(a_{e1}) $(a \cap b) \supset b = e$
(a_{e2}) $(c \supset a) \supset ((c \supset b) \supset (c \supset (a \cap b))) = e$

and $\mathbf{A}_e$ is also the the class of all algebras defined by the set D_e consisting of D_d together with these equations.

In view of Theorems 1_d, 2_d the two descriptions characterize the same classes of algebras $\mathbf{Mod}(D_e)$, and referring to the first description I may call them special e–frames. That $\mathbf{A}_e$ is contained in $\mathbf{Mod}(D_e)$ can be seen by deducing the new equations, replacing e by, say, $a \supset a$. Actually, it suffices the the terms d, said to be equal to e, are such that $\blacktriangle \Longrightarrow d$ is derivable because then also $g \Longrightarrow d$ is derivable for any g. Thus for instance

$$\frac{u, u \to w \Longrightarrow u \qquad v, u, u \to w \Longrightarrow v}{u, u \to w, u \to v \Longrightarrow v} \qquad \frac{u, u \to v \Longrightarrow u \qquad w, u, u \to v \Longrightarrow w}{u, u \to w, u \to v \Longrightarrow w}$$

$$u, u \to w, u \to v \Longrightarrow v \wedge w$$

$$u \to w . u \to v \Longrightarrow u \to v \wedge w$$

$$u \to v \Longrightarrow (u \to w) \to (u \to v \wedge w)$$

$$\Longrightarrow (u \to v) \to ((u \to w) \to (u \to v \wedge w))$$

It remains to be shown is that the special e–frames belong to $\mathbf{A}_e$.

Let A be an algebra with operations $\supset$ and $\cap$; let Q be a subset of A. The pair A,Q shall be called an e–*frame* if with $\supset$ it is a d–frame and if there holds for all a,b,c:

(f_{eo}) $(a \cap b) \supset a$ is in Q,
(f_{e1}) $(a \cap b) \supset b$ is in Q,
(f_{e2}) $(c \supset a) \supset ((c \supset b) \supset (c \supset (a \cap b)))$ is in Q.

The algebras A of e–frames with only the operations $\supset$, $\cap$ and for which Q consists of one element e only are precisely the *special* e–frames mentioned above.

The properties proved for d–frames remain valid for e–frames. The equivalence $\approx$ becomes a congruence relation also for $\cap$. Because $a \lesssim e$ and $b \lesssim f$ imply $a \cap b \lesssim e$ by (f_{eo}) and $a \cap b \lesssim f$ by (f_{e1}) whence $a \cap b \lesssim e \cap f$ by (f_{e2}); the reverse inequality follows by symmetry. For special e–frames I find

(f15) $(\mu \supset^* a) \supset ((\mu \supset^* b) \supset (\mu \supset^* (a \cap b))) = e$.

If $m = 1$ then this is (a_{e2}). For $m>1$ it follows from

$$\begin{aligned}(\mu_o \supset^* a) &\leq (\mu_o \supset^* b) \supset (\mu_o \supset^* (a \cap b)) && \text{by inductive hypothesis}\\ \mu(0) \supset (\mu_o \supset^* a) &\leq \mu(0) \supset ((\mu_o \supset^* b) \supset (\mu_o \supset^* (a \cap b))) && \text{by (f4)}\\ &= (\mu(0) \supset (\mu_o \supset^* b)) \supset (\mu(0) \supset (\mu_o \supset^* (a \cap b))) && \text{by } (f_{d1}) .\end{aligned}$$

In the definition of e-frames (f_{e2}) may be replaced by

(f_{e2o}) $a \supset (b \supset (a \cap b))$ is in Q .

Because this holds in e-frames since $a \supset (b \supset a)$ and $a \supset (b \supset b)$ are in Q by (f_{do}) and (f1), further $(a \supset (b \supset a)) \supset ((a \supset (b \supset b)) \supset (a \supset (b \supset (a \cap b))))$ is in Q by (f15), and thus an application of (f_{mp}) gives (f_{e2o}). Conversely, it follows from (f_{e2o}) together with (f1) that $c \leq a \supset (b \supset (a \cap b))$, hence $c \supset (a \supset (b \supset (a \cap b)))$ is in Q, and then also $(c \supset a) \supset ((c \supset b) \supset (c \supset (a \cap b)))$ will be in Q by two applications of (f_{d1}) and (f_{mp}). – It follows that in the definition of special e-frames the equation (a_{e2}) may be replaced by

(a_{e2o}) $a \supset (b \supset (a \cap b)) = e$.

Now I complete the proof of Theorem 1_e by proving that the equations from R_e hold in all special e-frames. Again it suffices to show that, for every sequent $M \Rightarrow u$ of a K_tP_e-derivation D there holds

(*) $h \cdot M \supset^* h(u) = e$.

For instances

$$\frac{v, M \Rightarrow u}{v \wedge w, M \Rightarrow u}$$

of $(E\wedge)$ it follows from (a_{eo}) that $h(v) \cap h(w) \leq h(v)$ whence (f8) implies $h(v) \supset (h \cdot M \supset^* h(u)) \leq (h(v) \cap h(w)) \supset (h \cdot M \supset^* h(u))$ such that with the left also the right side will be equal to e. For instances of $(I\wedge)$

$$\frac{M \Rightarrow v \qquad M \Rightarrow w}{M \Rightarrow v \wedge w}$$

it follows from (f15) that together with the premisses also the conclusion will be transformed into e.

Replacing T_d, R_d, $\mathbf{A}_d$, F_d, p_d, d-frames and KP_d by T_e, R_e, $\mathbf{A}_e$, F_e, p_e, e-frames and KP_e there hold *without changes* to their proofs the

COROLLARY 1_e [verbally as Corollary 1_d]

THEOREM 2_e [verbally as Theorem 2_d]

[In particular, a filter F on an algebra A from $\mathbf{A}_e$ is defined as for an algebra from $\mathbf{A}_d$, but now A,F is an e-frame and so the congruence relation $\approx$

is also a congruence with respect to $\cap$. Still, a filter is closed with respect to $\cap$ as follows immediately from (a_{e2o}).]

As a surprising algebraic application of proof theoretical methods I notice that F_d is isomorphic to the subalgebra of F_e generated by the variables under the operation $\rightarrow$ alone. Because if v,w are formulas of T_e, containing only the connective $\rightarrow$, then $[v,w] \in R_e$ implies $[v,w] \in R_d$ since the derivations establishing $[v,w] \in R_e$ will use only formulas with the connective $\rightarrow$. Thus the map from T_d/R_d into T_e/R_e, sending the congruence class of v under R_d into that under R_e, is injective, and it also is homomorphic for the operation $\supset$.

The remainder of this section is devoted to another description of the e-algebras. Starting point is the observation that, being d–algebras with respect to $\supset$, the e–algebras are ordered by the relation $\leq$ such that $a \cap b$ is the *infimum* of a, b. Because $a \cap b$ is by (a_{eo}), (a_{e1}) a lower bound for a,b, and it follows from (a_{e2}) that $c \leq a$ and $c \leq b$ implies $c \leq a \cap b$.

Ordered sets in which any two elements have an infimum may be called (lower) *semi-lattice ordered*. The (lower) semi–lattice ordered sets $<E, \leq>$ stand in bijective correspondence to the *semilattices,* namely the algebras $<E, \cap>$ defined by the set SL consisting of equations

$$a \cap a = a \ ,$$
$$a \cap b = b \cap a \ ,$$
$$a \cap (b \cap c) = (a \cap b) \cap c \ .$$

The proof is a special case of that establishing the correspondence between lattice ordered sets and lattices. Thus the algebras A from $\mathbf{A}_e$ are semilattices with respect to their operation $\cap$. These semilattices, however, have much stronger properties.

A semilattice satisfying

for all b, c there exists a largest a with $a \cap b \leq c$

is called an RPC–semilattice. For every RPC–semilattice $A = <u(A), \cap>$ I define an operation $\supset$ by taking $b \supset c$ to be this largest (and therefore unique) element a. Thus

(A) $a \cap b \leq c$ if, and only if, $a \leq b \supset c$.

The algebra $<u(A), \cap, \supset>$ then is called the $\supset$*–expansion* of A. I now shall show the

THEOREM 3 The algebras in $\mathbf{A}_e$ are the $\supset$–expansions of their underlying semilattices, and every $\supset$–expansion of a semilattice is in $\mathbf{A}_e$. The class $\mathbf{A}_e$ can be defined by the following systems of equations:

(I) $a \cap a = a$, $a \cap b = b \cap a$, $a \cap (b \cap c) = (a \cap b) \cap c$ from SL and

(a_0) $a \supset a = b \supset b$
(a_3) $(a \cap b) \supset c = a \supset (b \supset c)$
(a_4) $(a \cap b) \supset b = a \supset a$
(a_5) $(a \supset b) \cap a = a \cap b$.

(II) The equations D_d together with (a_3) .

(III) The equations $a \cap b = b \cap a$ together with (a_0), (a_5) and

(a_6) $(a \supset b) \cap b = b$
(a_7) $a \supset (b \cap c) = (a \supset b) \cap (a \supset c)$.

The logical way to show that (a_3) holds in e–algebras is to show that for the free algebra T_e/R_e by deriving the corresponding sequents in KP_e:

$$\dfrac{\dfrac{\dfrac{x,y \Rightarrow y \quad x,y \Rightarrow y}{x,y \Rightarrow x \wedge y} \quad x,y,z \Rightarrow z}{(x \wedge y) \dashv z, x, y \Rightarrow z}}{\dfrac{(x \wedge y) \dashv z, x \Rightarrow y \dashv z}{(x \wedge y) \dashv z \Rightarrow x \dashv (y \dashv z)}}$$

and

$$\dfrac{\dfrac{x \wedge y \Rightarrow x \quad \dfrac{\dfrac{y \Rightarrow y}{x \wedge y \Rightarrow y} \quad x \wedge y, z \Rightarrow z}{x \wedge y, y \dashv z \Rightarrow z}}{x \dashv (y \dashv z), x \wedge y \Rightarrow z}}{x \dashv (y \dashv z) \Rightarrow (x \wedge y) \dashv z} \quad .$$

Translating these KP_e-derivations into equations for e–frames, I arrive at an equational proof:

$((a \cap b) \supset c) \supset (a \supset (b \supset c)) = a \supset (b \supset (((a \cap b) \supset c) \supset c)$ by (f7) twice
$= a \supset ((b \supset ((a \cap b) \supset c)) \supset (b \supset c))$ by (a_2)
$= a \supset (((b \supset (a \cap b)) \supset (b \supset c)) \supset (b \supset c))$ by (a_2)
$= ((a \supset (b \supset (a \cap b)) \supset (a \supset (b \supset c))) \supset (a \supset (b \supset c))$ by (a_2) twice
$= (e \supset (a \supset (b \supset c))) \supset (a \supset (b \supset c))$ by (a_{e2o})
$= (a \supset (b \supset c))) \supset (a \supset (b \supset c))$ by (a_1)
$= e$ by (a_0)

and

$(a \supset (b \supset c)) \supset ((a \cap b) \supset c)$
$= (a \cap b) \supset ((a \supset (b \supset c)) \supset c)$ by (f7)
$= (a \cap b) \supset (a \supset (b \supset c))) \supset ((a \cap b) \supset c)$ by (a_2)
$= (((a \cap b) \supset a) \supset ((a \cap b) \supset (b \supset c))) \supset ((a \cap b) \supset c)$ by (a_2)

$= (e \supset ((a \cap b) \supset (b \supset c))) \supset ((a \cap b) \supset c)$ by (a_{e0})
$= ((a \cap b) \supset (b \supset c)) \supset ((a \cap b) \supset c)$ by (a_1)
$= (((a \cap b) \supset b) \supset ((a \cap b) \supset c))) \supset ((a \cap b) \supset c)$ by (a_2)
$= (e \supset ((a \cap b) \supset c)) \supset ((a \cap b) \supset c)$ by (a_{e1})
$= ((a \cap b) \supset c) \supset ((a \cap b) \supset c)$ by (a_1)
$= e$ by (a_0) .

Thus (viii) holds in every A of $\mathbf{A}_e$, and the equivalence (A) is an immediate consequence of (viii). Hence A is the $\supset$-expansion of its underlying semilattice.

Next I show that every expansion A of an RPC-semilattice B satisfies the equations in D_e. If a is *any* element such that $c \leq a$, then certainly $a \cap c \leq c$; thus a possibly existing *largest* a with $a \cap c \leq c$ will necessarily be a largest of *all* elements in my semilattice. But such a largest element *does* exist and is, by definition, $c \supset c$; thus $c \supset c$ will be *the* largest element e of my semilattice, proving the equations (a_0) to hold. For this largest element e, I find $a \cap e = a$ for any a; hence the largest a such that $a \cap e \leq b$ will be b itself whence $b = e \supset b$, proving the equations (a_1). As a tool for the following, I prove for all b,c the equivalence

(B) $b \supset c = e$ if, and only if, $b \leq c$

Because it follows from (A) that $b \supset c = e$ is equivalent to e being the largest a such that $a \cap b \leq c$, and this now is equivalent to $a \cap b \leq c$ holding *for all* a in B. Taking $a = e$, there follows $b = e \cap b \leq c$; conversely, $b \leq c$ implies $a \cap b \leq c$ for all a.

In order to prove (a_2), I first show $a \supset (b \supset c) \leq (a \supset b) \supset (a \supset c)$ which by (A) is equivalent to $(a \supset (b \supset c) \cap (a \supset b) \leq a \supset c$ and equivalent to $(a \supset (b \supset c) \cap (a \supset b) \cap a \leq c$; write d for the left element. Then $d = d \cap a$ since $d \leq a$; also $d = d \cap a \leq b$ since $d \leq a \supset b$, hence $d = d \cap b$; finally $d = d \cap a \leq b \supset c$ since $d \leq a \supset (b \supset c)$, and thus $d = d \cap b \leq c$ what was to be shown. Secondly, I show $(a \supset b) \supset (a \supset c) \leq a \supset (b \supset c)$ which by (A) is equivalent to $((a \supset b) \supset (a \supset c)) \cap a \cap b \leq c$; write d now for the left element here. Then $d \leq (a \supset b) \supset (a \supset c)$ implies $d \cap (a \supset b) \cap a \leq c$, and so $d \cap (a \supset b) \leq c$ since $d \leq a$. But also $d \leq b$ will hold, so that $d = d \cap a \leq b$ whence $d \leq a \supset b$ and, therefore, $d = d \cap (a \supset b) \leq c$.

The equations (a_{e0}), (a_{e1}) follow by (B) from $a \cap b \leq a$, $a \cap b \leq b$. Finally, (a_{e2}) will hold if $c \supset a \leq (c \supset b) \supset (c \supset (a \cap b))$, and it follows from (A) that this is equivalent to $(c \supset a) \cap (c \supset b) \cap c \leq a \cap b$. Writing d for the left element, I find $d \leq c$ whence $d = d \cap c$, further $d \leq c \supset a$ and $d \leq c \supset b$ whence $d = d \cap c \leq a$ and $d \leq b$, and thus indeed $d \leq a \cap b$.

This concludes the proof of the first statements of Theorem 3. As for the axiom system (I), I first verify (a_5); this follows from (D_e) because $a \supset b \leq a \supset b$ implies $(a \supset b) \cap a \leq b$, and since $b \leq a \supset b$ by (f_{do}) this gives $(a \supset b) \cap a \leq b \cap a \leq (a \supset b) \cap a$ whence (a_5). Conversely, an algebra $A = \langle E, \cap, \supset \rangle$, satisfying (a_3) and such that $\langle E, \cap \rangle$ is a semilattice, will satisfy (A) if it contains an element e such that (B) holds. Since I assume (a_0), there is the unique element $e = a \supset a$. But $b \leq c$ implies $b \supset c = (b \cap c) \supset c = b \supset b = e$ via (bv^{1*}), and e is a largest element since $e \cap a = (a \supset a) \cap a = a \cap a = a$ by (a_5). Thus $b \supset c = e$ implies $b = e \cap b = (b \supset c) \cap b = b \cap c$ via (a_5), hence $b \leq c$.

The axioms from (II) have already been shown to hold in e–algebras. Conversely, I start by observing that (a_3) implies the associativity of $\cap$: given a,b,c, the following statements are equivalent for all d:

$$a \cap (b \cap c) \leq d \quad a \leq (b \cap c) \supset d \quad a \leq b \supset (c \supset d) \quad a \cap b < c \supset d \quad (a \cap b) \cap c \leq d .$$

Next, I derive (a_{e0}) and (a_{e1}): it follows from (2) that $(a \cap b) \supset a = a \supset (b \supset a) = e$ and it follows from (a_0) and (3) that $(a \cap b) \supset b = a \supset (b \supset b) = a \supset e = e$. Finally, I obtain (a_{e2}) by

$$\begin{aligned}
&(c \supset a) \supset ((c \supset b) \supset (c \supset (a \cap b)) \\
&= (c \supset a) \supset (c \supset (b \supset (a \cap b))) && \text{by } (a_2) \\
&= c \supset (a \supset (b \supset (a \cap b))) && \text{by } (a_2) \\
&= (c \cap a) \supset (b \supset (a \cap b)) && \text{by } (a_3) \\
&= ((c \cap a) \cap b) \supset (a \cap b) && \text{by } (a_3) \\
&= (c \cap (a \cap b)) \supset (a \cap b) && \text{by associativity} \\
&= c \supset ((a \cap b) \supset (a \cap b)) && \text{by } (a_3) \\
&= c \supset e && \text{by } (a_0) \\
&= e \quad .
\end{aligned}$$

Of the axioms in (III), (a_6) is a consequence of $b \leq a \supset b$ which follows from D_e. For (a_7), observe that (a_5) implies $(a \supset (b \cap c)) \cap a = a \cap (b \cap c)$ such that $(a \supset (b \cap c)) \cap a \leq b$ gives $a \supset (b \cap c) \leq a \supset b$, and $(a \supset (b \cap c)) \cap a \leq c$ gives $a \supset (b \cap c) \leq a \supset c$. Thus the left side in (a_7) lies below the right one. For the other inequality, $(a \supset b) \cap (a \supset c) \leq a \supset (b \cap c)$, it will suffice to show that $d = (a \supset b) \cap (a \supset c) \cap a$ satisfies $d \leq b \cap c$. But $d \leq a$ gives $d = d \cap a$, and $d \leq a \supset b$ gives $d = d \cap a \leq b$, $d \leq a \supset c$ gives $d \leq c$.

Conversely, let A be an algebra satisfying the stated equations, and write e for the element $a \supset a$ uniquely determined by (a_0). I define a relation $\leq$ by $a \leq b$ if $a = a \cap b$; the commutativity of $\cap$ implies that $\leq$ is antisymmetric. Further, $b \leq e$ for every b since $e \cap b = b$ by (a_6). Reflexivity of $\leq$ follows from $a = e \cap a = (a \supset a) \cap a = a \cap a$ by (a_5). I now verify (B): $b \leq c$, i.e. $b = b \cap c$, implies $e = b \supset b = b \supset (b \cap c) = (b \supset b) \cap (b \supset c) = e \cap (b \supset c) = b \supset c$ by (a_7); conversely, $b \supset c = e$ implies $b = e \cap b = (b \supset c) \cap b = b \cap c$ by (a_5) and so $b \leq c$. Next, I show the transitivity of $\leq$: if $a \leq b$ and $b \leq c$ then $a \supset b = e$ and $b \cap c = b$ whence $e = a \supset b = a \supset (b \cap c) = (a \supset b) \cap (a \supset c) = e \cap (a \supset c) = a \supset c$ by (a_7) and so $a \leq c$ by (B).

At this point, I have proved that $\leq$ is an order; now I shall show that $a \cap b$ is the infimum of a,b . As an auxiliary result, I observe that $a \cap b = e$ implies $a = e$: this follows from $e = a \supset e = a \supset (a \cap b) = (a \supset a) \cap (a \supset b) = e \cap (a \supset b) = a \supset b$, hence $a \leq b$, $a \cap b = a$ and thus $a = e$. Now I begin by showing that $a \cap b$ is a lower bound of both a and b: $e = (a \cap b) \supset (a \cap b) = ((a \cap b) \supset a) \cap ((a \cap b) \supset b)$

holds by (a_7), and so the auxiliary result implies $(a\cap b)\supset a = e$, i.e. $a\cap b \leq a$, from where $a\cap b \leq b$ by the commutativity of $\cap$. If c is a lower bound of a and of b then $e\leq e$ implies $e = e\cap e = (c\supset a)\cap(c\supset b) = c\supset(a\cap b)$ by (a_7) whence $c\leq a\cap b$. Thus $a\cap b$ *is* the infimum of a,b .

I know by now that $<u(A),\cap>$ is a semilattice with largest element e. In view of Part I, the proof will be completed if I can show that A satisfies (A) because then $<u(A),\cap>$ will be an RPC–lattice with the expansion A. Assume, therefore, $a\cap b\leq c$. It follows from (a_6) that $a = a\cap(b\supset a)$, hence $a\leq b\supset a$; thus $a\leq b\supset c$ will follow if I can show $b\supset a\leq b\supset c$, i.e. $b\supset a = (b\supset a)\cap(b\supset c)$. But this holds since $(a\cap b)\cap c = a\cap b$ implies $b\supset a = (b\supset a)\cap e = (b\supset a)\cap(b\supset b) = b\supset(a\cap b)$ by (a_7) whence $b\supset a = b\supset(a\cap b) = b\supset((a\cap b)\cap c) = (b\supset(a\cap b))\cap(b\supset c) = (b\supset a)\cap(b\supset c)$ once more by (a_7). On the other hand, assume $a\leq b\supset c$. Then $a\cap(b\supset c) = a$ whence $a\cap b = (a\cap(b\supset c))\cap b = a\cap(b\cap(b\supset c)) = a\cap(b\cap c)$ by (a_5), i.e. $a\cap b = (a\cap b)\cap c$ and thus $a\cap b\leq c$. – This concludes the proof of Theorem 3.

3. g–Algebras, g–Frames and RPC–Lattices

Let T_g be the term algebra generated by the variables in X under the operations $\rightarrow$, $\wedge$ and $\vee$. Let R_g be the relation of interdeducibility with respect to KP, and let $\mathbf{A}_g$ be the class of all models of R_g. There holds the

THEOREM 1_g $\mathbf{A}_g$ is the class of all algebras $A = <u(A),\supset,\cap,\cup>$ with a distinguished element e in A which, in addition to (a_{mp}), (a_{do}), (a_{d1}), (a_d), (a_{eo}), (a_{e1}), (a_{e2}) satisfy

(a_{go}) $a\supset(a\cup b) = e$
(a_{g1}) $b\supset(a\cup b) = e$
(a_{g2}) $(a\supset c)\supset((b\supset c)\supset((a\cup b)\supset c)) = e$

and $\mathbf{A}_g$ is also the the class of all algebras defined by the set D_g consisting of D_e together with these equations.

The proof is a simple extension of that of Theorem 1_e. The algebras satisfying the conditions of the Theorem I call *special* g–*frames.* A derivation securing the validity of (a_{g2}) in the algebras from $\mathbf{A}_g$ is

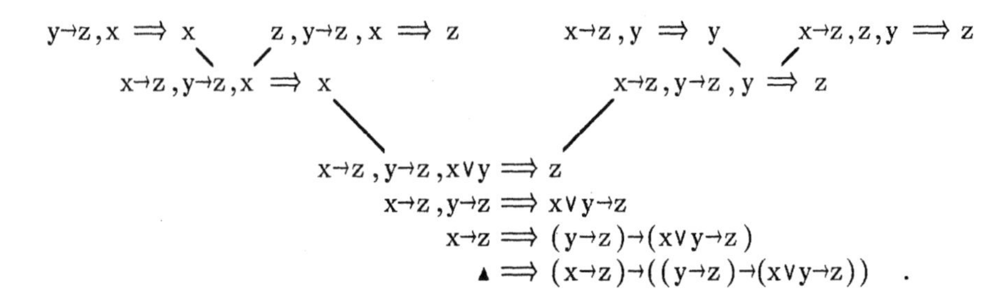

Let A be an algebra with operations $\supset$, $\cap$ and $\cup$; let Q be a subset of A. The pair A,Q shall be called an g-*frame* if with $\supset$ and $\cap$ it is an e-frame and if there holds for all a,b,c:

(f_{g0}) $a\supset(a\cup b)$ is in Q,
(f_{g1}) $b\supset(a\cup b)$ is in Q,
(f_{g2}) $(a\supset c)\supset((b\supset c)\supset((a\cup b)\supset c))$ is in Q.

The algebras A of g-frames with only the operations $\supset$, $\cap$, $\cup$ and for which Q consists of one element e only are precisely the *special* g-frames mentioned above.

The properties proved for e-frames remain in effect for g-frames. The equivalence $\approx$ becomes a congruence relation also for $\cup$. Because $e\lesssim a$ and $f\lesssim b$ imply $e\lesssim a\cup b$ by (f_{g0}) and $f\lesssim a\cup b$ by (f_{g1}) whence $e\cup f\lesssim a\cup b$ by (f_{g2}); the reverse inequality follows by symmetry. For special g-frames I find

$$(f16)\qquad (a\supset(\mu\supset^*c))\supset(b\supset(\mu\supset^*c))\supset((a\cup b)\supset(\mu\supset^*c))) = e\ .$$

Because (a_{g2}), namely $a\supset c\ \leq\ (b\supset c)\supset((a\cup b)\supset c)$, gives under (f4*) and (f9) first $\mu\supset^*(a\supset c)\ \leq\ (\mu\supset^*(b\supset c)\supset(\mu\supset^*((a\cup b)\supset\#)))$, and then applications of (f7*) will give the inequality belonging to (f16).

The proof of Theorem 1 will be completed by proving that the equations from R_g hold in all special g-frames. It suffices to show that, for every sequent $M\Longrightarrow u$ of a K_tP_g-derivation D there holds $h\cdot M\ \supset^*\ h(u) = e$. For instances of $(I\vee)$

$$\begin{array}{l} M\Longrightarrow v\\ M\Longrightarrow v\vee w\end{array}$$

(a_{g0}) gives $h(v)\leq h(v)\cup h(w)$ such that an application of (f4*) shows that together with the premiss also the conclusion is transformed into e.Foraninstance

$$\begin{array}{cc} v,M\Longrightarrow u & w,M\Longrightarrow u\\ \multicolumn{2}{c}{v\vee w,M\Longrightarrow u}\end{array}$$

of $(E\vee)$ an application of (f16) shows the corresponding result.

Replacing T_d, R_d, $\mathbf{A}_d$, F_d, p_d, d-frames and KP_d by T_g, R_g, $\mathbf{A}_g$, F_g, p_g, g-frames and KP there hold *without changes* to their proofs the

COROLLARY 1_g [verbally as Corollary 1_d]

THEOREM 2_g [verbally as Theorem 2_d]

As an algebraic application, I notice this time that F_d and F_e are isomorphic to the subalgebras of F_g generated by the variables under the operation $\rightarrow$ and the operations $\rightarrow$, $\wedge$ alone.

I now turn to the algebras from the class $\mathbf{A}_g$. It follows from (a_{g0}), (a_{g1}) that $a \cup b$ is an upper bound for a,b, and it follows from (a_{g2}) that if $a \leq c$ and $b \leq c$ then also $a \cup b \leq c$; thus $a \cup b$ is the *supremum* of a,b. Consequently, if $A = \langle E, \cap, \cup, \supset \rangle$ is in $\mathbf{A}_g$ then $\langle E, \cap, \cup \rangle$ is a lattice with the largest element e and satisfying again

(A) $a \cap b \leq c$ if, and only if, $a \leq b \supset c$.

Lattices of this kind are called RPC–lattices, and it follows from Theorem 3 that $\mathbf{A}_g$ consists precisely of the expansions of RPC–lattices. RPC–lattices are *distributive*. Because $(a \cap b) \cup (a \cap c) \leq a \cap (b \cup c)$ holds in any lattice, and so I only need to show that

$$a \cap (b \cup c) \leq (a \cap b) \cup (a \cap c) .$$

Indeed, $a \cap b \leq (a \cap b) \cup (a \cap c)$ and $a \cap c \leq (a \cap b) \cup (a \cap c)$ imply $b \leq a \supset ((a \cap b) \cup (a \cap c))$ and $c \leq a \supset h(a \cap b) \cup (a \cap c))$ whence $b \cup c \leq a \supset ((a \cap b) \cup (a \cap c))$ and thus $a \cap (b \cup c) \leq (a \cap b) \cup (a \cap c)$.

Every finite distributive lattice A is an RPC–lattice. Because since $a = c \cap b$ satisfies $a \cap b \leq c$ there are *some* a satisfying this condition, and if A is finite then there is at least one maximal such a, say a_0. But then for any such a it follows from $a_0 \cap b \leq c$ and $a \cap b \leq c$ that also $(a_0 \cup a) \cap b \leq c$ by distributivity. Since a_0 was maximal, there follows $a_0 \cup a = a_0$ and $a \leq a_0$. Thus a maximal a is already a largest one. – Unfortunately, it will become clear in a moment that the logically interesting RPC–lattices are anything but finite.

COROLLARY 2 The class $\mathbf{A}_g$ may be defined by adjoining, to any set defining $\mathbf{A}_e$, the equation

$$(a_8) \quad (a \cup b) \supset c = (a \supset c) \cap (b \supset c) .$$

In order to see that (a_8) holds in an algebra from $\mathbf{A}_g$, observe that given a,b,c, the following statements are equivalent for all d:

$d \leq (a \cup b) \supset c$
$d \cap (a \cup b) \leq c$
$(d \cap a) \cup (d \cap b) \leq c$

$d \cap a \leq c$ *and* $d \cap b \leq c$
$d \leq a \supset c$ *and* $d \leq b \supset c$
$d \leq (a \supset c) \cap (b \supset c)$.

Conversely, let $A = <E, \cap . \cup, \supset>$ be so that (a_8) holds and that $<E, \cap, \supset>$ is in $\mathbf{A}_e$. Taking c in (a_8) to be $a \cup b$, it follows that $e = (a \cup b) \supset (a \cup b) = (a \supset (a \cup b)) \cap (b \supset (a \cup b))$. But for the largest element e of a lattice it follows from $e = p \cap q$ that $p = e$ and $q = e$; hence $e = a \supset (a \cup b)$ and $e = b \supset (a \cup b)$. Thus $a \cup b$ is an upper bound of a,b, and if c is another upper bound of a,b then (a_8) implies $e = e \cap e = (a \supset c) \cap (b \supset c) = (a \cup b) \supset c$ whence $a \cup b \leq c$. It follows that $<E, \cap, \cup>$ is an RPC–lattice, and since A is its expansion, A will be in $\mathbf{A}_g$. – I mention a further identity which holds in $\mathbf{A}_g$:

(a_9) $\quad (a \supset c) \cup (b \supset c) \leq (a \cap b) \supset c$.

Because $a \supset c \leq (a \cap b) \supset c$ by (a_{eo}) and (f8) and $a \supset b \leq (a \cap b) \supset c$ by (a_{e1}) and (f8). Thus (a_9) follows from $p \cup q$ being the supremum of p,q.

THEOREM 4_g There are infinitely many formulas, containing only the two variables x, y, not two of which are interdeducible in KP.

The formulas containing only x,y form a subalgebra T_g^2 of T_g which, under the canonical homomorphism onto T_g/R_g, is mapped onto the algebra F_g^2 freely generated in $\mathbf{A}_g$. Thus it suffices to show that F_g^2 is infinite, and this will be the case if I can exhibit an infinite algebra N in $\mathbf{A}_g$ generated by two elements. Such an N with its largest element e is

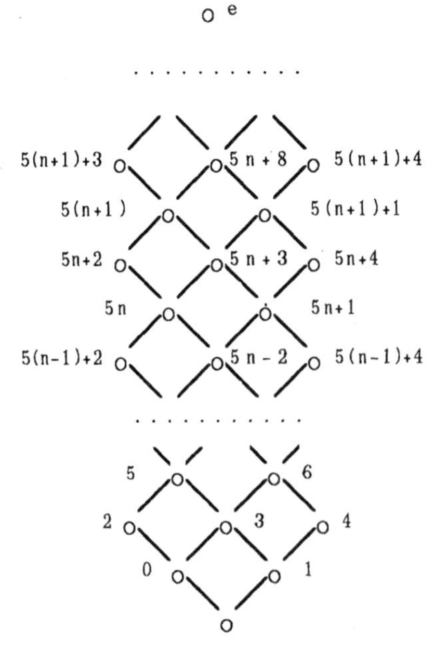

Here a_o, a_1 (indicated only by their indices) generate the algebra by

$$a_{5n} \cup a_{5n+1} = a_{5n+3} , \ a_{5n} \supset a_{5n+1} = a_{5n+4} , \ a_{5n+1} \supset a_{5n} = a_{5n+2} ,$$

$$a_{5n+2} \cup a_{5n+3} = a_{5n+5} , \ a_{5n+3} \cup a_{5n+4} = a_{5n+6} .$$

That the underlying lattice is distributive can be (seen from the diagram and) verified from these relations. It is not hard to verify that N actually *is* already the algebra F_g^2. The infinitely many formulas, pairwise not interdeducible, are $v_o = x$, $v_1 = y$

$$v_{5n+2} = v_{5n+1} \dashv v_{5n} , \ v_{5n+3} = v_{5n} \vee v_{5n+1}, \ v_{5n+4} = v_{5n} \dashv v_{5n+1} ,$$

$$v_{5n+5} = v_{5n+2} \vee v_{5n+3} , \ v_{5n+6} = v_{5n+3} \vee v_{5n+4} .$$

It may be mentioned that the algebra F_g^1 freely generated in $\mathbf{A}_g$ by *one* element x contains, besides x, only one more element, namely $e = x \supset x$.

4. m-Algebras, m-Frames and m-Lattices

Let $T = T_m$ be the term algebra generated by the variables in X under the operations $\dashv$, $\wedge$, $\vee$ and $\neg$. Let R_m be the relation of interdeducibility with respect to LM, and let $\mathbf{A}_m$ be the class of all models of R_m; the algebras in $\mathbf{A}_m$ I call m−*algebras.* There holds the

THEOREM 1_m $\mathbf{A}_m$ is the class of all algebras $A = <u(A), \supset, \cap, \cup, -> $ with a distinguished element e in A which, in addition to (a_{mp}), (a_{do}), (a_{d1}), (a_d), (a_{eo}), (a_{e1}), (a_{e2}), (a_{go}), (a_{g1}), (a_{g2}) satisfy

(a_{mo}) $\quad a \supset -b = b \supset -a$

and $\mathbf{A}_m$ is also the the class of all algebras defined by the set D_m consisting of D_g together with this equation.

The proof is a simple extension of that of Theorem 1_g. The algebras satisfying the conditions of the Theorem I call *special* m−*frames.* A derivation securing the validity of (a_{mo}) in the algebras from $\mathbf{A}_m$ is

$$\begin{array}{ccc} & & y , x \Longrightarrow y \\ y , x \Longrightarrow x & & \neg y , y , x \Longrightarrow \blacktriangle \\ & \diagdown \quad \diagup & \\ x \dashv \neg y , y , x \Longrightarrow \blacktriangle & & \\ x \dashv \neg y , y \Longrightarrow \neg x & & \\ x \dashv \neg y \Longrightarrow y \dashv \neg x & & \end{array}$$

Let A be an algebra with operations $\supset$, $\cap$, $\cup$ and $-$; let Q be a subset of A. The pair A,Q shall be called an m-*frame* if with $\supset$, $\cap$, $\cup$ it is an g-frame and if there holds for all a,b,c:

(f_{mo}) $\quad a\supset -b \leq b\supset -a$.

The algebras A of m-frames with only the operations $\supset$, $\cap$, $\cup$, $-$ and for which Q consists of one element e only are precisely the *special* m-frames mentioned above.

The properties proved for g-frames remain in effect for m-frames. (f_{mo}) implies $-a\supset -a \leq a\supset --a$, and since $-a\supset -a$ is in Q, it follows that

(f_{m1}) $\quad a \leq --a$.

Now $b \leq --b$ implies $a\supset b \leq a\supset --b$ by (f3); since $a\supset --b \leq -b\supset -a$ by (f_{mo}), transitivity gives

(f_{m2}) $\quad a\supset b \leq -b\supset -a$.

Thus the equivalence $\approx$ becomes a congruence relation also with respect to the operation $-$. For future use I mention

(f17) $\qquad b \leq (a\supset -b)\supset -a$

which follows from (f_{mo}) by (f7) and is, actually, equivalent to (f_{mo}).

(f_{m3}) $\quad -a \leq a\supset -b$.

This follows by (f_{mo}) from $-a \leq b\supset -a$ which is a consequence of (f_{do}).

The proof of Theorem 1 will be completed by proving that, for every homomorphism h from T into an A from $\mathbf{A}_m$, the LM-derivability of v $\Longrightarrow$ w implies $h(v) \leq h(w)$. Given an LM-derivation D, I shall prove as before that, for every sequent M $\Longrightarrow$ u in D there holds $h\cdot M\supset^* h(u) = e$ and, in addition, that for any sequent M $\Longrightarrow$ ▲ in D there holds $h\cdot M\supset^* -e = e$. For an instance of (I$\neg$)

$$\begin{array}{c} v,M \Longrightarrow \blacktriangle \\ M \Longrightarrow \neg v \end{array}$$

the premiss then goes into $h(v)\supset(h\cdot M\supset^* -e)$ which is $h\cdot M\supset^*(h(v)\supset -e)$ by (f7*). But $h(v)\supset -e = e\supset -h(v) = -h(v) = h(\neg v)$ by (f_{mo}) and (f6). Thus $h(v)\supset(h\cdot M\supset^* -e) = h\cdot M\supset^* h(\neg v)$, and this is the transform of the conclusion. For an instance of (E$\neg$)

$$\begin{array}{c} M \Longrightarrow v \\ \neg v,M \Longrightarrow \blacktriangle \end{array}$$

the conclusion goes into $-h(v)\supset(h\cdot M\supset^* -e)$ which is $h\cdot M\supset^*(-h(v)\supset -e)$ by (f7*). But $h(v) \leq (e\supset -h(v))\supset -e$ by (f17) whence $h(v) \leq -h(v)\supset -e$. Hence $h\cdot M\supset^* h(v) \leq h\cdot M\supset^*(-h(v)\supset -e)$ by (f4*), and if the the transform of the premiss is e then so is that of the conclusion.

Replacing T_d, R_d, $\mathbf{A}_d$, F_d, p_d, d–frames and KP_d by T_m, R_m, $\mathbf{A}_m$, F_m, p_m, m–frames and LM there hold *without changes* to their proofs the

COROLLARY 1_m [verbally as Corollary 1_d]

THEOREM 2_m [verbally as Theorem 2_d]

The algebras in $\mathbf{A}_m$ are (expansions of) RPC–lattices carrying the additional operation $-$. It follows from (f_{m2}) that the the operation $-$ inverts the order. Thus (f_{m1}) implies $-(--a) \leq -a$, but since also $-a \leq --(-a)$, it follows that

(1) $\quad -a = ---a$.

Now (f_{m3}) implies $-a \leq a \supset -e$, and (f_{mo}) implies $a \supset -e \leq e \supset -a = -a$ whence

(2) $\quad -a = a \supset -e$

which equation was the starting point when negation was introduced in Chapter **4**.

The algebras in $\mathbf{A}_m$ are in bijective correspondence with the pairs A, q consisting of algebras in $\mathbf{A}_g$ – or RPC–lattices – and elements q of A. Because given A and q, I can define an operation $-$ by $-a = a \supset q$. Then (a_{mo}) holds in consequence of (f7): $a \supset (b \supset q) = b \supset (a \supset q)$. If q is chosen as the largest element of A then the operation $-$ is the identity.

THEOREM 4_m There are infinitely many formulas, containing only the variable x, not two of which are interdeducible in LM.

The formulas containing only x form a subalgebra T^1 of T which, under the canonical homomorphism onto T/R_m, is mapped onto the algebra ${F_m}^1$ freely generated in $\mathbf{A}_m$. Thus it suffices to show that ${F_m}^1$ is infinite, and this will be the case if I can exhibit an infinite algebra N in $\mathbf{A}_m$ generated by one element. Such an N with a largest element e and a smallest element q can be generated from a_o with the relations

$$-a_o = a_1 \ , \quad -a_1 = a_2 \ , \quad -a_2 = a_1 \ , \quad -a_i = q \quad \text{for } i \geq 3 \ ,$$

$$a_{4n} \cup a_{4n+1} = a_{4n+3} \ , \quad a_{4n+1} \supset a_{4n} = a_{4n+2} \ ,$$

$$a_{4n+2} \cup a_{4n+3} = a_{4n+4} \ , \quad a_{4n+2} \supset a_{4n+3} = a_{4n+5} \ ,$$

as shown by the diagram

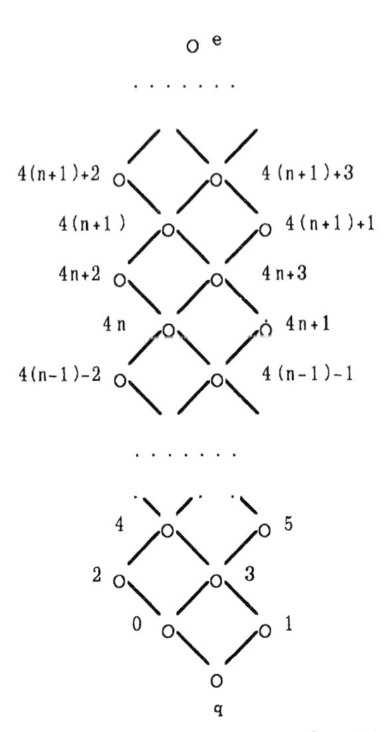

It is not hard to verify that F_m^1 arises from N by putting one additional element below q. The infinitely many formulas, pairwise not interdeducible, are $v_0 = x$, $v_1 = \neg x$

$$v_{4n+2} = v_{4n+1} \rightarrow v_{4n} \,,\quad v_{4n+3} = v_{4n} \vee v_{4n+1} \,,$$

$$v_{4n+4} = v_{4n+2} \vee v_{4n+3} \,,\quad v_{4n+5} = v_{4n+2} \rightarrow v_{4n+3} \,.$$

It follows from statement (1) that in an m–algebra A the elements a satisfying $a = --a$ are precisely the values of the operation $-$; they are called the *regular* elements of A. Together with (2), the equation (a_8) (from Corollary 2) implies

(3) $\quad -(a \cup b) = -a \cap -b$.

It follows that the infimum of two regular elements is regular again. Consider now the function ρ from A into A defined by

$$\rho a = --a \;;$$

it preserves the order since it is the composition of two order inverting maps. If c is regular, i.e. $\rho c = c$, then $a \leq c$ implies $\rho a \leq c$; thus ρa is the *smallest* regular element covering a. In particular, $\rho(a \cup b)$ is the smallest regular element covering both a and b :

The regular elements of A form a lattice REG(A) with $\inf(a,b) = a \cap b$ as in A and $\sup(a,b) = --(a \cup b) = -(-a \cap -b)$.

The supremum in REG(A) of ρa and ρb is $--(--a\cup--b)=-(---a\cap---b)=-(-a\cap-b)=--(a\cup b)=\rho(a\cup b)$ by repeated applications of (3); hence ρ is homomorphic for the supremum operations of A and REG(A), and I shall prove below in (11) that it is also homomorphic for the infimum operations. Since $a\cap b\leq a$ implies $-a\leq-(a\cap b)$ and since $-(a\cap b)$ is regular, I find

(4) $\quad -a\cup-b\leq--(-a\cup-b)\leq-(a\cap b)$.

The largest element e of A cannot be anything but regular since $e\leq--e$ implies $e=--e$. Since $a\leq e$ implies $-e\leq-a$, it follows that $-e$ is the *smallest* regular element of A. Now if a is regular then $-a$ is much more so; hence also $a\cap-a$ is regular. As A is an RPC–lattice, the equivalence (A) for these lattices shows that (f_{m3}) is equivalent to

(5) $\quad -a\cap a\leq-b$.

Thus, for regular a, the regular element $a\cap-a$ lies below any other regular element $-b$ and is, therefore, also the smallest regular element $-e$. This implies for *any* a

(6) $\quad -a\cap--a=-e$.

Since $a\leq--a$ implies $a\cap-a\leq--a\cap-a=-e$, it follows that $e=--e\leq-(a\cap-a)$ whence for *any* a

(7) $\quad -(a\cap-a)=e$.

In particular, $e=-(-a\cap--a)=--(a\cup-a)$ by (3) whence

(8) $\quad --(a\cup-a)=e$ *and* $-(a\cup-a)=-e$.

In particular, for regular a the regular supremum of a and $-a$ is e, and since the (regular) infimum of a and $-a$ is $-e$ by (5), I obtain

> The lattice REG(A) has the largest element e and the smallest element $-e$, and $-a$ is a complement of a in this lattice.

The two observations

(9) $\quad d\cap a\leq-e$ if, and only if, $d\leq-a$

(10) $\quad$ if $a\cap b\leq-e$ then $--a\cap b\leq-e$

are preparatory for (11). The first of them is just (2), rewritten with the equivalence (A) for RPC–lattices. The second holds since $a\cap b\leq-e$ implies $b\leq-a$ whence $--a\cap b\leq--a\cap-a=-e$. Now I shall prove

(11) $\quad --a\cap--b=--(a\cap b)$.

It follows from $a\cap b\leq a$ that $--(a\cap b)\leq--a\cap--b$. Conversely, since $a\cap b\cap-(a\cap b)\leq-e$ by (5), it follows by two applications of (10) that $--a\cap--b\cap-(a\cap b)\leq-e$ whence $--a\cap--b\leq--(a\cap b)$ by (9).

Now (11) this implies that the map ρ is homomorphic for the infimum operation also. Hence ρ is a homomorphism of lattices, and since A was distri–

butive, the homomorphic image REG(A) of A will be distributive as well. Finally, ρ is homomorphic for the complement operation by (1). Thus

REG(A) is a Boolean lattice, and ρ is a homomorphism from A onto REG(A) for the lattice and complement operations.

Finally, I can characterize the congruence relation induced by ρ on A in the manner described for Boolean algebras in Chapter **1.4**. I define the set DENS(A) of *dense* elements of A to consist of all d such that $\rho d = e$; evidently, DENS(A) is a *filter* in the sense defined for Boolean algebras. The desired characterization then is

(12) $\rho a = \rho b$ if, and only if, there exists d in DENS(A) such that $-(d \cap a) = -(d \cap b)$.

If such an element d is given then it follows from $\rho d = e$ that $\rho a = \rho d \cap \rho a = \rho(d \cap a) = --(d \cap a) = --(d \cap b) = \rho(d \cap b) = \rho b$ since ρ is homomorphic. Conversely, if $\rho a = \rho b$ then $d = (a \cup -a) \cap (b \cup -b)$ is dense since $a \cup -a$ and $b \cup -b$ are dense by (8). Now $--a = --b$ implies $-a = -b$ whence $a \cap -b = a \cap -a$, hence $a \cap -b \leq a \cap b$ by (5), and so $-(a \cap b) \leq -(a \cap -b)$. Since $a \cap d = a \cap (a \cup -a) \cap (b \cup -b) = a \cap (b \cup -b) = (a \cap b) \cup (a \cap -b)$, also $-(a \cap d) = -(a \cap b) \cap -(a \cap -b)$ by (3), hence $-(a \cap d) = -(a \cap b)$, and since by symmetry also $-(b \cap d) = -(a \cap b)$, I arrive at $-(a \cap d) = -(b \cap d)$. – I collect the results proven now in the

THEOREM 5_m For A in $\mathbf{A}_m$ the set REG(A) of regular elements is a Boolean lattice under the induced operations $\cap$ and $-$ and with the supremum operation $--(a \cup b) = -(-a \cap -b)$.

The map ρ is homomorphic from A onto the Boolean lattice REG(A) for lattice and complement operations, and the congruence relation induced by ρ is characterized by (12) .

This Theorem is due to GLIVENKO 29. Analyzing its proof, the reader will notice that it remains in effect for algebras A which are distributive lattices with a largest element e and a smallest element 0 and such that, for any a, there exists a largest d such that $d \cap a = 0$ which is chosen as $-a$. The element $-a$ then is called a *pseudocomplement* of a and the lattices mentioned are called *pseudocomplemented.* The lattices in $\mathbf{A}_m$ may not contain a smallest element 0, but now for every b there exists, for any a, the largest d such that $d \cap a = b$, namely $d = a \supset b$. For this reason, $a \supset b$ is also called the *relative pseudocomplement* of a for b, and the RPC-lattices are the *relatively pseudocomplemented* lattices.

Let p_m be the natural homomorphism from T onto T/R_m and let ρ now be the homomorphism from T/R_m onto $REG = REG(T/R_m)$. I then find that

(13) The following formulas are LM–interdeducible

$\neg\neg(x \wedge y)$ and $\neg\neg x \wedge \neg\neg y$
$\neg\neg(x \vee y)$ and $\neg(\neg x \wedge \neg y)$.

This follows from $p_m(\neg\neg(x \wedge y)) = --(p_m(x) \cap p_m(y)) = \rho(p_m(x) \cap p_m(y)) = \rho \cdot p_m(x) \cap \rho \cdot p_m(y) = p_m(\neg\neg x) \cap p_m(\neg\neg y) = p_m(\neg\neg x \wedge \neg\neg y)$ by (11) and from $p_m(\neg\neg(x \vee y)) = --(p_m(x) \cup p_m(y)) = -(-p_m(x) \cap -p_m(y)) = -(p_m(\neg x) \cap p_m(\neg y) = p_m(\neg(\neg x \wedge \neg y))$ by (3). The explicit LM–derivations I presented as **4**.(1)–(4).

I conclude this section with a construction which will be used in order to provide counter examples. Let A be an RPC–lattice and let B be an algebra in $\mathbf{A}_m$, both with largest elements e_A and e_B, and assume that $-e_B$ is the smallest element of B (e.g. if B is a Boolean algebra). Define C by putting B on top of A so that e_A becomes the unique lower neighbour of $-e_B$. It is clear then that C is a lattice (and it may be left to the reader to verify that C is distributive). On C I shall define the operation $\supset_c$ in such a way that C becomes an RPC–lattice:

(A) $c \supset_c c'$ is the largest d such that $d \cap c \leq c'$.

If c, c' are both in B then I set $c \supset_c c' = c \supset_b c'$ whence (A) follows from the same property for $\supset_b$. If c is in A and c' is in B then $c < c'$, and I set $c \supset_c c' = e_B$. If c is in B and c' is in A then $c' < c$, and since $c' < -e_B \leq d \cap c$ for any d in B and $d = d \cap c$ for any d in A, I may set $c \supset_c c' = c'$. If c, c' are both in A and $c \supset_a c' < e_A$ then I set $c \supset_c c' = c \supset_A c'$; and if $c \supset_a c' = e_A$ then I set $c \supset_c c' = e_B$. Thus (A) holds in any case. It remains to define the operation $-$, and for the elements from B I keep the value inherited from B, while for the elements a from A I set $-a = e_B$. I now shall verify (f_{mo}) which is inherited from B if both arguments a,b there belong to B. If a is from A and b from B then $a \supset_c -b = e_B$ since also $-b$ is in B, and $b \supset_c -a = b \supset_c e_B = e_B$ by definition of $-a$; for a from B and b from A the argument is symmetric. Finally, if both a,b are from A, then $a \supset_c -b = a \supset_c e_B = e_B$ and analogously $b \supset_c -a = e_B$. Thus C is in $\mathbf{A}_m$.

In the next section I shall refer to C as *the (counter) example* obtained by *putting* B *on top of* A. A striking particular case is that obtained by putting two 1–element algebras on top of each other. If a and e are their elements then $-a = e = -e$ whence $-a \cap a < -e$. Thus (6) cannot be improved.

5. i–Algebras, i–Frames and Heyting Algebras

Let T be the term algebra as before. Let R_i be the relation of interdeducibility with respect to LJ, and let $\mathbf{A}_i$ be the class of all models of R_i ; the algebras in $\mathbf{A}_i$ are called *Heyting algebras.* There holds the

THEOREM 1_i $\mathbf{A}_i$ is the class of all algebras $A = \langle u(A), \supset, \cap, \cup, - \rangle$ with a distinguished element e in A which, in addition to (a_{mp}), (a_{do}), (a_{d1}), (a_d), (a_{eo}), (a_{e1}), (a_{e2}), (a_{go}), (a_{g1}), (a_{g2}), (a_{mo}) satisfy

$$(a_{io}) \quad -a \supset (a \supset b) = e \ ,$$

and $\mathbf{A}_i$ is also the the class of all algebras defined by the set D_i consisting of D_m together with this equation.

The proof is a simple extension of that of Theorem 1_m. The algebras satisfying the conditions of the Theorem I call *special* i–*frames.* A derivation securing the validity of (a_{io}) in the algebras from $\mathbf{A}_i$ is

$$\begin{array}{r} x \Longrightarrow x \\ \neg x, x \Longrightarrow \blacktriangle \\ \neg x, x \Longrightarrow y \\ \neg x \Longrightarrow x \rightarrow y \ . \end{array}$$

Let A be an algebra with operations $\supset$, $\cap$, $\cup$ and $-$; let Q be a subset of A. The pair A,Q shall be called an i–*frame* if with $\supset$, $\cap$, $\cup$ it is an m-frame and if there holds for all a,b,c:

$(f_{io}) \quad -a \leq a \supset b$.

The algebras A of i–frames with only the operations $\supset$, $\cap$, $\cup$, $-$ and for which Q consists of one element e only are precisely the *special* i–frames mentioned above.

The properties proved for m–frames remain in effect for i–frames, and the equivalence $\approx$ remains a congruence relation.

The proof of Theorem 1 will be completed by proving that, for every homomorphism h from T into an A from $\mathbf{A}_i$, the LJ–derivability of $v \Longrightarrow w$ implies $h(v) \leq h(w)$. Given an LJ-derivation D, I shall prove as before that, for every sequent $M \Longrightarrow u$ in D there holds $h \cdot M \supset^* h(u) = e$ and for any sequent $M \Longrightarrow \blacktriangle$ in D there holds $h \cdot M \supset^* -e = e$. There is only the new case of an instance of (AIN) :

$$\begin{array}{l} M \Longrightarrow \blacktriangle \\ M \Longrightarrow u \ . \end{array}$$

Since $-e \leq e \supset h(u)$ by (f_{io}), $h \cdot M \supset^* -e \leq h \cdot M \supset^* h(u)$ follows from (f4*), and so together with the premiss also the conclusion will be mapped to e.

Replacing T_d, R_d, $\mathbf{A}_d$, F_d, p_d, d–frames and KP_d by T_i, R_i, $\mathbf{A}_i$, F_i, p_e, i-frames and LJ there hold *without changes* to their proofs the

COROLLARY 1_i [verbally as Corollary 1_d]

THEOREM 2_i [verbally as Theorem 2_d]

The algebras in $\mathbf{A}_i$ also can be defined as those algebras A in $\mathbf{A}_m$ in which $-e$ is a *smallest* element of A. Because their defining equation (a_{io}) is, again by the equivalence (A) for RPC–lattices, equivalent to $-a \cap a \leq b$ for all a,b, and taking a to be regular, I find $-a \cap a = -e$ by (6) whence $-e \leq b$ for all b. Thus (6) now improves to

(14) $\quad -a \cap a = -e$.

for every a. Thus the Heyting algebras may be described as the expansions of RPC–lattices A containing a *smallest element* 0.

An element a in a Heyting algebra is said to be *decidable* if $a \cup -a = e$. Actually, a is decidable already if there exists a complement b of a, i.e. $a \cup b = e$, $a \cap b = 0$, because the latter equation implies by (9) that $b \leq -a$ whence $e = a \cup b = a \cup -a$. Decidable elements a are regular, because $a = a \cup 0 = a \cup (-a \cap --a) = (a \cup -a) \cap (a \cup --a) = e \cap (a \cup --a) = a \cup --a$ implies $--a \leq a$.

LEMMA 2 The decidable elements of a Heyting algebra A form a Boolean sublattice of A and of REG(A).

Clearly, together with a also $-a$ is decidable; let now a and b be decidable. Then (3) implies $(a \cup b) \cup -(a \cup b) = (a \cup b) \cup -a \cap -b = (a \cup b \cup -a) \cap (a \cup b \cup -b) = e \cap e = e$; thus $a \cup b$ is decidable and regular. Hence also $-a \cup -b$ is regular, and so (3) implies $-(a \cap b) = -(--a \cap --b) = --(-a \cup -b) = -a \cup -b$ and $(a \cap b) \cup -(a \cap b) = (a \cap b) \cup (-a \cup -b) = (a \cup -a \cup -b) \cap (b \cup -a \cup -b) = e \cap e = e$.

Besides the *algebraic* semantics given by its associated class of algebras, the calculus LJ permits another kind of semantical value structures discovered by KRIPKE 65; for a detailed discussion cf. DUMMETT 77, RAUTENBERG 79 and TROELSTRA – VAN DALEN 88.

It follows from the Corollaries 1_x that the algebraic consequence operator $cn^{\mathbf{A}}$ defined with respect to the entire *class* $\mathbf{A}_x$ of algebras coincides with with the consequence operator cn^F defined by the single algebra F_x free in $\mathbf{A}_x$. In the case of classical logic we know that the infinite algebra F_x may be replaced here by a very small finite one; it was observed by GÖDEL 32 that for LJ, as for any of the earlier calculi admitting **JI**→, it leads to a contradiction if I assume that there is a finite algebra A with a distinguished element e such that (x) $a \supset a = e$, $e \cap e = e$, $e \cup a = a \cup e = e$ and that (y) at least the derivability of $\blacktriangle \Longrightarrow v$ is equivalent to $h(v) = e$ for every homomorphism from T into A. Because assume such A to exist and let n be number of its elements; let v be the formula

$$\mathbb{W} < (x_i \rightarrow x_j) \wedge (x_j \rightarrow x_i) \mid i,j \leq n,\ i \neq j > .$$

Then any homomorphism h into A will map at least two different of the n+1 variables x_i, $i \leq n$, onto the same a in A . If x_p, x_q are such then

$h((x_p \rightarrow x_q) \wedge (x_q \rightarrow x_p)) = (a \supset a) \cap (a \supset a) = e \cap e = e$ by (x), hence also $h(v) = e$ by (x). As this holds for every h, it follows from (y) that $\blacktriangle \Longrightarrow v$ must be derivable, and then it follows from **JI**v that at least one formula $(x_i \rightarrow x_j) \wedge (x_j \rightarrow x_i)$ must be derivable. But neither LJ nor any of the previous calculi can produce such derivation.

While no single finite algebra A in $\mathbf{A}_i = \mathbf{A}$ satisfies $cn^A = cn^{\mathbf{A}}$, JASKOWSKI 36 has observed that there is a sequence of finite algebras J_n in $\mathbf{A}_i$ such that $cn^{\mathbf{J}} = cn^{\mathbf{A}}$ for the set $\mathbf{J}$ of these J_n; here J_0 is the Boolean algebra *2* and J_{n+1} is obtained from the (n+1)-th power $(J_n)^{n+1}$ of J_n by putting one new element on top of it. A proof of these facts can be found in ROSE 53 ; for a semantical proof cf. RAUTENBERG 79 .

THEOREM 6 None of the four operations $\supset, \cap, \cup, -$ of Heyting algebras is equationally definable by the others.

If one of these operations were defined in terms of the three others then a subset closed with respect to the these others would have to be closed for the fourth. So it suffices to find a counterexample for each of the operations, and such were provided by MCKINSEY 39 whose first two examples I use below. The subset {1,2} of the Jaskowski algebra J_1:

1
2
3

is closed with respect to $\cup, \cap$ and $\supset$ (since $1 \supset 2 = 2$) but not for $-$ since $-2 = 3$. The subset {1,2,3,5} of the Heyting algebra

1
4
2 3
5

is closed with respect to $\cap, \supset$ (since $2 \supset 3 = 3$, $3 \supset 2 = 2$) and $-$ (since $-2 = 3$, $-3 = 2$), but not for $\cup$. Consider finally the Heyting algebra N_6 obtained by simplifying the algebra N of Theorem 4_m:

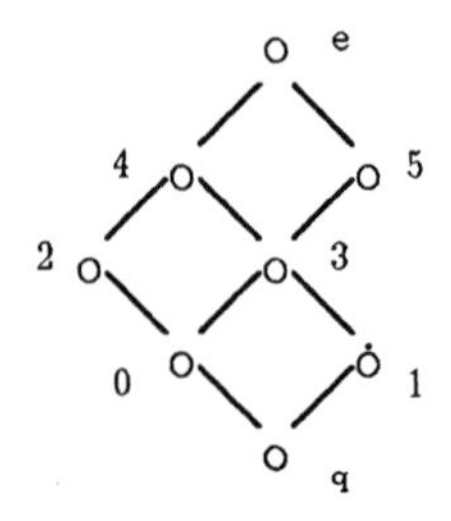

Then the subset of elements different from 5 is closed with respect to $\cap$, $\cup$ and $-$ (since $-0=1, -1=2, -2=1, -i=q$ for $i \geq 3$) but not for $\supset$ since $2\supset 3 = 5$. And the subset of elements different from 0 is closed with respect to $\cup$, $\supset$ (since only $e\supset 0=0$) but not for $\cap$.

THEOREM 4_i There are infinitely many formulas, containing only the variable x, not two of which are interdeducible in LJ.

The algebra N constructed in Theorem 4_m is a Heyting algebra, and it has been observed by NISHIMURA 60 that it is the free Heyting algebra F_i^1 with one generator.

It follows from (8) that every element $a\cup -a$ is dense; thus in a Heyting algebra I have

(15) d is dense if, and only if, there exists an a such that $d = a\cup -a$

since $-d = -e$ implies $d\cup -d = d\cup -e = d$. As a consequence, I find for any a

(16) $--a\supset a$ is dense.

This will follow from (15), together with the fact that elements laying above a dense element must be dense themselves, if I can show that $a\cup -a \leq --a\supset a$. But this is a consequence of $a \leq --a\supset a$ which holds by (d0), and of $--a \leq -a\supset a$ which holds by (f_{i0}). There must, of course, be a proof of (16) avoiding the auxiliary use of the operation $\cup$, and it can be found by translating the LJ–derivation **4**.(8) of the formula corresponding to (16):

$$
\begin{aligned}
& a \leq --a\supset a \\
& a \cap -(--a\supset a) \leq (--a\supset a) \cap -(--a \cap a) = -e \\
& -(--a\supset a) \leq -a \\
& -(--a\supset a) \cap --a \leq -a \cap --a \leq -e \\
& -(--a\supset a) \cap --a \leq a \\
& -(--a\supset a) \leq --a \supset a \\
& -(--a\supset a) = -(--a\supset a) \cap (--a\supset a) = -e
\end{aligned}
$$

A counterexample showing that (16) does not hold in algebras from $\mathbf{A}_m$ is obtained by putting a 2–element Boolean algebra with elements e, $-e$ on top of a 1–element algebra with the element a: then $-a = e$, $--a = -e$ whence $--a\supset a = a$. It also refutes $--a\supset --b \leq --(a\supset b)$ with $b = -e$ since then $--b = b$ and $--a\supset --b = e$ but $a\supset b = a$ whence $--(a\supset b) = -e$.

The characterization of the congruence relation induced by the homomorphism ρ, stated in (12), may now be simplified to read

(12_i) $\rho a = \rho b$ if, and only if, there exists d in DENS(A) such that $d\cap a = d\cap b$.

If such a d is given, it follows from $\rho d = e$ that $\rho a = \rho d \cap \rho a = \rho(d\cap a) = \rho(d\cap b) = \rho b$ since ρ is homomorphic. Conversely, if $\rho a = \rho b$ then $d =$

$(a\cup -a)\cap(b\cup -b)$ was dense, and $--a=--b$ again implies $-a=-b$ whence $a\cap -b=a\cap -a=-e$ and $a\cap d=a\cap(a\cup -a)\cap(b\cup -b)=a\cap(b\cup -b)=(a\cap b)\cup(a\cap -b)=a\cap b$, but by symmetry also $b\cap d=a\cap b$. - Finally, I find

the map ρ from A onto REG(A) is homomorphic also for the operation $\supset$.

This follows from **4**.(6)-(7) where it was shown that

(17) The formulas $\neg\neg(x\to y)$ and $\neg\neg x\to\neg\neg y$ are LJ-interdeducible

THEOREM 5_i For A in $\mathbf{A}_i$ the set REG(A) of regular elements is a Boolean lattice under the induced operations $\cap$ and $-$ and with the supremum operation $--(a\cup b)$.

The map ρ is homomorphic from A onto the Boolean lattice REG(A) for $\cap$, $\cup$, $\supset$ and $-$, and the congruence relation induced by ρ is characterized by condition (12_i).

If h is a homomorphism for $\cap$, $\cup$, $\supset$ and $-$ from A into a Boolean lattice B then there is a unique homomorphism f from REG(A) to B such that $h=f\cdot\rho$.

The last observation follows from the fact that $\rho r=\rho s$ implies $h(r)=h(s)$, and this is so because $d=a\cup -a$ implies $h(d)=e_B$ in the Boolean lattice B whence $d\cap r=d\cap s$ implies $h(r)=h(d\cap r)=h(d\cap s)=h(s)$.

Let p_i be the natural homomorphism from T onto T/R_i and let ρ now be the homomorphism from T/R_i onto $REG=REG(T/R_i)$. A formula r in T is called *regular* if $p_i(r)$ belongs already to REG; clearly, all formulas of the form $\neg v$ are regular. Also, r is regular if, and only if, $p_i(r)=\rho\cdot p_i(r)=p_i(\neg\neg r)$, hence if $\neg\neg r\Longrightarrow r$ is LJ-derivable. It follows from the above that together with v and w also $v\wedge w$ and $v\to w$ are regular; hence the set NEG of formulas defined by

$\neg x$ is in NEG for every variable x,
if v, w are in NEG then so are $\neg v$, $v\wedge w$ and $v\to w$

will contain only regular formulas (in TROELSTRA - VAN DALEN 88 it is called the *negative fragment* of (intuitionistic) formulas). Obviously, NEG consists of all formulas not containing the connective $\vee$ and containing variables only prefixed by $\neg$. Also NEG is a full set of representatives for REG: for every formula v there is a formula $v^\curlyvee$ in NEG such that $p_i(v^\curlyvee)=\rho\cdot p_i(v)$, namely

$$x^\curlyvee=\neg\neg x\,,\quad (\neg v)^\curlyvee=\neg v^\curlyvee\,,$$
$$(v\to w)^\curlyvee=v^\curlyvee\to w^\curlyvee\,,\quad (v\wedge w)^\curlyvee=v^\curlyvee\wedge w^\curlyvee\,,\quad (v\vee w)^\curlyvee=\neg(\neg v^\curlyvee\wedge\neg w^\curlyvee)$$

because e.g. $\rho\cdot p_i(v\to w)=\rho\cdot p_i(v)\supset\rho\cdot p_i(w)=p_i(v^\curlyvee)\supset p_i(w^\curlyvee)=p_i(v^\curlyvee\to w^\curlyvee)$, $\rho\cdot p_i(v\vee w)=--(p_i(v)\cup p_i(w))=-(-p_i(v)\cap -p_i(w))=-(-\rho\cdot p_i(v)\cap -\rho\cdot p_i(w))$

$= -(-p_i(v^\gamma) \cap -p_i(w^\gamma)) = p_i(\neg(\neg v^\gamma \wedge \neg w^\gamma))$. Induction on the complexity of r in NEG yields explicit LJ-derivations of $\neg\neg r \Longrightarrow r$:

$$r \Longrightarrow \neg\neg r$$
$$\neg\neg\neg r, r \Longrightarrow \blacktriangle$$
$$\neg\neg\neg r \Longrightarrow \neg r$$

$$r\wedge s \Longrightarrow r$$
$$\neg\neg(r\wedge s) \Longrightarrow \neg\neg r \qquad \neg\neg r \Longrightarrow r \qquad\qquad \neg\neg(r\wedge s) \Longrightarrow \neg\neg s \qquad \neg\neg s \Longrightarrow s$$
$$r\wedge s \Longrightarrow s$$
$$\neg\neg(r\wedge s) \Longrightarrow r \qquad\qquad \neg\neg(r\wedge s) \Longrightarrow s$$
$$\neg\neg(r\wedge s) \Longrightarrow r\wedge s$$

$$b \Longrightarrow b$$
$$a \Longrightarrow a \qquad \neg b, b \Longrightarrow \blacktriangle$$
$$a, \neg b, a\to b \Longrightarrow \blacktriangle$$
$$a, \neg b \Longrightarrow \neg(a\to b)$$
$$a, \neg b, \neg\neg(a\to b) \Longrightarrow \blacktriangle$$
$$a, \neg\neg(a\to b) \Longrightarrow \neg\neg b \qquad \neg\neg b \Longrightarrow b$$
$$a, \neg\neg(a\to b) \Longrightarrow b$$
$$\neg\neg(a\to b) \Longrightarrow a\to b \quad .$$

6. c-Algebras, c-Frames and Boolean Algebras

Let T be the term algebra as before. Let R_c be the relation of interdeducibility with respect to LK, and let $\mathbf{A}_c$ be the class of all models of R_c; the algebras in $\mathbf{A}_c$ I shall call c-*algebras*. There holds the

THEOREM 1_c $\mathbf{A}_c$ is the class of all algebras $A = \langle u(A), \supset, \cap, \cup, - \rangle$ with a distinguished element e in A which, in addition to (a_{mp}), (a_{do}), (a_{d1}), (a_d), (a_{eo}), (a_{e1}), (a_{e2}), (a_{go}), (a_{g1}), (a_{g2}), (a_{mo}), (a_{io}) satisfy

(a_{co}) $\quad (-a \supset a) \supset a = e$

and $\mathbf{A}_c$ is also the the class of all algebras defined by the set D_c consisting of D_i together with this equation.

The proof is a simple extension of that of Theorem 1_i. The algebras satisfying the conditions of the Theorem I call *special* c-*frames*. A derivation securing the validity of (a_{co}) in the algebras from $\mathbf{A}_c$ is

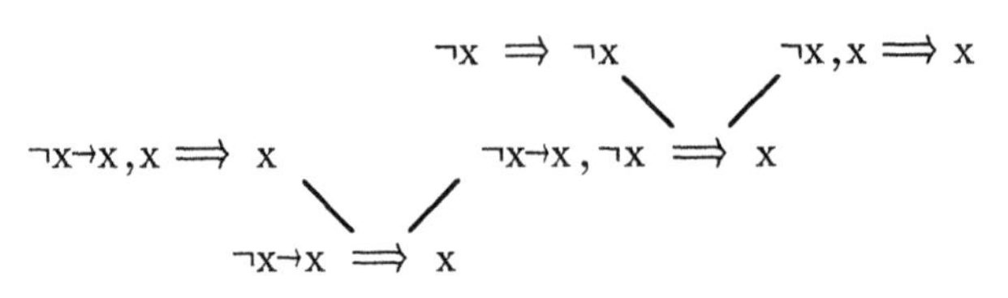

making use of the rule (K_0).

Let A be an algebra with operations $\supset$, $\cap$, $\cup$ and $-$; let Q be a subset of A. The pair A,Q shall be called a c-*frame* if with $\supset$, $\cap$, $\cup$ it is an i-frame and if there holds for all a,b,c:

(f_{co}) $\qquad -a\supset a \leq a$.

The algebras A of c-frames with only the operations $\supset$, $\cap$, $\cup$, $-$ and for which Q consists of one element e only are precisely the *special* c-frames mentioned above. The equation (a_{io}) from D_i implies $--a\supset(-a\supset a) = e$, $--a \leq -a\supset a$, and so (f_{co}) gives

(f_{c1}) $\qquad --a \leq a$.

Thus (f_{m1}) implies that $a = --a$ holds in the algebras A from $\mathbf{A}_c$, and (a_{mo}) can be written as

(18) $\qquad a\supset b = -b\supset -a$.

I shall also need

(f18) $\qquad (\mu\supset^*(b\supset c)) \cap (\mu\supset^*(-b\supset c)) = \mu\supset^* c$.

This holds if μ consists of one element a only since

$$
\begin{aligned}
&(a\supset(b\supset c)) \cap (a\supset(-b\supset c)) \\
&\quad = a\supset((b\supset c) \cap (-b\supset c)) && \text{by } (a_7) \\
&\quad = a\supset((-c\supset -b) \cap (-c\supset b)) && \text{by } (18) \\
&\quad = a\supset(-c\supset(-b\cap b)) && \text{by } (a_7) \\
&\quad = a\supset(-c\supset -e) && \text{by } (14) \\
&\quad = a\supset(e\supset c) && \text{by } (18) \\
&\quad = a\supset c \ .
\end{aligned}
$$

The general case follows by induction:

$$
\begin{aligned}
&(\mu\supset^*(b\supset c)) \cap (\mu\supset^*(-b\supset c)) \\
&\quad = (\mu(0)\supset(\mu_0\supset^*(b\supset c))) \cap \mu(0)\supset(\mu\supset^*(-b\supset c))) \\
&\quad = \mu(0)\supset(\mu_0\supset^*(b\supset c))) \cap (\mu_0\supset^*(-b\supset c))) && \text{by } (a_7) \\
&\quad = \mu(0)\supset(\mu_0\supset^* c) = \mu\supset^* c \ .
\end{aligned}
$$

The properties proved for i-frames remain in effect for c-frames, and the equivalence $\approx$ remains a congruence relation.

The proof of Theorem 1_c will be completed as before, proving for an LK-derivation D: for every sequent $M \Rightarrow u$ in D there holds $h\cdot M\supset^* h(u)=e$

and, for any sequent $M \Longrightarrow \blacktriangle$ in D, there holds $h \cdot M \supset^* -e = e$. There is only the new case of an instance of (K_0):

$$\frac{w, M \Longrightarrow v \qquad \neg w, M \Longrightarrow v}{M \Longrightarrow v}$$

The premisses go into $h(w) \supset (h \cdot M \supset^* h(v)) = h \cdot M \supset^* (h(w) \supset h(v)$ and into $h \cdot M \supset^* (-h(w) \supset h(v)$, and if both these elements are e then by (f18) so is $h \cdot M \supset^* h(v)$.

Replacing T_d, R_d, $\mathbf{A}_d$, F_d, p_d, d–frames and KP_d by T_c, R_c, $\mathbf{A}_c$, F_c, p_c, c–frames and LK there hold *without changes* to their proofs the

COROLLARY 1_c [verbally as Corollary 1_d]

THEOREM 2_c [verbally as Theorem 2_d]

It follows from $a = --a$ that the homomorphism ρ in Theorem 5_i becomes the identity from A onto REG(A), i.e. A = REG(A), and (8) improves to

$$a \cap -a = -e \quad \textit{and} \quad a \cup -a = e \ .$$

It also follows from $a = --a$ that the map $-$ is an order reversing involution on A which, therefore, will transform suprema into infima and vice versa. Thus (4) improves to $-a \cup -b = -(a \cap b)$. Of course, all of this also follows from A = REG(A) and the fact that reduct of REG(A) under $\cap$, $\cup$ and $-$ is a Boolean lattice with e and $-e$ as its largest and smallest elements.

In an RPC–lattice the element $b \supset c$ is the largest a satisfying $a \cap b \leq c$. On a given lattice A, therefore, there can be *at most* one operation f(b,c) making it into an RPC–lattice. If A is Boolean then there *is* such an operation, namely $f(b,c) = -b \cup c$: this follows from $(-b \cup c) \cap b = (-b \cap b) \cup (c \cap b) = c \cap b \leq c$ and the observation that $a \cap b \leq c$ implies $a \leq -b \cup a = (-b \cup a) \cap (-b \cup b) = -b \cup (a \cap b) \leq -b \cup c$. Thus in A from $\mathbf{A}_c$ there holds

(19) $\quad b \supset c = -b \cup c$.

Thus every Boolean lattice is an RPC–lattice, and I can expand it by (19) to an algebra $A_B = \langle u(B), \cap, \cup, \supset \rangle$ which belongs to $\mathbf{A}_g$. It is in $\mathbf{A}_m$ since (a_{mo}) holds as $a \supset -b = -a \cup -b = -b \cup -a = b \supset -a$. It is in $\mathbf{A}_i$ since (a_{io}) holds as $-a \supset (a \supset b) = --a \cup (a \supset b) = --a \cup -a \cup b = --a \cup -a = e$. It finally is in $\mathbf{A}_c$ since $(-a \supset a) \supset a = -(-a \supset a) \cup a = -(--a \cup a) \cup a = -a \cup a = e$. Thus every Boolean lattice comes from an algebra in $\mathbf{A}_c$:

THEOREM 5_c $\mathbf{A}_c$ consists of the expansions by (19) of the complemented lattices underlying Boolean algebras.

It was the class of these algebras which, under the name **B**, was used in Chapter 1.10 to define the Boolean valued consequence operation $cn^{\mathbf{B}}$ of which it then was shown there that it coincides with the classical 2-valued consequence operation cn. Thus Theorem 2_c now implies the

COROLLARY 3 A sequent $M \Longrightarrow v$ is LK-derivable if, and only if, $v \in cn(C)$.

If the connection with the classes $\mathbf{A}_m$ and $\mathbf{A}_i$ is not of interest and if, instead of LK, the multiple calculus MK is considered, then it is easily seen that the class of algebras associated to the calculus MK is that of [the expansions of the lattices underlying] Boolean algebras. On the one hand, it is straightforward to derive the Boolean equations in MK, e.g. distributivity

$$x,y \Longrightarrow x \qquad x,y \Longrightarrow y \qquad\qquad x,z \Longrightarrow x \qquad x,z \Longrightarrow z$$
$$x,y \Longrightarrow x\wedge y \qquad\qquad x,z \Longrightarrow x\wedge z$$
$$x,y \Longrightarrow (x\wedge y)\vee(x\wedge z) \qquad\qquad x,z \Longrightarrow (x\wedge y)\vee(x\wedge z)$$
$$x,y\vee z \Longrightarrow (x\wedge y)\vee(x\wedge z)$$
$$x,x\wedge(y\vee z) \Longrightarrow (x\wedge y)\vee(x\wedge z)$$
$$x\wedge(y\vee z),x\wedge(y\vee z) \Longrightarrow (x\wedge y)\vee(x\wedge z)$$
$$x\wedge(y\vee z) \Longrightarrow (x\wedge y)\vee(x\wedge z) \quad ,$$

the inequalities $b\cap -b \leq a$ and $a \leq b\cup -b$

$$x \Longrightarrow x \qquad\qquad y,x \Longrightarrow x$$
$$x,\neg x \Longrightarrow \blacktriangle \qquad\qquad y \Longrightarrow x,\neg x$$
$$x,\neg x \Longrightarrow y \qquad\qquad y \Longrightarrow x\vee\neg x, x\vee\neg x$$
$$x\wedge\neg x \Longrightarrow y \qquad\qquad y \Longrightarrow x\vee\neg x$$

and the equation (19)

$$x \Longrightarrow x \qquad\qquad\qquad x \Longrightarrow x$$
$$x,\neg x \Longrightarrow \blacktriangle \qquad\qquad\qquad \blacktriangle \Longrightarrow x,\neg x$$
$$x,\neg x \Longrightarrow y \qquad x,y \Longrightarrow y \qquad\qquad \blacktriangle \Longrightarrow x,\neg x\vee y \qquad y \Longrightarrow y$$
$$\qquad\qquad\qquad\qquad\qquad\qquad y \Longrightarrow \neg x\vee y$$
$$x,\neg x\vee y \Longrightarrow y \qquad\qquad\qquad x\rightarrow y \Longrightarrow \neg x\vee y \quad .$$
$$\neg x\vee y \Longrightarrow x\rightarrow y$$

On the other hand, for every sequent $M \Longrightarrow N$ occurring in an MK-derivation and for every homomorphism h from T into a Boolean algebra B, I find

$$(^{**}) \qquad \bigcap h(M) \leq \bigcup h(N)$$

where $\bigcap\vartheta = 1_B$ and $\bigcup\vartheta = 0_B$ for the empty sequence ϑ. This is clear for axioms of MK since in that case M and N have a common member x whence $\bigcap h(M) \leq h(x) \leq \bigcup h(N)$. That (**) is preserved under structural rules of

MK follows from trivial Boolean computations, and for the logical rules I use in the case of

(I→) $h(v) \cap \bigcap h(M) \leq h(w) \cup \bigcup h(N)$ implies

$$\bigcap h(M) \leq -h(v) \cup \bigcap h(M) = -h(v) \cup (h(v) \cap \bigcap h(M)) \leq -h(v) \cup h(w) \cup \bigcup h(N) = (h(v) \supset h(w)) \cup \bigcup h(N)$$

(E→) $\bigcap h(M) \leq h(v) \cup \bigcup h(N)$ implies $-h(v) \cap \bigcap h(M) \leq -h(v) \cap \bigcup h(N)$, thus $h(w) \cap \bigcap h(M) \leq \bigcup h(N)$ implies

$$(h(v) \supset h(w)) \cap \bigcap h(M) = (-h(v) \cup h(w)) \cap \bigcap h(M)$$
$$\leq (-h(v) \cap \bigcap h(M)) \cup (h(w) \cap h(M)) \leq (-h(v) \cap \bigcup h(N)) \cap \bigcup h(N) = \bigcup h(N).$$

(I¬) $h(v) \cap \bigcap h(M) \leq \bigcup h(N)$ implies

$$\bigcap h(M) \leq -h(v) \cup \bigcap h(M) = -h(v) \cup (h(v) \cap \bigcap h(M)) \leq -h(v) \cup \bigcup h(N)$$

(E¬) $\bigcap h(M) \leq h(v) \cup \bigcup h(N)$ implies

$$-h(v) \cap \bigcap h(M) \leq -h(v) \cap (h(v) \cup \bigcup h(N)) = -h(v) \cap \bigcup h(N) \leq \bigcup h(N) .$$

The cases of the rules for $\wedge$ and $\vee$ are trivial.

If in the above LK is taken in place of MK then the sequents $y \Longrightarrow x \vee \neg x$ and $x \rightarrow y \Longrightarrow \neg x \vee y$ require the use of a rule such as (K_0) since they cannot be derived in LJ :

$$\frac{\dfrac{x,y \Longrightarrow x}{x,y \Longrightarrow x\vee\neg x} \qquad \dfrac{\neg x,y \Longrightarrow \neg x}{\neg x,y \Longrightarrow x\vee\neg x}}{y \Longrightarrow x\vee\neg x}\,(K_0) ,$$

$$\frac{\dfrac{\dfrac{x \Longrightarrow x \qquad y \Longrightarrow y}{x, x\rightarrow y \Longrightarrow y}}{x, x\rightarrow y \Longrightarrow \neg x\vee y} \qquad \dfrac{\dfrac{\neg x \Longrightarrow \neg x}{\neg x, x\rightarrow y \Longrightarrow \neg x}}{\neg x, x\rightarrow y \Longrightarrow \neg x\vee y}}{x\rightarrow y \Longrightarrow \neg x\vee y}\,(K_0) .$$

On the other hand, I have work again with (*) instead of (**), and there the one element case of (f18) follows easily from (19) and distributivity

$$(a\supset(b\supset c)) \cap (a\supset(-b\supset c)) = (-a\cup(-b\cup c)) \cap (-a\cup(b\cup c)) = -a\cup((-b\cup c)\cap(b\cup c))$$
$$= -a\cup((-b\cap b)\cup c) = -a\cup c .$$

7. Translations from Classical into Intuitionistic Logic

LK–derivations are particular LJ–derivations. Hence the congruence classes of T under R_i are contained in those under R_c, and so there is a homomorphism h from T/R_i into T/R_c such that $h \cdot p_i = p_c$ for the natural homomorphisms p_i and p_c onto T/R_i and T/R_c. It follows from Theorem 5_i that h factors as $h = f \cdot \rho$ through $\mathbf{REG} = \mathbf{REG}(T/R_i)$ whence $f \cdot \rho \cdot p_i = p_c$.

LEMMA 3 f is an isomorphism from $\mathbf{REG}(T/R_i)$ onto T/R_c .

As T is generated by the variables x, **REG** is generated by the elements $\rho p_i(x) = --p_i(x) = p_i(\neg\neg x)$. Now f maps these to the $p_c(x)$, and so f is certainly surjective. But T/R_c is the free Boolean algebra generated by the elements $p_c(x)$, and so there exists a homomorphism g_1 from T/R_c to **REG** mapping $p_c(x)$ to $\rho p_i(x)$. Thus $g_1 \cdot f$ is the the identity on the generators of **REG** and, therefore, on **REG** itself. Consequently, f is injective and so an isomorphism.

It follows that $p_c(v) = p_c(w)$ implies $p_i(\neg\neg v) = \rho p_i(v) = \rho p_i(w) = p_i(\neg\neg w)$: if v,w are LK-interdeducible then $\neg\neg v, \neg\neg w$ are LJ-interdeducible. But more can be said. A *translation* shall be a map τ from T into itself such that

(t) v and v^τ are LK-interdeducible,
(tt) v^τ and $\neg\neg v^\tau$ are LJ-interdeducible, i.e. $p_i(v^\tau) = p_i(\neg\neg v^\tau) = \rho p_i(v^\tau)$.

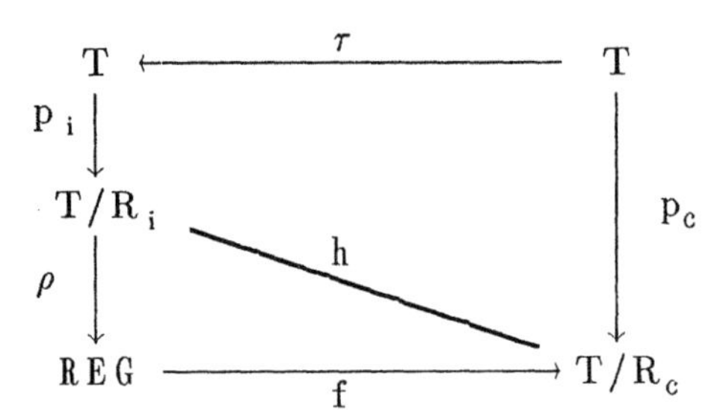

LEMMA 4 $M \Longrightarrow v$ is LK-derivable if, and only if, $M^\tau \Longrightarrow v^\tau$ is LJ-derivable.

Here M^τ is $<m_k{}^\tau \mid k<m-1>$ for $M = <m_k \mid k<m-1>$. The proof follows from the following chain of implications:

$M \Longrightarrow v$ is LK-derivable ,
$M^\tau \Longrightarrow v^\tau$ is LK-derivable by (t)
$p_c \cdot M^\tau \supset^* p_c(v^\tau) = e$ in T/R_c by Corollary 5_c
$h \cdot p_i \cdot M^\tau \supset^* h(p_i(v^\tau)) = e$ in T/R_c since $h \cdot p_i = p_j$
$f \cdot \rho \cdot p_i \cdot M^\tau \supset^* f(\rho(p_i(v^\tau))) = e$ in T/R_c since $h = f \cdot \rho$
$f(\rho \cdot p_i \cdot M^\tau \supset^* \rho(p_i(v^\tau))) = e$ in T/R_c since f is homomorphic
$\rho \cdot p_i \cdot M^\tau \supset^* \rho(p_i(v^\tau)) = e$ in T/R_i since f is isomorphic
$p_i \cdot M^\tau \supset^* p_i(v^\tau) = e$ in T/R_i by (tt)
$M^\tau \Longrightarrow v^\tau$ is LJ-derivable by Corollary 5_i .

A first example of a translation is the trivial map ϑ, defined explicitly as $v^\vartheta = \neg\neg v$; here (t) is obvious, and (tt) follows from the fact that v^ϑ starts with a negation. A second example is the KOLMOGOROV map κ defined recursively as

$$x^{\kappa} = \neg\neg x \ , \quad (\neg v)^{\kappa} = \neg v^{\kappa} \ ,$$
$$(v \to w)^{\kappa} = \neg\neg(v^{\kappa} \to w^{\kappa}) \ , \quad (v \wedge w)^{\kappa} = \neg\neg(v^{\kappa} \wedge w^{\kappa}) \ , \quad (v \vee w)^{\kappa} = \neg\neg(v^{\kappa} \vee w^{\kappa}) \ .$$

It follows (t) by a straightforward induction, e.g. if $p_c(v) = p_c(v^{\kappa})$ and $p_c(w) = p_c(w^{\kappa})$ then $p_c(v \vee w) = p_c(v) \cup p_c(w) = p_c(v^k) \cup p_c(w^k) = p_c(v^k \vee w^k) = \neg\neg p_c(v^k \vee w^k)$. It follows (tt) from the fact that v^{κ} is of the form $\neg v_0$ such that $p_i(v^k) = p_i(\neg v_0) = -p_i(v_0) = \rho(-p_i(v_0))$.

A third example is the GÖDEL map γ, mentioned already in connection with the subset NEG of regular formulas after Theorem 5_i, and defined recursively as

$$x^{\gamma} = \neg\neg x \ , \quad (\neg v)^{\gamma} = \neg v^{\gamma} \ ,$$
$$(v \to w)^{\gamma} = v^{\gamma} \to w^{\gamma} \ , \quad (v \wedge w)^{\gamma} - v^{\gamma} \wedge w^{\gamma} \ , \quad (v \vee w)^{\gamma} = \neg(\neg v^{\gamma} \wedge \neg w^{\gamma}) \ .$$

Again (t) follows by a straightforward induction, and (tt) follows for $\to$ and $\wedge$ again by induction, for $\neg$ and $\vee$ form the fact the γ-image begins with a negation.

Given a translation τ, the proof of Lemma 4 provides a rough way to transform an LK-derivation of $M \Longrightarrow v$ into an LJ-derivation of $M^{\tau} \Longrightarrow v^{\tau}$, to pursue it requires to transform algebraic computations from T/R_c to T/R_i and these then back into LJ-derivations. Consequently, direct translations of derivations themselves are desirable. In the following I shall distinguish the four classical calculi LK_0, LK_t, LK_n, LK_d depending on the rules (K_0), (K_t), (K_n), (K_d):

LEMMA 5 For each of the four calculi LK_x and for each of the maps ϑ, κ, γ there is an operator $\mathbf{G}_x^{\varphi}$ transforming an LK_x-derivation D of $M \Longrightarrow u$ into an LJ-derivation of $M^{\varphi} \Longrightarrow u^{\varphi}$.

The function *length*(G_x^{κ}(D)) depends linearly on *length*(D) for $x = n, d$, and the function *length*(G_0^{κ}(D)) depends linearly on *length*(D) when computed in the calculus obtained from LJ by adding the admissible rule (M_0). In all other cases, the lengths may grow exponentially due to the occurrence of cuts.

I begin by observing that each of my maps sends LK-axioms into reflexive LJ-axioms. Structural rules and the rules $(I\neg)$, $(E\neg)$, (AIN) translate immediately. The translations of logical rules I consider first in the case of κ , making use of the inversion rules $\mathbf{JI}\neg$ and $\mathbf{JS}\neg\neg$ and of the fact that all images u^{κ} under κ are of the form $\neg t$ for some t:

(I→)

$$\begin{array}{rcll} v^K, M^K & \Longrightarrow & w^K & \\ M^K & \Longrightarrow & v^K \rightarrow w^K & \\ \neg(v^K \rightarrow w^K), M^K & \Longrightarrow & \blacktriangle & \\ M^K & \Longrightarrow & \neg\neg(v^K \rightarrow w^K) & \end{array}$$

(E→)

$$M^K \Longrightarrow v^K \qquad w^K, M^K \Longrightarrow u^K$$

$$\begin{array}{rcll} v^K \rightarrow w^K, M^K & \Longrightarrow & u^K & \\ \neg u^K, v^K \rightarrow w^K, M^K & \Longrightarrow & \blacktriangle & \\ \neg u^K, M^K & \Longrightarrow & \neg(v^K \rightarrow w^K) & \\ \neg\neg(v^K \rightarrow w^K),\ \neg\neg t, M^K & \Longrightarrow & \blacktriangle & u^K = \neg t \\ \neg\neg(v^K \rightarrow w^K),\ t, M^K & \Longrightarrow & \blacktriangle & \mathbf{JS}\neg\neg \\ \neg\neg(v^K \rightarrow w^K),\ M^K & \Longrightarrow & u^K & \neg t = u^K \end{array}$$

(I∧)

$$M^K \Longrightarrow v^K \qquad M^K \Longrightarrow w^K$$

$$\begin{array}{rcll} M^K & \Longrightarrow & v^K \wedge w^K & \\ M^K & \Longrightarrow & \neg\neg(v^K \wedge w^K) & \end{array}$$

(E∧)

$$\begin{array}{rcll} v^K, M^K & \Longrightarrow & u^K & \\ v^K \wedge w^K, M^K & \Longrightarrow & u^K & \\ v^K \wedge w^K, \neg u^K, M^K & \Longrightarrow & \blacktriangle & \\ \neg\neg(v^K \wedge w^K), \neg\neg t, M^K & \Longrightarrow & \blacktriangle & u^K = \neg t \\ \neg\neg(v^K \wedge w^K), t, M^K & \Longrightarrow & \blacktriangle & \mathbf{JS}\neg\neg \\ \neg\neg(v^K \wedge w^K), M^K & \Longrightarrow & u^K & \end{array}$$

(I∨)

$$\begin{array}{rcll} M^K & \Longrightarrow & v^K & \\ M^K & \Longrightarrow & v^K \vee w^K & \\ \neg(v^K \vee w^K), M^K & \Longrightarrow & \blacktriangle & \\ M^K & \Longrightarrow & \neg\neg(v^K \vee w^K) & \end{array}$$

(E∨)

$$v^K, M^K \Longrightarrow u^K \qquad w^K, M^K \Longrightarrow u^K$$

$$\begin{array}{rcll} v^K \vee w^K, M^K & \Longrightarrow & u^K & \\ \neg\neg(v^K \vee w^K), \neg\neg t, M^K & \Longrightarrow & \blacktriangle & u^K = \neg t \\ \neg\neg(v^K \vee w^K), t, M^K & \Longrightarrow & \blacktriangle & \mathbf{JS}\neg\neg \\ \neg\neg(v^K \vee w^K), M & \Longrightarrow & u^K & \end{array}$$

For the map ϑ I use the interdeducibilities (13), (17):

(I→)

$$\begin{array}{rcll} \neg\neg v, M^\vartheta & \Longrightarrow & \neg\neg w & \\ v, M^\vartheta & \Longrightarrow & \neg\neg\neg t & \mathbf{JS}\neg\neg \text{ and } w = \neg t \\ \neg\neg\neg\neg t, v, M^\vartheta & \Longrightarrow & \blacktriangle & \\ t, v, M^\vartheta & \Longrightarrow & \blacktriangle & \mathbf{JS}\neg\neg \\ v, M^\vartheta & \Longrightarrow & w & \neg t = w \\ M^\vartheta & \Longrightarrow & v \rightarrow w & \\ \neg(v \rightarrow w), M^\vartheta & \Longrightarrow & \blacktriangle & \\ M^\vartheta & \Longrightarrow & \neg\neg(v \rightarrow w) & \end{array}$$

(E→) (17)

$$M^{\vartheta} \Rightarrow \neg\neg v \qquad \neg\neg w, M^{\vartheta} \Rightarrow \neg\neg u$$
$$\neg\neg(v \to w) \Rightarrow \neg\neg v \to \neg\neg w \quad \text{CUT} \quad \neg\neg v \to \neg\neg w, M^{\vartheta} \Rightarrow \neg\neg u$$
$$\neg\neg(v \to w), M^{\vartheta} \Rightarrow \neg\neg u$$

(I∧)

$$M^{\vartheta} \Rightarrow \neg\neg v \qquad M^{\vartheta} \Rightarrow \neg\neg w \qquad (13)$$
$$M^{\vartheta} \Rightarrow \neg\neg v \wedge \neg\neg w \quad \text{CUT} \quad \neg\neg v\ \neg\neg w \Rightarrow \neg\neg(v \wedge w)$$
$$M^{\vartheta} \Rightarrow \neg\neg(v \wedge w)$$

(E∧) (13)

$$\neg\neg v, M^{\vartheta} \Rightarrow \neg\neg u$$
$$\neg\neg(v \wedge w) \Rightarrow \neg\neg v \wedge \neg\neg w \quad \text{CUT} \quad \neg\neg v \wedge \neg\neg w, M^{\vartheta} \Rightarrow \neg\neg u$$
$$\neg\neg(v \wedge w), M^{\vartheta} \Rightarrow \neg\neg u$$

(I∨)

$$M^{\vartheta} \Rightarrow \neg\neg v$$
$$\neg v, M^{\vartheta} \Rightarrow \blacktriangle \qquad \mathbf{JI\to}$$
$$\neg v \wedge \neg w, M^{\vartheta} \Rightarrow \blacktriangle \qquad (13)$$
$$M^{\vartheta} \Rightarrow \neg(\neg v \wedge \neg w) \quad \text{CUT} \quad \neg(\neg v \wedge \neg w) \Rightarrow \neg\neg(v \vee w)$$
$$M^{\vartheta} \Rightarrow \neg\neg(v \vee w)$$

(E∨)

$$\neg\neg v, M^{\vartheta} \Rightarrow \neg\neg u \qquad \neg\neg w, M^{\vartheta} \Rightarrow \neg\neg u$$
$$\mathbf{JI\neg},\ \mathbf{JS\neg\neg} \qquad v, \neg u, M^{\vartheta} \Rightarrow \blacktriangle \qquad w, \neg u, M^{\vartheta} \Rightarrow \blacktriangle$$
$$\neg u, M^{\vartheta} \Rightarrow \neg v \qquad \neg u, M^{\vartheta} \Rightarrow \neg w$$
$$(13) \qquad \neg u, M^{\vartheta} \Rightarrow \neg v \wedge \neg w$$
$$\neg\neg(v \vee w) \Rightarrow \neg(\neg v \wedge \neg w) \quad \text{CUT} \quad \neg(\neg v \wedge \neg w), M^{\vartheta} \Rightarrow \neg\neg u$$
$$\neg\neg(v \vee w), M^{\vartheta} \Rightarrow \neg\neg u \quad .$$

For the map γ the rules for → and ∧ translate trivially. For ∨ I cannot assume that u^{γ} is of the form $\neg t$; instead, I have to use that the images under γ are regular formulas:

(I∨)

$$M^{\gamma} \Rightarrow v^{\gamma}$$
$$\neg v^{\gamma}, M^{\gamma} \Rightarrow \blacktriangle$$
$$\neg v^{\gamma} \wedge \neg w^{\gamma}, M^{\gamma} \Rightarrow \blacktriangle$$
$$M^{\gamma} \Rightarrow \neg(\neg v^{\gamma} \wedge \neg w^{\gamma})$$

(E∨)

$$v^{\gamma}, M^{\gamma} \Rightarrow u^{\gamma} \qquad w^{\gamma}, M^{\gamma} \Rightarrow u^{\gamma}$$
$$v^{\gamma}, \neg u^{\gamma}, M^{\gamma} \Rightarrow \blacktriangle \qquad w^{\gamma}, \neg u^{\gamma}, M^{\gamma} \Rightarrow \blacktriangle$$
$$\neg u^{\gamma}, M^{\gamma} \Rightarrow \neg v^{\gamma} \qquad \neg u^{\gamma}, M^{\gamma} \Rightarrow \neg w^{\gamma}$$
$$\neg u^{\gamma}, M^{\gamma} \Rightarrow \neg v^{\gamma} \wedge \neg w^{\gamma}$$
$$\neg(\neg v^{\gamma} \wedge \neg w^{\gamma}), M^{\gamma} \Rightarrow \neg\neg u^{\gamma} \quad \text{CUT} \quad \neg\neg u^{\gamma} \Rightarrow u^{\gamma}$$
$$\neg(\neg v^{\gamma} \wedge \neg w^{\gamma}), M^{\gamma} \Rightarrow u^{\gamma}$$

Turning to the rules specific for LK, (K_0) translates under ϑ and κ with help of (M_0) as

$$\begin{array}{rcll}
\neg\neg w, M^{\vartheta} \Rightarrow \neg\neg u & & \neg\neg\neg w, M^{\vartheta} \Rightarrow \neg\neg u & \\
M^{\vartheta} & \Rightarrow & \neg\neg\neg\neg u & \\
\neg\neg\neg\neg\neg u, M^{\vartheta} & \Rightarrow & \blacktriangle & \\
\neg u, M^{\vartheta} & \Rightarrow & \blacktriangle & \mathbf{JS}\neg\neg \\
M^{\vartheta} & \Rightarrow & \neg\neg u &
\end{array}$$

$$\begin{array}{rcll}
w^{\kappa}, M^{\kappa} \Rightarrow v^{\kappa} & & \neg w^{\kappa}, M^{\kappa} \Rightarrow v^{\kappa} & \\
M^{\kappa} & \Rightarrow & \neg\neg v^{\kappa} & \\
\neg\neg\neg t, M^{\kappa} & \Rightarrow & \blacktriangle & v^{\kappa}=\neg t \\
\neg t, M^{\kappa} & \Rightarrow & \blacktriangle & \\
M^{\kappa} & \Rightarrow & v^{\kappa} &
\end{array}$$

while under γ another cut with $\neg\neg v^{\gamma} \Rightarrow v^{\gamma}$ becomes necessary. (K_t) translates under ϑ with a cut from the derivation of $\blacktriangle \Rightarrow \neg(\neg w \wedge \neg\neg w)$ established as **4**.(5)

$$\begin{array}{c}
(13) \\
\neg(\neg w \wedge \neg\neg w) \Rightarrow \neg\neg(w \vee \neg w) \qquad \neg\neg(w \vee \neg w) \\
\blacktriangle \Rightarrow \neg(\neg w \wedge \neg\neg w) \qquad \neg(\neg w \wedge \neg\neg w), M^{\vartheta} \Rightarrow \neg\neg v \\
M^{\vartheta} \Rightarrow \neg\neg v
\end{array}$$

and similarly under κ and γ. (K_n) translates under ϑ and κ as

$$\begin{array}{rcll}
\neg\neg\neg v, M^{\vartheta} & \Rightarrow & \neg\neg v & \\
\neg v, \neg\neg\neg v, M^{\vartheta} & \Rightarrow & \blacktriangle & \mathbf{JI}\neg \\
\neg v, \neg v, M^{\vartheta} & \Rightarrow & \blacktriangle & \mathbf{JS}\neg\neg \\
\neg v, M^{\vartheta} & \Rightarrow & \blacktriangle & \\
M^{\vartheta} & \Rightarrow & \neg\neg v &
\end{array}
\qquad
\begin{array}{rcll}
\neg v^{\kappa}, M^{\kappa} & \Rightarrow & v^{\kappa} & \\
\neg v^{\kappa}, \neg\neg t, M^{\kappa} & \Rightarrow & \blacktriangle & v^{\kappa}=\neg t \\
\neg v^{\kappa}, t, M^{\kappa} & \Rightarrow & \blacktriangle & \mathbf{JS}\neg\neg \\
\neg\neg t, t, M^{\kappa} & \Rightarrow & \blacktriangle & \\
t, t, M^{\kappa} & \Rightarrow & \blacktriangle & \mathbf{JS}\neg\neg \\
t, M^{\kappa} & \Rightarrow & \blacktriangle & \\
M^{\kappa} & \Rightarrow & v^{\kappa} &
\end{array}$$

and (K_d) translates as

$$\begin{array}{rcll}
M^{\vartheta} & \Rightarrow & \neg\neg\neg\neg v & \\
\neg\neg\neg\neg\neg\neg\neg t, M^{\vartheta} & \Rightarrow & \blacktriangle & v = \neg t \\
t, M^{\vartheta} & \Rightarrow & \blacktriangle & \mathbf{JS}\neg\neg \\
M^{\vartheta} & \Rightarrow & v &
\end{array}
\qquad
\begin{array}{rcll}
M^{\kappa} & \Rightarrow & \neg\neg v^{\kappa} & \\
\neg\neg\neg\neg t, M^{\kappa} & \Rightarrow & \blacktriangle & v^{\kappa}=\neg t \\
t, M^{\kappa} & \Rightarrow & \blacktriangle & \mathbf{JS}\neg\neg \\
M^{\kappa} & \Rightarrow & v^{\kappa} & .
\end{array}$$

References

A.Diego: Sobre algebras de Hilbert. Tesis Buenos Aires 1961. Also: Notas di Mat. **12** Bahia Blanca 1965 . French translation: Sur les algebres de Hilbert. Paris 1966

M.Dummett: Elements of Intuitionism. Oxford 1977

V.Glivenko: Sur quelques points de la logique de M.Brouwer. Bull.Acad.Roy.Belgique **15** (1929) 183–188

K.Gödel: Zum intuitionistischen Aussagenkalkul. Anzeiger Akad.Wiss.Wien **69** (1932) 65–66

K.Gödel: Eine Interpretation des intuitionistischen Aussagenkalkuls. Ergeb.math.Kolloq. **4** (1933) 39–40

St.Jaskowski: Recherches sur les système de la logique intuitioniste. Actes du Congr.Int.de Philosophie Sci. **6** Paris 1936, 58–61 . Translation in McCall 67 .

S.Kripke: Semantical analysis of intuitionistic logic. in: J.Crossley and M.A.Dummett eds.: Formal Systems and Recursive Functions. Amsterdam 1965

St.McCall: Polish Logic. Oxford 1967

J.C.C.McKinsey: Proof of the independence of the primitive symbols of Heyting's calculus of propositions. JSL 4 (1939) 155–158

I.Nishimura: On formulas of one variable in intuitionistic propositional calculus. J.S.L. **25** (1960) 327–331

W.Rautenberg: Klassische und nichtklassische Aussagenlogik. Braunschweig 1979

G.F.Rose: Propositional calculus and realizability. Trans.A.M.S. 75 (1953) 1–19

A.S.Troelstra – D. van Dalen: Constructivism in Mathematics I . Amsterdam 1988

Chapter 7. Calculi of Formulas

Sequents, the objects of the sequent calculi studied so far, were conceived as *deductive situations*: the rules of such calculi have premisses and conclusions which are deductive situations. Viewed in this way, the sequent calculi do *not* begin with a description of how to deduce but teach how to get new deductive situations from given ones. So they describe derivations of *deductive situations from deductive situations*, and only if they start from trivial deductive situations of the form $x \Longrightarrow x$ becomes it possible to *implicitly* read off actual deductions *of formulas from formulas*.

The earlier analyses of the concepts of deducibility and logical consequence, starting with FREGE 79, § 6, all had concerned themselves with deductions of *formulas from formulas*. The characteristic property of these *calculi of formulas* was the use of the rule of *modus (ponendo) ponens*, saying that from v and $v \rightarrow w$ one may conclude upon w. Clearly, this rule has the same inconstructive disadvantages as has the cut rule for sequent calculi: given a formula w, there is no indication from which formula v, together with $v \rightarrow w$, it may have been deduced.

When discussing calculi of formulas I shall speak again of *rules*, but the reader be advised to notice that the objects of my rules now will formulas and not sequents or deductive situations.

Let T be an algebra of formulas (or terms). A general notion of *consequence* will determine to every set M of formulas the set cp(M) of *consequences of* M. Thus cp will be a map from subsets of T to subsets of T, and it shall be assumed to have the properties of a *closure operation:*

$M \subseteq cp(M)$	extensitivity
if $M \subseteq N$ then $cp(M) \subseteq cp(N)$	monotonicity
$cp(cp(M)) = cp(M)$	transitivity .

Moreover, it should be *formal* in the sense that, if v is in cp(M) and if a substitution g is made for the variables occurring in these formulas, then also the result g(v) of v under this substitution shall be a consequence of the formulas g(M) which result from M under this substitution. I therefore define a *consequence operation on* T to be a closure operation on (the underlying set of) T which is *invariant*

$$g(cp(M)) \subseteq cp(g(M))$$

for every endomorphism g of T. In that case I shall write $M \vdash v$ instead of $v \epsilon cp(M)$; thus the invariance condition may be expressed as

$$\text{if } M \vdash v \text{ then } g(M) \vdash g(v) .$$

In particular, for the empty set 0 the set cp(0) will be closed with respect to all endomorphisms of T, and instead of $0 \vdash v$ I shall simply write $\vdash v$.

If T contains the operation $\rightarrow$ then two formulas v,w are called *interdeducible* if there holds $\vdash$ v$\rightarrow$w as well as $\vdash$ w$\rightarrow$v. The class **A** of algebras *associated* to a consequence operation cp shall be the equational class defined by the equations [v,w] such that v and w are interdeducible. In all my examples, the relation V of interdeducibility will be an invariant congruence relation on T, and algebra T/V is called the *Lindenbaum algebra* of cp.

A consequence operation cp may *admit* or *respect* certain *rules*. For instance, if T contains $\rightarrow$ and if for two variables x,y there holds

(MP) x, x$\rightarrow$y $\vdash$ y

(and thus v,v$\rightarrow$w $\vdash$ w for any formulas v,w) then cp will be said to *admit the rule* (MP), and consequence operations admitting (MP) will be called *modus ponens* or MP-*operations*. More generally, a *rule* shall be a pair, consisting of a finite (possibly empty) sequence μ of formulas, its *premisses*, and a single formula u as its *conclusion*; I shall, for the moment, denote such a rule by $\mu \Vdash$ u, and the length of μ I will call its arity. Conclusions of rules of arity 0 are called *axioms*. I define that a consequence operation cp *admits* the rule $\mu \Vdash$ u if $\mu \vdash$ u (and thus $g \cdot \mu \vdash g(u)$ for every endomorphism g of T).

Given a set $\mathcal{R}$ of rules, a consequence operation cp is called the *calculus defined by* $\mathcal{R}$ if it is the *smallest* consequence operation admitting the rules of $\mathcal{R}$; I shall prove that such a smallest (and therefore unique) cp *always exists*.

To begin with I will discuss the special case that $\mathcal{R}$ contains (MP) as its only rule of positive arity; I then denote by AX the set of all axioms given by the 0-ary rules in $\mathcal{R}$ and call cp the *pure* MP-*calculus defined by* AX. Let M be any set of formulas; in order to describe cp(M), I introduce the notion of a *pure* MP-*deduction* S from M and the axioms AX. By this I understand a tree together with a function assigning a formula to each of its nodes such that

every maximal node e carries an element from M or the endomorphic image of an axiom,

every non-maximal node e has exactly two upper neighbours, e carries a formula q and the upper neighbours carry formulas p and p$\rightarrow$q.

A deduction then is said to be a *deduction from* M *of* the formula carried by its minimal node. I define cp(M) to be the set of formulas which have deductions from M. In order to see that cp is a closure operation, observe that M $\subseteq$ cp(M) holds since the elements of M have deductions consisting of one-node trees, and M $\subseteq$ N implies cp(M) $\subseteq$ cp(N) since every deduction from M is a fortiori a deduction from D. As for cp(cp(M)) $\subseteq$ cp(M), observe that for v in cp(cp(M)) there is a deduction S of v the maximal nodes of which either carry axioms or formulas w in cp(M). If w occurs at a maximal node e_w then I find a deduction S_w of w from M; implanting S_w on top of e_w, I obtain a deduction of v from M. Finally, if g is an endomorphism of T then it follows from $g(p \rightarrow q) = g(p) \rightarrow g(q)$ that a deduction of v from M turns into a deduction of g(v) from g(M) if, at each node, I replace the formula assigned to it by its image under g. This proves that cp is an MP-operation. Consider now any consequence operation cp' admitting the rules of my particular $\mathcal{R}$; for the elements u in AX this means that they are in cp'(0). But then they are also in any cp'(M), meaning that the formulas at the maximal nodes of a pure MP-deduction S from M are in cp'(M), and since cp' admits (MP), induction in the tree of S shows that the formulas at each of its nodes are in cp'(M). It follows that $cp(M) \subseteq cp'(M)$, and thus cp is the smallest consequence operation admitting the rules in $\mathcal{R}$.

If $\mathcal{R}$ is an arbitrary set of rules then the definition of an *$\mathcal{R}$-deduction* S from M is an immediate generalization: it shall be a tree together with a function assigning to each of its nodes (i) a formula, (ii) a rule from $\mathcal{R}$, (iii) an endomorphism r of T such that

every maximal node e is either assigned an element from M, or the rule assigned to e is of arity 0 with an axiom u, and e is assigned the formula r(u) for the assigned endomorphism r,

every non-maximal node e is assigned a rule $\mu \Vdash u$ of positive arity m and has exactly m upper neighbours which are assigned the formulas $r \cdot \mu$, and e is assigned the formula r(u) for the assigned endomorphism r.

Defining cp(M) to be the set of all formulas which have $\mathcal{R}$-deductions from M, the proof that cp is the smallest consequence operation admitting $\mathcal{R}$ carries over immediately.

[The existence of the smallest consequence operation admitting a set $\mathcal{R}$ of rules can be obtained also by arguments about algebras. Every pair R,g, consisting of a rule $R: \mu \Vdash u$ of arity m in $\mathcal{R}$ and of an endomorphism g of T, defines an operation f of arity m on u(T): if α is a sequence in T of length m then

$$f(\alpha) = g(u) \text{ if } \alpha = g \cdot \mu \quad \text{and } f(\alpha) = \alpha(0) \text{ otherwise.}$$

Under these operations f the set of u(T) becomes an algebra A. Consider now a consequence operation cp' admitting the rule R, and let α be a sequence in M. If $f(\alpha) = \alpha(0)$ then $f(\alpha)$ is trivially in cp'(M); if $f(\alpha) = g(u)$ for $\alpha = g \cdot \mu$ then $\alpha \vdash g(u)$ implies that $f(\alpha)$ is in cp'(M). Consequently, if cp' admits the rules of $\mathcal{R}$ then the set $[M]^A$ generated by M in A will be contained in cp'(M). On the other hand, the map cp defined by $cp(M) = [M]^A$ *is* a closure operation which *is* invariant and which *does* admit each of the rules of $\mathcal{R}$; hence cp is the smallest consequence operation admitting the rules $\mathcal{R}$.]

I now shall use the properties of frames developed in the last Chapter. It is no accident that the name (f_{mp}), of the first property defining frames, uses the same two letters which occur in the name of the rule (MP). Let T_d be the algebra of formulas with the connective $\rightarrow$ only. If cp is an MP-operation cp such that the two formulas

(d0) $x\rightarrow(y\rightarrow x)$
(d1) $(x\rightarrow(y\rightarrow z))\rightarrow((x\rightarrow y)\rightarrow(x\rightarrow z))$

belong to cp(0) then T_d, cp(0) is a d-frame. Thus the statements proved for d-frames can be read as saying that various other formulas belong to cp(0); for instance by **6**.(f8) the formula

(d2) $(x\rightarrow y)\rightarrow((y\rightarrow z)\rightarrow(x\rightarrow z))$.

will be in cp(0). The proofs of those statements were not always obvious, and the reader may have wondered how some apparently inperspicuous arguments possibly might have been discovered. The answer, and actually a recipe of how to find such proofs, will become clear immediately.

I shall say that an (arbitrary) MP-operation cp is *deductive* if always

(D) $M,v \vdash w$ implies $M \vdash v\rightarrow w$

where, again, M,v abbreviates $M\cup\{v\}$; this is nothing but the rule (I$\rightarrow$) for deduction situations $M \vdash u$. The property (D) will be called the *deduction principle,* and that (D) holds is also expressed by saying that the *deduction theorem* holds. If cp is any MP-operation then every deducibility

(*) $\vdash m_0\rightarrow(m_1\rightarrow(\ldots\rightarrow(m_{n-2}\rightarrow m_{n-1})\ldots))$

gives rise to the deducibility

(**) $m_0, m_1, \ldots, m_{n-2} \vdash m_{n-1}$,

said to be the *rule belonging to the axiom* (*). Conversely, if cp is deductive then (**) implies (*), and (*) then is said to be the *axiom belonging to the rule* (**). Assume now that cp *is* deductive and consider again the task to prove (d2). For each of the following lines

$$\vdash (x\rightarrow y)\rightarrow((y\rightarrow z)\rightarrow(x\rightarrow z))$$
$$x\rightarrow y \vdash (y\rightarrow z)\rightarrow(x\rightarrow z)$$
$$x\rightarrow y,\ y\rightarrow z \vdash x\rightarrow z$$
$$x\rightarrow y,\ y\rightarrow z,\ x \vdash z$$

the deducibility of any but the last line will follow by (D) from the deducibility of the line below it. But the deducibility of the last line follows immediately by two applications of (MP), deducing first y from x and $x\rightarrow y$ and then z from y and $y\rightarrow z$. This illustrates the ease with which deductions may be found for deductive MP-operations. So far, of course, I have not exhibited a single deductive MP-operation, and so it is time to prove the

LEMMA 1 For a deductive MP–operation the rule

(d1R) $x \rightarrow y \; , \; x \rightarrow (y \rightarrow z) \vdash x \rightarrow z$

is admissible and the formulas (d0), (d1) are deducible. Conversely, a pure MP–calculus is deductive if (d0), (d1) are deducible, or if (d0) is deducible and (d1R) is admissible.

First, the formula (d0) is deducible and (d1R) is admissible if (D) holds:

$$\begin{array}{r}x,y \vdash x \\ x \vdash y \rightarrow x \\ \vdash x \rightarrow (y \rightarrow x)\end{array}$$

and

$$\begin{array}{rll} & x, x \rightarrow y, x \rightarrow (y \rightarrow z) \vdash y & \text{(MP)} \\ & x, x \rightarrow y, x \rightarrow (y \rightarrow z) \vdash y \rightarrow z & \text{(MP)} \\ & y, y \rightarrow z \vdash z & \text{(MP)} \\ \text{transitivity of cp}: & x, x \rightarrow y, x \rightarrow (y \rightarrow z) \vdash z & \\ & x \rightarrow y, x \rightarrow (y \rightarrow z) \vdash x \rightarrow z & \text{(MP)} \end{array}$$

Also, if (d1R) is admissible then the last deduction extends to a deduction of (d1). In an MP–calculus deducing (d1) with length m there is a deduction H of length 2+m showing the admissibility of (d1R):

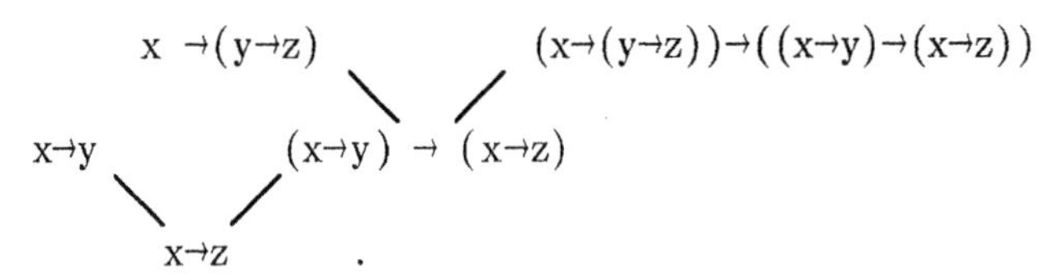

Finally, every MP–calculus deducing (d0) and admitting (d1R) deduces $x \rightarrow x$. Because let u be a deducible formula, e.g. (d0). Then

1.	$\vdash u$	by choice of u
2.	$\vdash u \rightarrow (x \rightarrow u)$	by (d0)
3.	$\vdash x \rightarrow u$	by 1, 2 and (MP)
4.	$\vdash x \rightarrow (u \rightarrow x)$	by (d0)
	$\vdash x \rightarrow x$	by (d1R) .

The remainder of the proof will follow from the

LEMMA 2 Let cp be a pure MP–calculus deducing (d0) and admitting (d1R). There is an operator which transforms a cp–deduction S for $M,v \vdash w$ into a cp–deduction T for $M \vdash v \rightarrow w$. If k_0 is 1 plus the maximum of the lengths of cp–deductions of (d0) and $v \rightarrow v$, and if k_1 is the length of a fixed cp–deduction H of $x \rightarrow z$ from $x \rightarrow y$ and $x \rightarrow (y \rightarrow z)$, then $\mathit{length}(T) \leq k_1 \cdot \mathit{length}(S) + k_0$.

By recursion on the depth $|e|$ of nodes e of S, I shall construct for every e a cp–deduction T_e such that if e carries the formula q then T_e deduces $v \dashv q$ from M , and $length(T_e) \leq k_1 \cdot |e| + k_0$.

Once this will have been done, I shall set $T = T_r$ with r the root of S.

Consider first a maximal node e of S carrying a formula q. If q is an axiom of cp or an element of M then it follows from (d0) that there is a c-deduction G ending with a node g carrying $q \dashv (v \dashv q)$. I define T_e by putting G on top of e, making g into an upper neigbour of e, adding a second upper neighbour f of e carrying q, and proceeding by (MP) to $v \dashv q$. T_e then has length k_0. If q is v then I define T_e to be the cp-deduction of $v \dashv v$ obtained by substituting v for x in the above deduction of $x \dashv x$ where, of course, the steps using $\vdash$ u , (d0) and (d1R) will be replaced by the actual cp-deductions. By choice of k_0 again T_e has this number as length.

If e is not maximal then, (MP) being the only rule, there are two upper neighbours f and g of e which carry p, $p \dashv q$ while e carries q. Let II be a fixed cp-deduction of length k_1, deriving $x \dashv z$ from $x \dashv y$ and $x \dashv (y \dashv z)$; thus H under substitution becomes a deduction H' of length k_1 deriving $v \dashv q$ from $v \dashv p$ and $v \dashv (p \dashv q)$. By inductive hypothesis, there are c-deductions T_f of $v \dashv p$ and T_g of $v \dashv (p \dashv q)$. Implanting T_f, T_g at the maximal nodes of H' which carry their results, I obtain the desired deduction T_e of $v \dashv q$. The length of T_e is bound by the larger of $length(T_g) + k_1$ and $length(T_f) + k$ by k_1. Since $|g| \leq |e| - 1$, $|f| \leq |e| - 1$ there follows from $length(T_g) \leq k_1 \cdot |g| + k_0$ and $length(T_f) \leq k_1 \cdot |f| + k_0$ that $length(T_e) \leq k_1 \cdot (|e| - 1) + k_1 + k_0 = k_1 \cdot |e| + k_0$.

If both (d0), (d1) are axioms of cp then k_1 is 2. There is another possibility to apply Lemma 2 in certain situations, namely that to consider, in addition to cp, the calculus $cp^{\S}$ obtained by adding (d1R) as a defining rule. Then the transformation in Lemma 2 can be viewed as mapping cp-deductions into $cp^{\S}$-deductions, and in that case k_1 is 1.

It follows from Lemma 1 that there is a *smallest* deductive MP-operation cd, viz. the pure MP-calculus defined by the axioms (d0), (d1).

I now can formulate the recipe of how to find proofs for properties of d-frames which may be formulated, say, as inequalities $t \precsim s$. If such an inequality is to hold for all d-frames, then it will hold, in particular, for the d-frame determined by the calculus cd on the term algebra T_d. Thus I shall look for a deduction for $\vdash t \dashv s$ in cd. As cd is deductive, I search for a chain of deducibilities, connected by the deduction principle, starting with $\vdash t \dashv s$ and ending with a deducibility which has a derivation in cd. Repeated application of the algorithm described in Lemma 2 then produces a deduction T for $\vdash t \dashv s$ in cd. Finally, induction in the tree of T shows that, for *any* d-frame A, Q, the formulas carried by the nodes of S are contained in the set Q.

1. Modus Ponens Calculi for Positive Logic

Let cd be the pure MP-calculus defined by (d0), (d1) on the algebra T_d of formulas containing the connective $\rightarrow$ only. The cd-deductions provide a formal notion of proof for statements on d-frames. In analogy to Theorem **6**.1, Corollary **6**.1_d and Theorem **6**.2_d there holds the

THEOREM 1_d The class associated to cd is $\mathbf{A}_d$. The pair T_d,cd(0) is a d-frame, and the relation V_d of cd-interdeducibility coincides with the relation R_d of KP_d-interdeducibility. For every finite M and every v it is equivalent that

$v \in cd(M)$

$M \Longrightarrow v$ is KP_d-derivable

for every A in $\mathbf{A}_d$ and for for every homomorphism h from T_d in A: if $h(m) = e_A$ for every m in M then $h(v) = e_A$.

It follows from the definitions that T_d,cd(0) is a c-frame. Thus V_d is the congruence relation determined by that frame – though instead of using **6**.(f4),(f8) those formulas now may be proved easily with the deduction theorem. Also with the deduction theorem I deduce the equations D_d defining $\mathbf{A}_d$; thus $\mathbf{Mod}(V_d) \subseteq \mathbf{A}_d$. For the reverse inclusion I show that, for every frame A,Q, if there is a derivation S establishing $\vdash v \rightarrow w$ then $h(v) \leq h(w)$ holds for every homomorphism from T_d into A; this, again, follows immediately by induction in the tree of S. Since cd is invariant as a calculus, V_d is an invariant congruence relation, and so $\mathbf{Mod}(V_d) = \mathbf{A}_d = \mathbf{Mod}(R_c)$ implies $V_d = R_c$. – The class associated to cd was shown to be $\mathbf{A}_d$ by DIEGO 61 who invented the equation **6**.(a_d).

By definition of T_d/V_d the distinguished element of that algebra is cd(0), and by Corollary **6**.1_d the distinguished element of T_d/R_d is the set of all v such that $\blacktriangle \Longrightarrow v$ is KP_d-derivable. Hence $T_d/V_d = T_d/R_d$ implies that $v \in cd(0)$ if, and only if, $\blacktriangle \Longrightarrow v$ is KP_d-derivable. But then the following statements are equivalent:

(x_0) $m_0, m_1, \ldots, m_{k-1} \Longrightarrow u$ is derivable in KP_d
(x_1) $\blacktriangle \Longrightarrow m_0 \rightarrow (m_1 \rightarrow (\ldots \rightarrow (m_{k-1} \rightarrow u)\ldots))$ is derivable in KP_d
(x_2) $m_0 \rightarrow (m_1 \rightarrow (\ldots \rightarrow (m_{k-1} \rightarrow u)\ldots)) \in cd(0)$
(x_3) $u \in cd(\{m_0, m_1, \ldots, m_{k-1}\})$

The semantical completeness result it but a restatement of Theorem **6**.2_d (and a direct proof, not making use of the connection with KP_d, is even simpler and will be carried out below in the variant Theorem 1_{f1}).

It follows from the above that, for any KP_d-derivation D with an endsequent $M \Longrightarrow u$, there exist cd-derivations S for $M \vdash u$ and vice versa. Starting from S establishing (x_3), the deduction theorem will give a deduction establishing (x_2), and for calculi as simple as the propositional ones considered here a proof search may produce a derivation of (x_1) and then one of (x_0). Conversely, a derivation D of (x_0) will give a derivation of (x_1), but in view of the inconstructive character of the rule (MP) there is no systematic procedure to then find a deduction of (x_2). However, I shall prove the

THEOREM 2_d There is an operator **P** transforming cd-deductions S of $M \vdash u$ into KP_dC-derivations **P**(S) of $M \Longrightarrow u$, and *length*(**P**(S)) $\leq$ *length*(S) + k for a suitable k.

There is an operator **Q** transforming KP_dC-derivations D of $M \Longrightarrow u$ into cd-deductions **Q**(D) of $M \vdash u$, and the length of **Q**(D) is bounded by a linear function of *length*(D).

Let S be a cd-deduction of $M \vdash u$. In the tree of S I replace at every node e the formula v_e at e by the sequent $M \Longrightarrow v_e$ which I shall call D(e). By recursion in the tree of S, I construct for every e a KP_d-derivation D_e of $M \Longrightarrow u$. Consider first the case that e is maximal; if v_e is in M then $M \Longrightarrow v_e$ is a reflexive axiom, and if v_e is one of the axioms (d0), (d1) then I know how to find a KP_d-derivations already of $\blacktriangle \Longrightarrow v_e$. So D_e is known for all these maximal e, and the length of such D_e has an upper bound k depending only on (the complexity of the formulas in) M. If e is not maximal then it has upper neighbours f, g such that $v_f = p$, $v_g = p \rightarrow q$, $v_e = q$; hence there are D_f and D_g deriving $M \Longrightarrow p$ and $M \Longrightarrow p \rightarrow q$. The inversion rule for $(I\rightarrow)$ produces from D_g a KP_d-derivation H of $M, p \Longrightarrow q$. Connecting D_f and H by (CUT), I arrive at a KP_dC-derivation D_e of D(e). Finally, I set $\mathbf{P}(D) = D_r$ for the root of S. Clearly *length*(**P**(S)) $\leq$ *length*(S) + k holds with the above k.

Let now D be a KP_dC-derivation of $M \Longrightarrow u$; I shall first construct an operator $\mathbf{Q}_0$ which is not linear. If D has length 0 then u must be a variable contained in M; thus also $\mathbf{Q}_0(D)$ is trivial. If D has positive length then it ends with an instance either of $(I\rightarrow)$ or of $(E\rightarrow)$ or ends with a cut. In the first case, u is $p \rightarrow q$ and there is a (shorter) subderivation H of the premiss $M, p \Longrightarrow q$; to H I find the derivation $\mathbf{Q}_0(H)$ of $M, p \vdash q$, and Lemma 2 produces from $\mathbf{Q}_0(H)$ the derivation D*. In the second case, M is $v \rightarrow w, N$ and there are (shorter) subderivations G of $N \Longrightarrow v$ and H of $w, N \Longrightarrow u$ for which I have derivations $\mathbf{Q}_0(G)$ of $N \vdash v$ and $\mathbf{Q}_0(H)$ of $w, N \vdash u$. I form $\mathbf{Q}_0(D)$ by taking the maximal nodes e of $\mathbf{Q}_0(H)$ which carry the formula w, and give each of them one new upper neighbour carrying $v \rightarrow w$ and a second upper neighbour which is the end of a copy of $\mathbf{Q}_0(G)$:

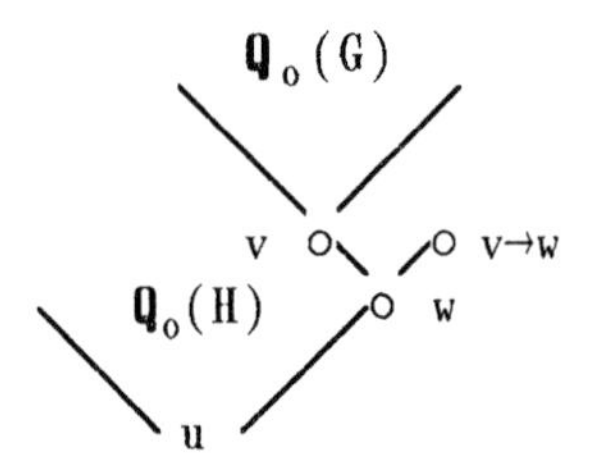

In the third case, if M ⟹ u arises under a cut from M ⟹ v and v,M ⟹ u then I implant a deduction of M ⊢ v at those maximal nodes of a deduction of v,M ⊢ u which carry the formula v.

The length of $\mathbf{Q}_0(D)$ is bounded by $2^n + n \cdot k$ for $n = \mathit{length}(D)$. Because in the first case there holds $\mathit{length}(\mathbf{Q}_0(D)) \leq 2 \cdot \mathit{length}(\mathbf{Q}_0(H)) + k$, in the second case there holds $\mathit{length}(\mathbf{Q}_0(D)) \leq \mathit{length}(\mathbf{Q}_0(H)) + \mathit{length}(\mathbf{Q}_0(G)) + 1$ and in the third case I have $\mathit{length}(\mathbf{Q}_0(D)) \leq \mathit{length}(\mathbf{Q}_0(H)) + \mathit{length}(\mathbf{Q}_0(G))$ for the derivations of H and G of the premisses of the cut.

In order to construct an operator **Q** with linear growth I follow an idea of GORDEEV and begin by transforming D into a $cd^§$-deduction $D^§$ for the calculus $cd^§$ with (d1R) as an additional rule; from $D^§$ I define **Q**(D) replacing the instances of (d1R) by their cd-deductions. As these have length 2, **Q**(D) will have at most twice the length of $D^§$, and so it will suffice to see that the length of $D^§$ is bounded by a linear function in *length*(D). In the first case, I form $D^§$ as I formed $\mathbf{Q}_0(D)$, but as a $cd^§$-deduction its length now is bound by $\mathit{length}(\mathbf{Q}(H)) + k$. In the second case, I apply Lemma 2 to the deduction $H^§$ of w,N ⊢ u obtaining a $cd^§$-deduction $H^†$ of N ⊢ w→u. I then form $D^§$ as

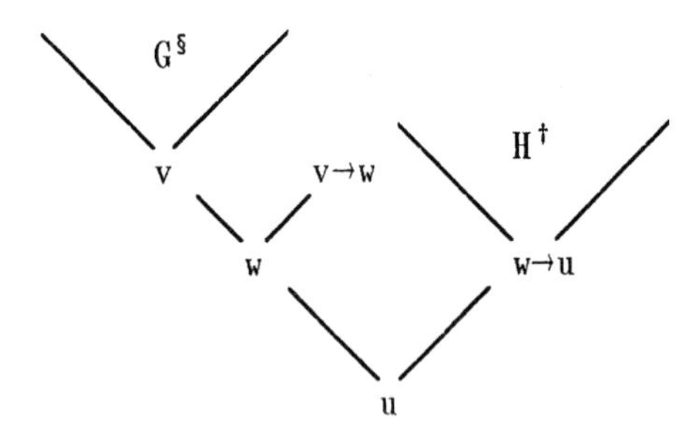

whence

$$\begin{aligned} \mathit{length}(D^§) &\leq 2 + \mathit{max}(\mathit{length}(G^§), \mathit{length}(H^†)) \\ &\leq 2 + k + \mathit{max}(\mathit{length}(G^§), \mathit{length}(H^§)) . \end{aligned}$$

In the third case, let G and H be the subderivations of M ⟹ v and of v,M ⟹ u . Applying Lemma 2 to $H^§$, I find a deduction $H^†$ of M ⊢ v→u and form $D^§$ as

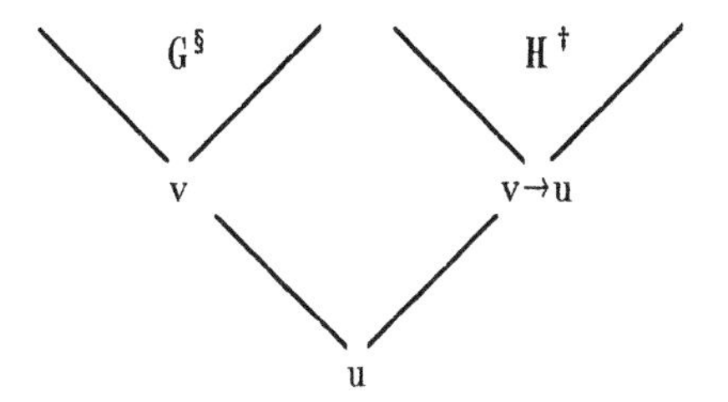

whence $length(D^{\S}) \leq 1+k+max(length(G^{\S}), length(H^{\S}))$. Induction therefore shows that $length(D^{\S}) \leq 2 \cdot length(D) + k \cdot (1+length(D))$.

It also is possible to use the translation employed in the proof of Theorem **6**.2; such an operator $\mathbf{Q}_1$ then will work in two steps: (1) a K_tP_dC-derivation D of $M \Longrightarrow u$ is transformed into a cd-deduction $D^{\S}$ of $M \rightarrow^* u$ and (2) from $D^{\S}$ a cd-deduction $\mathbf{Q}_1(D)$ for $M \vdash u$ is obtained by applying (MP). The construction of $D^{\S}$ is by recursion on D and makes use of statements proved for frames in the previous Chapter and, therefore, holding in particular for the frame T_d, cd(0). If M is a sequence of formulas then there are cd-deductions for

(FRP) $\vdash (M \rightarrow^* u) \rightarrow (M \cdot \pi \rightarrow^* u)$ by **6**.(f11),

(FW) $\vdash ((M \rightarrow^* u) \rightarrow ((v,M) \rightarrow^* u)$ by **6**.(f12),

(FRC) $\vdash ((v,M) \rightarrow^* u) \rightarrow ((v,v,M) \rightarrow^* u)$ by **6**.(f13),

(FI→) $\vdash ((v,M \rightarrow^* w) \rightarrow (M \rightarrow^* (v \rightarrow w))$ by **6**.(f7*),

and by **6**.(f12) there is a cd-deduction of

(FE→) $M \rightarrow^* v \vdash ((w,M) \rightarrow^* u) \rightarrow ((v \rightarrow w,M) \rightarrow^* u)$.

[A direct proof can start from a deduction of $a \rightarrow ((b \rightarrow c) \rightarrow ((a \rightarrow b) \rightarrow c))$ which can be found making use of (D). This would produce a deduction of

$$(M \rightarrow^* v) \rightarrow (((M \rightarrow^* w) \rightarrow (M \rightarrow^* u)) \rightarrow (((M \rightarrow^* v) \rightarrow (M \rightarrow^* w)) \rightarrow (M \rightarrow^* u))$$

and thus a deduction

$$(M \rightarrow^* v) \vdash (((M \rightarrow^* w) \rightarrow (M \rightarrow^* u)) \rightarrow (((M \rightarrow^* v) \rightarrow (M \rightarrow^* w)) \rightarrow (M \rightarrow^* u)).$$

Making use of **6**.(f9*), I deduce from the latter formula

$$(M \rightarrow^* (w \rightarrow u)) \rightarrow ((M \rightarrow^* ((v \rightarrow w) \rightarrow u))$$

and from that by **6**.(f7*)

$$(w \rightarrow (M \rightarrow^* u)) \rightarrow ((v \rightarrow w) \rightarrow (M \rightarrow^* u)) \quad .]$$

Finally, there is a cd-deduction of

(FCUT) $M \rightarrow^* v \; , \; (v,M) \rightarrow^* u \vdash M \rightarrow^* u$

since **6**.(f7*), (f9*) deduce from the formula $(v,M) \rightarrow^* u = v \rightarrow (M \rightarrow^* u)$ first $M \rightarrow^* (v \rightarrow u)$ and then $(M \rightarrow^* v) \rightarrow (M \rightarrow^* u)$.

Now the construction of $D^§$ is clear, since deductions named after rules ensure that the transformed conclusion of such rule can be deduced from the transformed premiss(es), employing at most one instance of (MP) *and* these well established deductions. Thus the length of $D^§$ is that of D plus the lengths of the established deductions. These lengths, however, depend only on the number of formulas in the respective M. In K_tP_dC this number may increase at most by 1 when going from a conclusion to its premisses, hence it is bounded by $w(D) \cdot \mathit{length}(D)$ where $w(D)$ is the number of formulas in the antecedent of the endsequent of D. Consequently, the length of $D^§$ is bounded by $w(D) \cdot \mathit{length}(D) \cdot \mathit{length}(D)$, and the change from $D^§$ to $\mathbf{Q}_1(D)$ then will add another $w(D)$ steps.

LEMMA 3 cd is also the pure MP-calculus defined by the axioms

(d0) $x \to (y \to x)$
(d2) $(x \to y) \to ((y \to z) \to (x \to z))$
(d3) $(x \to (x \to y)) \to (x \to y)$

as well as the pure MP-calculus defined by the axioms

(d0) $x \to (y \to x)$
(d4) $(y \to z) \to ((x \to y) \to (x \to z))$
(d3) $(x \to (x \to y)) \to (x \to y)$
(d5) $(x \to (y \to z)) \to (y \to (x \to z))$.

That the formulas of these axioms are cd-deducible follows by obvious applications of (D) and (MP). Thus the pure MP-calculi cd_1 and cd_2, defined by the first and the second axiom system, have axioms belonging to cd(0), and since they employ the same rule as does cd, every deduction in cd_1 or cd_2 can be expanded into a cd-deduction. Conversely, I do *not* know yet whether cd_1 or cd_2 are deductive; I *do* have, however, for cd_1 the rule *belonging* to (d2)

(CR) $x \to y,\ y \to z \vdash x \to z$,

also called the *chain rule*. Also, since

$\vdash (y \to z) \to ((x \to y) \to (x \to z)$ by (d4)
$\vdash (y \to z) \to ((x \to y) \to (x \to z)) \to (x \to y) \to ((y \to z) \to (x \to z))$ by (d5)

it follows with (MP) that (d2) is derivable in cd_2. Consequently, cd_1 is also contained in cd_2. Thus it remains to be shown that cd is contained in cd_1, and this will be the case if I can derive (d1R) in cd_1. I begin by showing that the axiom belonging to (MP)

(AMP) $\vdash x \to ((x \to y) \to y)$

is derivable in cd_1 :

1. $\vdash x\to((x\to y)\to x)$ by (d0)
2. $\vdash ((x\to y)\to x)\to((x\to y)\to((x\to y)\to y)$ by (d2)
3. $\vdash x\to((x\to y)\to((x\to y)\to y)$ from 1,2 by (CR)
4. $\vdash ((x\to y)\to((x\to y)\to y))\to((x\to y)\to y)$ substituting $x\to y$ for x in (d3)
5. $\vdash x\to((x\to y)\to y)$ from 3,4 by (CR) .

From this I conclude

1. $\vdash y\to((y\to z)\to z)$ by (AMP)
2. $x\to y,\ y\to((y\to z)\to z)\vdash x\to((y\to z)\to z)$ by (CR)
3. $x\to y\vdash x\to((y\to z)\to z$ by 1 , 2
4. $\vdash x\to(y\to z)\to((y\to z)\to z)\to(x\to z)$ by (d2)
5. $x\to(y\to z)\vdash((y\to z)\to z)\to(x\to z)$ by 4
6. $x\to y,\ x\to(y\to z)\vdash x\to((y\to z)\to z),\ ((y\to z)\to z)\to(x\to z)$
7. $x\to((y\to z)\to z),((y\to z)\to z)\to(x\to z)\vdash x\to(y\to z)$ by (CR) [by 4,5
8. $x\to y,\ x\to(y\to z)\vdash x\to(x\to z)$ by 6 and 7
9. $x\to(x\to z)\vdash x\to z$ by (d3)
10. $x\to y,\ x\to(y\to z)\vdash x\to z$ by 8 and 9.

This ends the proof of Lemma 3.

For the algebra T_e with the connectives $\to$ and $\wedge$ I define the pure MP-calculus ce by any set of axioms for cd plus the axioms

(e0) $x\wedge y\to x$
(e1) $x\wedge y\to y$
(e2) $(z\to x)\to((z\to y)\to(z\to x\wedge y))$;

for the algebra T_g with the connectives $\to$, $\wedge$ and $\vee$ I define the pure MP-calculus cg by any set of axioms for ce plus the axioms

(g0) $x\to x\vee y$
(g1) $y\to x\vee y$
(g2) $(x\to z)\to((y\to z)\to(x\vee y\to z))$.

In the first case, the pair T_e, ce(0) becomes an E-frame, in the second T_g, cg(0) becomes an F-frame. Since in E-frames the condition **6**.(f_{e2}) can be replaced by **6**.(f_{e2_o}), here the axioms (e2) can be replaced by

($e2_0$) $x\to(y\to x\wedge y)$

and the proof of **6**.(f_{e2_o}) provides MP-deductions establishing this equivalence.

Defining V_e and V_g by the same conditions as V_d, and replacing cd, $\mathbf{A}_d$, T_d, V_d, R_d, d-frame, KP_d in Theorem 1_d by ce, $\mathbf{A}_e$, T_e, V_e, R_e, e-frame, KP_e and by cg, $\mathbf{A}_g$, T_g, V_g, R_g, g-frame, KP, there hold without changes to their proofs

THEOREM 1_e [verbally as Theorem 1_d]

THEOREM 1_g [verbally as Theorem 1_d]

Replacing cd and KP_dC by ce, KP_eC and cg, KPC there also hold

THEOREM 2_e [verbally as Theorem 2_d]

THEOREM 2_g [verbally as Theorem 2_d] .

Here the construction of **P** has to be supplement by KP_e– and KP–derivations of the new axioms; for (g2) this was done during the proof of Corollary 6.2, and the other cases I can leave to the reader. For the construction of **Q** I again use the auxiliary calculi $ce^§$, $cg^§$ with the additional rule (d1R). If D ends with an instance of (I∧) with subderivations G and H leading to the premisses M ⟹ v and M ⟹ w then I form $D^§$ as

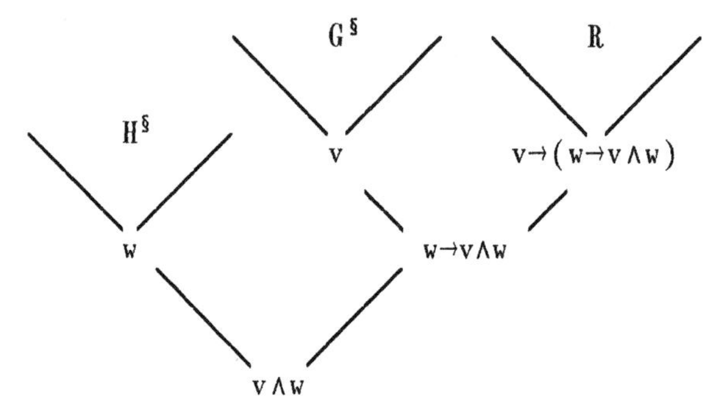

where R is a deduction of $(e2_0)$ of length r; thus

$$length(D^§) \leq 2 + max(length(G^§), length(H^§), r) .$$

If D ends with (E∧) and H is the subderivation of, say, the premiss v,M ⟹ u then Lemma 2 transforms $H^§$ into a deduction $H^†$ of M ⊢ v→u, and I form $D^§$ with help of (e0) as

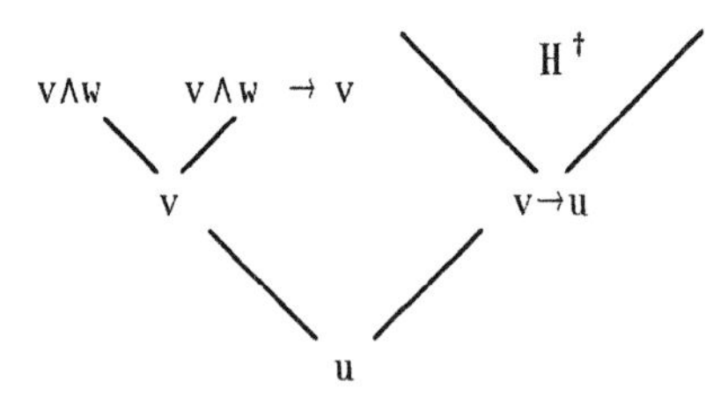

whence $length(D^§) \leq 1 + max(1, length(H^†)) = k+1+length(H^§)$. If D ends with (I∧) and H is the subderivation of, say, the premiss M ⟹ v then I form $D^§$ with help of (g0) from $H^§$ such that $length(D^§) = 1 + length(H^§)$. If D ends with an instance of (E∨) with subderivations G and H leading to the premisses v,M ⟹ u and w,M ⟹ u then I first apply Lemma 2 to $G^§$, $H^§$ obtaining $G^†$: M ⊢ v→u and $H^†$: M ⊢ w→u , and then I form $D^§$ with help of (g2) as

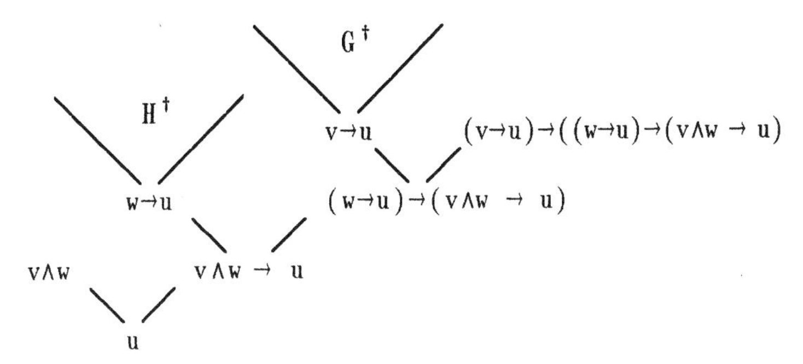

whence $length(D^§) \leq 3+k+max(length(G^§), length(H^§))$. Induction therefore shows that $length(D^§) \leq (3+r)\cdot length(D) + k\cdot(1+length(D))$.

Deductions in a pure modus ponens calculus cp will, unless they consist of an axiom alone, always make use of formulas containing the connective →. Let H be a subset of the set of propositional connectives → , ∧ ,∨ , ¬ ; cp will be called *separated* with respect to the connectives in H if the following holds: if M ⊢ v can be deduced in cp and if M, v contain only connectives from H then a cp-deduction of M ⊢ v can be constructed which uses only formulas containing → and the connectives from H. It follows from Theorem 2_g that cg is separated for each of the subsets of {→,∧,∨} . Because a cgdeduction S of M ⊢ v can first be transformed into a KP-derivation S* which, by the subformula principle, uses only formulas containing the connectives from H. It then follows from the constructions, performed in the proofs of Theorem 2_d and its extensions, that the cp-deduction of M ⊢ v formed from S* uses formulas which, apart from →, contain only connectives from H. – In SCHMIDT 60, § 117, it is shown that ce is separated for {→,∧} *without* making use of the correspondence between deductions and derivations.

2. Modus Ponens Calculi for Minimal and for Intuitionistic Logic

Let T be a term algebra containing the positive connectives and ¬ . Let cm be the pure modus ponens calculus defined by a set of axioms for the positive calculus cg together with the axiom

(m0) $(x\to\neg y)\to(y\to\neg x)$.

Defining V_m by the same conditions as V_d, and replacing cd, $\mathbf{A}_d$, T_d, V_d, R_d, d-frame, KP_d in Theorem 1_d by cm, $\mathbf{A}_m$, T, V_m, R_m, m-frame, LM, there holds without changes to its proof

THEOREM 1_m [verbally as Theorem 1_d].

If a_0 is a fixed axiom, there also holds

THEOREM 2_m There is an operator **P** transforming cm–deductions S of M $\vdash$ u into LMC–derivations **P**(S) of M $\Longrightarrow$ u , and *length*(**P**(S)) $\leq$ *length*(S) + k for a suitable k.

There is an operator **Q** transforming KP_dC–derivations D of M $\Longrightarrow$ u , M $\Longrightarrow$ ▲ into cm–deductions **Q**(D) of M $\vdash$ u or M $\vdash$ $\neg a_0$, and the length of **Q**(S) is bounded by a linear function of *length*(S).

For the construction of **P** I use the LM–derivation of (a_{mp}) given in the proof of Theorem **6**.1_m. For the construction of **Q** I use again the calculus $cm^§$ with the additional rule (d1R). Observe that the only way a sequent M $\Longrightarrow$ ▲ can arise as conclusion is by an instance of (E$\neg$); so consider the case that D ends with such and let H be the subderivation of the premiss M $\Longrightarrow$ v . It follows from **6**.(f_{m3}) that there is a cm–deduction of $\neg x \rightarrow (x \rightarrow \neg a_0)$ of a fixed length r_m. I then form $D^§$ as

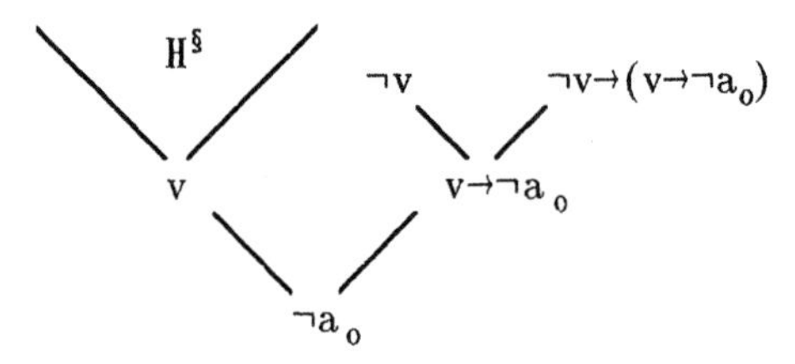

whence *length*($D^§$) $\leq$ 1 + *max*(*length*($H^§$), r_m+1). If D ends with an instance of (I$\neg$) and the conclusion M $\Longrightarrow$ $\neg v$ then let H be the subderivation of the premiss v,M $\Longrightarrow$ ▲ . Thus $H^§$ deduces v,M $\vdash$ $\neg a_0$, and Lemma 2 produces $H^†$ deducing M $\vdash$ $v \rightarrow \neg a_0$. Making use of the axiom a_0 and the axiom (m0), I form $D^§$ as

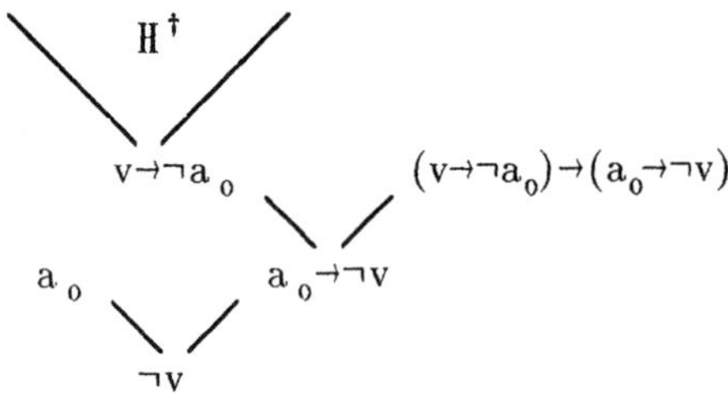

whence *length*($D^§$) $\leq$ 2+k + *length*($H^§$).

It follows from this Theorem that the calculus cm is separated for each subset H of the connectives $\rightarrow, \wedge, \vee, \neg$. Because a cm–deduction S of M $\vdash$ v can first be transformed into an LM–derivation S of M $\Longrightarrow$ v which, by the subformula principle, uses only formulas containing connectives from set set H of connectives occurring in M $\vdash$ v . If H does not contain $\neg$ then S is a KP–derivation which can be transformed into a cg–derivation of M $\vdash$ v using only connectives from H. If H does contain $\neg$ then S may contain instances of (I$\neg$) or (E$\neg$). Transforming S into an cm–derivation, this may contain, besides the formulas from S, also the formula $\neg a_0$. But if a_0 is, for instance, (d0) then $\neg a_0$ will contain only the connectives $\neg$ and $\rightarrow$.

LEMMA 4 The calculus cm may be defined by adjoining, to any set of axioms for the positive calculus cg, one of the following four axiom systems:

(A) (m1) $x \rightarrow \neg\neg x$ (m2) $x \rightarrow y \rightarrow \neg y \rightarrow \neg x$

(B) (m3) $\neg x \rightarrow (x \rightarrow \neg y)$ (m4) $(x \rightarrow \neg x) \rightarrow \neg x$

(C) (m2) (m4)

(D) (m5) $(x \rightarrow \neg y) \rightarrow ((x \rightarrow y) \rightarrow \neg x)$.

cm-deductions of (m1)-(m3) can be taken the proofs of 6.(f_{m1})-(f_{m3}) for M-frames. In order to derive (m4), let a_0 be in cd(0) and proceed as follows:

$\neg x \vdash x \rightarrow \neg a_0$	by (m3)
$x \rightarrow \neg x \vdash x \rightarrow (x \rightarrow \neg a_0)$	by (d4)
$x \rightarrow (x \rightarrow \neg a_0) \vdash (x \rightarrow x) \rightarrow (x \rightarrow \neg a_0)$	by (d1)
$x \rightarrow (x \rightarrow \neg a_0) \vdash x \rightarrow \neg a_0$	
$x \rightarrow (x \rightarrow \neg a_0) \vdash a_0 \rightarrow \neg x$	by (m0)
$x \rightarrow (x \rightarrow \neg a_0) \vdash \neg a_0$	by choice of a_0
$x \rightarrow \neg x \vdash \neg x$	by (CR)
$\vdash (x \rightarrow \neg x) \rightarrow \neg x$	by the deduction principle .

So the axioms from (A), (B), (C) are derivable by cm. Assume now the system (A); then $\vdash y \rightarrow \neg\neg y$ from (m1) implies $\vdash (\neg\neg y \rightarrow \neg x) \rightarrow (y \rightarrow \neg x)$ by (d2), and since $\vdash (x \rightarrow \neg y) \rightarrow (\neg\neg y \rightarrow \neg x)$ by (m2), the formula (m0) becomes derivable by (CR). Thus (A) defines cm. Assume next (C). Since

$x \vdash \neg x \rightarrow x$	by (d0)
$x \vdash \neg x \rightarrow \neg\neg x$	by (m2)
$x \vdash \neg\neg x$	by (m4)

the deduction principle gives a derivation of (m1). Thus (C) is equivalent to (A). Now consider (B). Then

$\neg y \vdash y \rightarrow \neg x$	by (m3)
$x \rightarrow \neg y \vdash x \rightarrow (y \rightarrow \neg x)$	by (d4)
$x \rightarrow \neg y \vdash (x \rightarrow y) \rightarrow (x \rightarrow \neg x)$	by (d1)
$(x \rightarrow y) \rightarrow (x \rightarrow \neg x) \vdash (x \rightarrow y) \rightarrow \neg x$	by (m4) and (d4)
$x \rightarrow \neg y \vdash (x \rightarrow y) \rightarrow \neg x$	by (CR)

and the deduction principle gives a derivation of (m5). Thus (B) implies (D), but clearly (D) defines cm since

$\vdash$ y→(x→y) by (d0)
$\vdash$ ((x→y)→¬x)→(y→¬x) by (d2)

and so (m5) together with (CR) implies (m0).

Together with cm, also the four modus ponens calculi $cm_a - cm_d$, defined by the axiom systems (A)–(D), will be separated for every subset of the connectives →, ∧, ∨, ¬ . Because the deductions, given above, of each of the axiom systems from the others, employ only the connectives → and ¬ if a_0 is chosen as (d0), say. Thus every cm_a–derivation, for instance, may be transformed into a cm–derivation containing the same connectives, and vice versa.

Let ci be the pure modus ponens calculus defined on T by a set of axioms for the minimal calculus cm together with the axiom

(i0) $\neg x \to (x \to y)$.

Defining V_i by the same conditions as V_d, and replacing cd, $\mathbf{A}_d$, T_d, V_d, R_d, d–frame, KP_d in Theorem 1_d by ci, $\mathbf{A}_i$, T, V_i, R_i, i–frame, LJ, there holds without changes to its proof

THEOREM 1_i [verbally as Theorem 1_d].

There also holds

THEOREM 2_i [verbally as Theorem 2_m]

In order to construct **P** I need an LJ–proof of (i0), and that can be found in the proof of Theorem **6**.1_i. For the construction of **Q**, consider an LJ–derivation D ending with an instance of (AIN), extending a subderivation H of M ⟹ ▲ to M ⟹ v . I then define $D^§$ from $H^§$ and the axioms a_0 and (i0) as

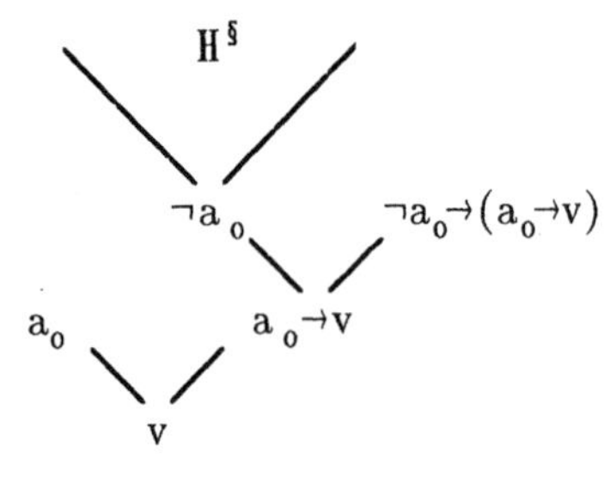

whence $length(D^§) \leq 2 + length(H^§)$.

Also the calculus ci is separated for each subset H of the connectives →, ∧, ∨, ¬ . Because let S be a ci–deduction of M $\vdash$ v , transformed into an LJ–derivation S* of M ⟹ v . If the set H of connectives occurring in M, v does not contain ¬ then S* does not contain instances of (E¬). Consequent–

ly, S^* cannot contain sequents $N \Longrightarrow \blacktriangle$ and, therefore, no instances of the rule (AIN). Thus S^* is a KP-derivation containing only connectives from H and may be transformed into a cg-deduction of $M \vdash v$ with the same property. And if H does contain $\neg$ then the only additional formula appearing in the transform of S^* will be $\neg a_0$, containing only $\neg$ and $\rightarrow$.

LEMMA 5 The calculus ci may be defined by adjoining to any set of axioms for the minimal calculus cm the axiom

(i1) $\neg x \wedge x \rightarrow y$

and by adjoining to any set of axioms for the positive calculus cg either both the two axioms (m4) and (i0) or both the two axioms (m3) and

(i2) $\neg\neg x \rightarrow (\neg x \rightarrow x)$.

Here (i1) is just a reformulation of (i0) in view of the equivalence (A) for RPC-lattices. Replacing y in (i0) by $\neg y$, I obtain (i2); hence the axioms (m4) and (i0) imply the axiom system (B) for cm from Lemma 4. The axiom (i2) is a particular instance of (i0); however, together with (m3) this instance is equivalent to (i0) over the axioms of cg since

$x, \neg x \vdash \neg\neg y$	by (m3)
$\neg\neg y \vdash \neg y \rightarrow y$	by (i2)
$x, \neg x \vdash \neg y \rightarrow y$	by (CR)
$x, \neg x \vdash \neg y$	by (m3)
$x, \neg x \vdash y$	by (MP)

from where the deduction principle leads to (i0). – It follows from these proofs that also the pure modus ponens calculi defined by these new axiom systems will be separated for each subset of $\rightarrow$, $\wedge$, $\vee$, $\neg$. Again, it follows from $T/V_i = F_i$ that, for any formula v, the deducibility of $\vdash v$ for ci is equivalent to $p_i(v) = e$ in F_i. Thus each of the various axiom systems for ci gives, simultaneously, a system of defining equations for the class $\mathbf{A}_i$.

3. Modus Ponens Calculi for Classical Logic

Let T be a term algebra containing the positive connectives and $\neg$. Let cc be the pure modus ponens calculus defined by a set of axioms for the intuitionistic calculus ci together with the axiom

(c0) $(\neg x \rightarrow x) \rightarrow x$.

Defining V_c by the same conditions as V_d, and replacing cd, $\mathbf{A}_d$, T_d, V_d, R_d, d-frame, KP_d in Theorem 1_d by cc, $\mathbf{A}_c$, T, V_c, R_c, c-frame, LK, there holds without changes to its proof

THEOREM 1_c [verbally as Theorem 1_d].

I next deduce in cc

(c1) $\neg\neg x \to x$

from (i2) and (c0) by (CR) in analogy to **6**.(f_{c1}).

(c2) $x \vee \neg x$

since
$\vdash x \to x\vee\neg x$
$\vdash \neg(x\vee\neg x) \to \neg x$ by (m2)
$\vdash \neg(x\vee\neg x) \to x\vee\neg x$
$\vdash \neg(x\vee\neg x) \to \neg\neg(x\vee\neg x)$ by (m2)
$\vdash (\neg(x\vee\neg x) \to \neg\neg(x\vee\neg x)) \to \neg\neg(x\vee\neg x)$ by (m4)
$\vdash \neg\neg(x\vee\neg x)$
$\vdash x\vee\neg x$ by (c1)

(c3) $(\neg x\to y) \to (\neg y\to x)$

since
$\vdash (\neg x\to y) \to (\neg y\to\neg\neg x)$ by (m2)
$\vdash (\neg y\to\neg\neg x) \to (\neg y\to x)$ by (c1), (d4)

(c4) $(\neg x\to\neg y) \to (y\to x)$

since
$\vdash (\neg x\to\neg y) \to (\neg\neg y\to x)$ by (c3)
$\vdash (\neg\neg y\to x) \to (y\to x)$ by (c1), (d4)

(c5) $(x\to y) \to ((\neg x\to y)\to y)$

since
$\vdash (\neg y\to\neg x) \to ((\neg y\to x) \to \neg\neg y)$ (m5)
$\vdash (x\to y) \to ((\neg y\to x) \to \neg\neg y)$ by (m2)
$\vdash ((\neg y\to x) \to \neg\neg y) \to ((\neg x\to y) \to \neg\neg y)$ by (c3), (d2)
$\vdash (x\to y) \to ((\neg x\to y) \to \neg\neg y)$ by (CR)
$\vdash ((\neg x\to y) \to \neg\neg y) \to ((\neg x\to y) \to y)$ by (c1), (d4)
$\vdash (x\to y) \to ((\neg x\to y) \to y)$ by (CR) .

I now can prove

THEOREM 2_c [verbally as Theorem 2_d]

For the construction of **P**, I need an LK-proof of (dc0) and that can be found in the proof of Theorem **6**.1_c. For the construction of **Q**, let D be an LK-derivation ending with an instance of (K_0) and consider the calculus $cc^{\S}$ using (dR1). From the subderivations H and G ending with $w,M \Longrightarrow v$ and $\neg w,M \Longrightarrow v$ I define $D^{\S}$ as

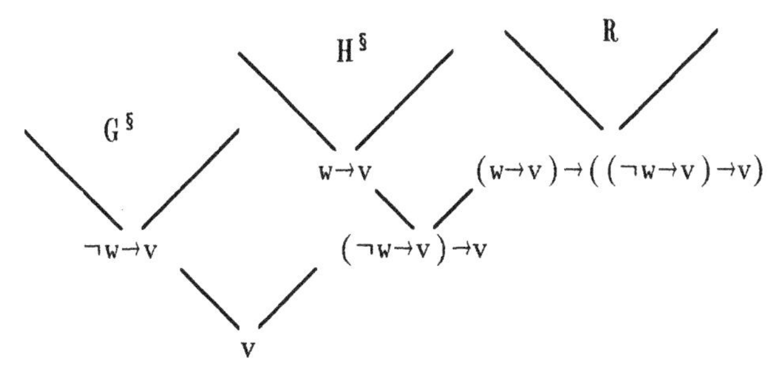

whence

$$length(\mathrm{D}^{\S}) \leq 2 + max(length(\mathrm{G}^{\S}),\ length(\mathrm{H}^{\dagger}),\ \mathrm{r})$$

with the length r of a deduction R of (c5). Induction therefore shows that $length(\mathrm{D}^{\S}) \leq 2 \cdot length(\mathrm{D}) + \mathrm{r}$.

I collect some more properties of the calculus cc in the

LEMMA 6 The calculus cc may be defined by any set of axioms for

(A) ci plus the axiom (cp) $((x \to y) \to x) \to x$

(B) ci plus the axiom (cpn) $((x \to \neg x) \to x) \to x$

(C) cm plus the axiom (c1)

(D) cg plus the axiom (c4)

(E) cg plus the axioms (i0) and (c0) .

(F) cm plus the axiom (c2) .

If X is one of A to F then cc_x shall be the calculus defined in (X). I begin by proving the equivalence of cc with cc_a and cc_b and this I start by deducing (cp) for cc:

$$\begin{array}{rll} \neg x \vdash & x \to y & \text{by (i0)} \\ (x \to y) \to x \vdash & \neg x \to x & \text{by (d2)} \\ (x \to y) \to x \vdash & x & \text{by (c0)} \end{array}$$

from where (cdp). Also, the instance (cpn) of (cp) gives (c0) since

$$\begin{array}{rll} x \to \neg x \vdash & \neg x & \text{by (m4)} \\ ((x \to \neg x) \to x) \to x \vdash & (\neg x \to x) \to x & \text{by (d2) twice} \\ \vdash & (\neg x \to x) \to x & \text{by (cpn)} \ . \end{array}$$

Thus cc contains cc_a, cc_a contains cc_b, and cc_b again contains cc. Next, cc_c contains cc since

$$\begin{array}{rll} x \vdash & \neg\neg x & \text{by (m1)} \\ \neg x \to x \vdash & \neg x \to \neg\neg x & \text{by (d4)} \\ \neg x \to \neg\neg x \vdash & \neg\neg x & \text{by (m4)} \end{array}$$

$$\neg x \to x \vdash \neg\neg x \qquad \text{by (CR)}$$
$$\neg x \to x \vdash x \qquad \text{by (c1)}$$

whence (c0) by the deduction principle. As for (D), (c4) permits to derive (i0), (c1) and (m0):

$$\neg x \vdash \neg y \to \neg x \qquad \text{by (d0)}$$
$$\neg x \vdash x \to y \qquad \text{by (c4)}$$
$$\vdash \neg x \to (x \to y) \qquad \text{(i0)}$$

$$\neg\neg x \vdash \neg x \to \neg y \qquad \text{by (i0)}$$
$$\neg\neg x \vdash y \to x \qquad \text{by (c4)}$$
$$\neg\neg x \vdash \neg\neg x \to x \qquad \text{specializing y}$$
$$\vdash \neg\neg x \to (\neg\neg x \to x)$$
$$\vdash \neg\neg x \to x \qquad \text{(c1) by (d1)}$$

$$x \to \neg y \vdash \neg\neg x \to \neg y \qquad \text{by (c1)}$$
$$x \to \neg y \vdash y \to \neg x \qquad \text{by (c4)}$$
$$\vdash (x \to \neg y) \to (y \to \neg x) \qquad \text{(m0)}$$

Thus cc_c is contained in cc_d whence cc is contained in cc_d. As for (E) it remains to be shown that cc in contained in cc_e, and since (m3) is an instance of (i0), Lemma 4 (B) shows that it will suffice to derive (m4) in cc_e:

$$\neg\neg x \vdash x \qquad \text{by (c1) which follows from (c0) and the instance (i2) of (i0)}$$
$$(\neg\neg x \to \neg x) \to \neg x \vdash (x \to \neg x) \to x \qquad \text{by (d2) twice}$$
$$\vdash (\neg\neg x \to \neg x) \to \neg x \qquad \text{by (c0)}$$
$$\vdash (x \to \neg x) \to x \qquad \text{by (MP)} \ .$$

Finally, for (F) it remains to be shown that cc is contained in cc_f, and this I do by deriving (c0) from (c2) with help of cc_f:

$$\vdash z \to x \vee z \qquad \text{by (g1)}$$
$$y \to z \vdash y \to x \vee z \qquad \text{by (d4)}$$
$$\vdash x \to x \vee z \qquad \text{by (g0)}$$
$$\vdash (x \to x \vee z) \to ((y \to x \vee z) \to (x \vee y \to x \vee z) \qquad \text{by (g2)}$$
$$\vdash (y \to x \vee z) \to (x \vee y \to x \vee z) \qquad \text{by (MP)}$$
$$y \to z \vdash x \vee y \to x \vee z \qquad \text{by (CR)}$$
$$\neg x \to x \vdash x \vee \neg x \to x \vee x \qquad \text{specializing y,z}$$
$$\vdash (x \to x) \to ((x \to x) \to (x \vee x \to x)) \qquad \text{by (g2)}$$
$$\vdash x \vee x \to x \qquad \text{by (MP)}$$
$$\neg x \to x \vdash x \qquad \text{by (CR)}$$
$$\vdash (\neg x \to x) \neg x \ .$$

It follows from the proof of Theorem 2_c that cc can also be defined as ci plus (c5). Another definition of cc is that as cg plus the axioms (m1) and (c3), because (c3) with $y = \neg x$ implies (c1).

I have shown in Chapter 5 that the calculus MC of classical logic permits to derive positive formulas which are underivable in positive (or intuitionistic) logic. Hence also cc shares this undesirable property of progressiveness.

Let cx be one of the pure MP-calculi cd, ce or cg on T_d, T_e or T_g, and let cxp be cx plus the *Peirce axiom* (cp). Since (cp) is cc-deducible, every cxp-deduction can be expanded into a cc-deduction.

Consider now a cc-deduction S of M $\vdash$ u where u and the formulas of M belong to T_x, and let x be one of d, e or g. By Theorem 2_x then **P**(S) is an LK-derivation D of M $\Longrightarrow$ u which by Lemma 5.5 can be transformed into a KPPC-derivation D' = $\mathbf{P}_p$(D) of M $\Longrightarrow$ u . The formulas in D' contain only the connectives occurring in M and u and so belong to T_x; hence D' is derivation of the calculus KP_xPC obtained from KP_x by adding (K_p). Observe also that there is no reason to eliminate cuts from D'. The operator **Q** from Theorem 2_x transforms KP_xC-derivations into cx-deductions, and now I shall extended it to an operator $\mathbf{Q}_p$ sending KP_xPC-derivations D' into cxp-deductions.

Here I use again the calculus $cxp^§$ with (d1R), and if D' ends with an instance of (K_p) loading to M $\longrightarrow$ u from a subderivation II of the premiss u$\rightarrow$w,M $\Longrightarrow$ u then I apply Lemma 2 to the deduction $H^§$ of u$\rightarrow$w,M $\vdash$ u , obtaining a $cxp^§$-deduction $H^†$ of M $\vdash$ (u$\rightarrow$w)$\rightarrow$u . I then form $(D')^§$ as

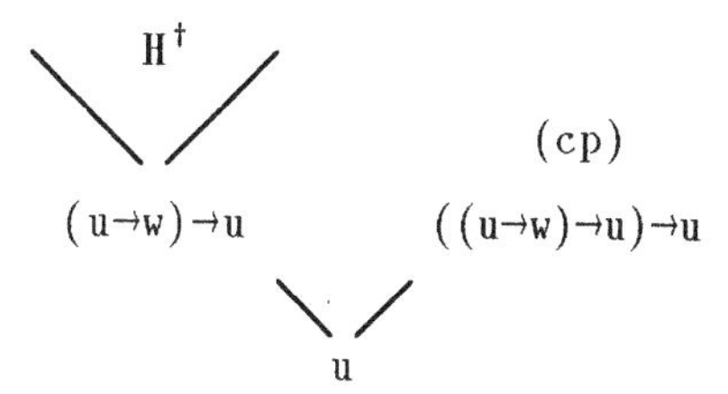

and replacing instances of (d1R) I obtain $\mathbf{Q}_p$(D) from $D^§$. All this started from the cc-deduction S which so has been transformed into the cxp-deduction $\mathbf{Q}_p(\mathbf{P}_p(\mathbf{P}(S)))$. Thus I have shown the

THEOREM 3 A deducibility M $\vdash$ u with positive formulas from T_x is cc-deducible if, and only if, it is cxp-deducible.

For cg I can prove this directly without making use of the transformations to and from sequential calculi. To this end, let cc be defined by the axiom system (D) of Lemma 6, i.e. as cg plus (c4). Given a cc-deduction S of M $\vdash$ u, let z be a new propositional variable and let ν be the map which is the identity on variables, is homomorphic for positive connectives and sends $\neg$v to $v^\nu\rightarrow z$. Let S^ν arise from S replacing every formula v by v^ν. Then instances of (MP) are preserved as are instances of positive axioms; an instance $(\neg v\rightarrow\neg w)\rightarrow(w\rightarrow v)$ of (c4) becomes

$$((v^{\nu}\to z)\to(w^{\nu}\to z)) \to (w^{\nu}\to v^{\nu}) \ .$$

Thus S^{ν} is a cg-deduction of u from M and these formulas. But such a formula has a cdp-deduction from $z\to v^{\nu}$:

$$\begin{array}{rll} & \vdash ((v^{\nu}\to z)\to v^{\nu}) \to v^{\nu} & \text{by (cp)} \\ & \vdash (w^{\nu}\to ((v^{\nu}\to z)\to v^{\nu}) \to (w^{\nu}\to v^{\nu}) & \text{by (d4)} \\ & \vdash ((v^{\nu}\to z) \to (w^{\nu}\to v^{\nu})) \to (w^{\nu}\to v) & \text{by (d5), (d2)} \\ z\to v^{\nu} & \vdash (w^{\nu}\to z) \to (w^{\nu}\to v^{\nu}) & \text{by (d4)} \\ z\to v^{\nu} & \vdash ((v^{\nu}\to z)\to(w^{\nu}\to z)) \to ((v^{\nu}\to z)\to(w^{\nu}\to v^{\nu})) & \text{by (d4)} \\ z\to v^{\nu} & \vdash ((v^{\nu}\to z)\to(w^{\nu}\to z)) \to (w^{\nu}\to v^{\nu}) & \text{by (CR).} \end{array}$$

So if $v_0, \dots, v_{k-1}$ are the formulas occurring in (c4)-instances of S then these cdp-deductions can be implanted into S^{ν}, changing it into a cgp-deduction $S^{\nu *}$ of

$$z\to v_0{}^{\nu}, \ \dots \ , z\to v_{k-1}{}^{\nu}, \ M \vdash u \ .$$

Let τ be the map which replaces z in all formulas of $S^{\nu *}$ by u. Then $S^{\nu *}$ is transformed into a cgp-deduction $U_k = S^{\nu * \tau}$ of

$$u\to v_0{}^{\nu\tau}, \ \dots \ , u\to v_{k-2}{}^{\nu\tau}, \ u\to v_{k-1}{}^{\nu\tau}, \ M \vdash u \ .$$

By the deduction theorem, U_k gives rise to a cgp-deduction of

$$u\to v_0{}^{\nu\tau}, \ \dots \ , u\to v_{k-2}{}^{\nu\tau}, \ M \vdash (u\to v_{k-1}{}^{\nu\tau}) \to u \ .$$

such that the (cp)-instance $((u\to v_{k-1}{}^{\nu\tau})\to u) \to u$ leads to a cgp-deduction U_{k-1} of

$$u\to v_0{}^{\nu\tau}, \ \dots \ , u\to v_{k-2}{}^{\nu\tau}, \ M \vdash u \ .$$

Repeating this (k−1) times, I arrive at a cgp-deduction U_0 of $M \vdash u$.

Theorem 3 is due to TARSKI who, in place of (cp), made use of the axiom $((x\to y)\to z)\to((x\to z)\to z)$; the simpler axiom (cp) was proposed by BERNAYS, cf. ŁUKASIEWICZ–TARSKI 30. In presence of (cp), the axiom (d2) for cd may be replaced by the axiom (d2) since then (d3) becomes deducible such that Lemma 3 may be applied:

$$\begin{array}{rll} & \vdash (x\to(x\to y)) \to (((x\to y)\to y) \to x\to y) & \text{by (d2)} \\ ((x\to(x\to y))\to x\to y) \to x\to y & \vdash (x\to(x\to y)) \to x\to y & \text{by (d2)} \\ & \vdash (x\to(x\to y)) \to x\to y & \text{by (cp) .} \end{array}$$

While in intuitionistic logic none of the connectives $\to, \wedge, \vee, \neg$ can be defined from the others (by Theorem 6.6), matters are different for classical logic since in every Boolean algebra there hold the identities

$$a \supset b = -a \cup b = -(a \cap -b) \ , \ a \cap b = -(a \supset -b) = -(-a \cup -b) \ , \ a \cup b = a \supset -b = -(-a \cap -b)$$

and so each of the positive connectives $\to$, $\wedge$, $\vee$ may be classically defined in terms of any of the others together with $\neg$. Consequently, classical deducibility may also be defined on term algebras containing only one of these positive connectives together with $\neg$. A pure modus ponens calculus for formulas containing only $\neg$ and $\vee$, for instance, may be obtained simply by translating cc : (MP) becomes

$$x\,,\ \neg x \vee y \vdash y$$

and the axioms are translated in the same manner: (d0) into $\neg x \vee (\neg y \vee x)$ and so on. While such obvious translations will result in rather cumbersome axiom systems, much simpler systems have been found, for instance WHITEHEAD-RUSSELLs 10 pure modus ponens calculus cc_{wr} for $\neg, \vee$ with $p \to q$ an abbreviation for $\neg p \vee q$ and with the axioms

$$x \vee x \to x$$
$$x \to x \vee y$$
$$x \vee y \to y \vee x$$
$$(x \to y) \to ((z \vee x) \to (z \vee y))\ .$$

They are easily deduced in cc, and conversely the (translations of the) axioms of cc can be deduced in cc_{wr} (cf. e.g. SCHMIDT 60). Of more interest are the modus ponens calculi for the connectives $\to$ and $\neg$ which I will discuss next.

Let T_n be a term algebra with the connectives $\to$ and $\neg$. Let cc_{fl} be the pure MP-calculus defined by the axioms

(d0) $\quad x \to (y \to x)$

(d1) $\quad (x \to (y \to z)) \to ((x \to y) \to (x \to z))$

(c4) $\quad (\neg x \to \neg y) \to (y \to z)$;

cc_{fl} is known as the FREGE-ŁUKASIEWICZ calculus. I shall first present the reduction of cc_{fl} to the calculus cc already known:

THEOREM 4 A deducibility $M \vdash u$ with formulas from T_n is cc-deducible if, and only if, it is cc_{fl}-deducible.

Once this will have been shown, I shall discuss a more direct proof of the completeness of cc_{fl} for the Boolean valued semantical consequence operation.

The one direction of the theorem is obvious since by Lemma 6 (D), a cc_{fl}-deduction is also a cc-deduction. In order to prove the other direction, let τ be the map from the term algebra T, generated by X for $\to, \wedge, \vee, \neg$, to T_n which is the identity on variables, is homomorphic for $\to$ and $\neg$, and which satisfies

$$(u \wedge w)^\tau = \neg(u^\tau \to \neg w^\tau)\ \ ,\ \ (u \vee w)^\tau = \neg u^\tau \to w^\tau\ .$$

Of course, this definition is motivated by the identities holding in Boolean algebras. Assume now that I have found cc_{fl}-deductions S_{e0}, S_{e1}, S_{e2} of the formulas

(te0) $\neg(x\to\neg y)\to x$
(te1) $\neg(x\to\neg y)\to y$
(te2) $(z\to x) \to ((z\to y)\to(z\to\neg(x\to\neg y)))$

and S_{go}, S_{g1}, S_{g2} for the formulas

(tg0) $x \to \neg x\to y$
(tg1) $y \to \neg x\to y$
(tg2) $(x\to z) \to ((y\to z)\to((\neg x\to y)\to z)))$.

Let U be a cc–deduction of M $\vdash$ u where I can assume cc to be defined as in Lemma 6 (D), i.e. with the axioms (d0),(d1),(e0),(e1),(e2),(g0),(g1),(g2),(c4). Replacing the formulas in U by their images under τ, M and u remain unchanged, instances of (d0),(d1),(c4) remain such instances, and instances of (e0),(e1),(e2),(g0),(g1),(g2) become instances of (te0),(te1),(te2),(tg0),(tg1),(tg2) for which I then implant the deductions S_{eo}, S_{e1}, S_{e2}, S_{go}, S_{g1}, S_{g2} . In this way, I have constructed a cc_{f1}–deduction of M $\vdash$ u.

In order to motivate how to find the deductions S_{eo}, S_{e1}, S_{e2}, S_{go}, S_{g1}, S_{g2}, it may be convenient to use the language of frames. Let A be an algebra with operations $\supset$ and $-$; let Q be a subset of A. The pair A,Q shall be called a fl–*frame* if with $\supset$ it is a d–frame and if there holds for all a,b

(f_{f1}) $(-a\supset -b) \supset (b\supset a)$ is in Q .

On an fl–frame I define an operation $\cap$ by $a\cap b = -(a\supset -b)$. The following cc_{f1}–deductions are motivated by the indicated corresponding inequalities for fl–frames.

	$\vdash \neg x \to x\to\neg y$	by (m3)	
	$\vdash \neg(x\to\neg y) \to \neg\neg x$	by (m2)	
	$\vdash \neg\neg x \to x$	by (c1)	
1.	$\vdash \neg(x\to\neg y) \to x$	(te0)	$a\cap b\leq a$
	$\vdash z\to\neg z \to \neg z$	by (m4)	
	$\vdash \neg\neg z \to \neg(z\to\neg z)$	by (m2)	
	$\vdash z \to \neg\neg z$	by (m1)	
2.	$\vdash z \to \neg(z\to\neg z)$	by (CR)	$c\leq c\cap c$
	$\vdash y\to\neg y \to x\to\neg y$	by (m0)	
3.	$\vdash \neg(x\to\neg y) \to \neg(y\to\neg x)$	by (m1)	$a\cap b\leq b\cap a$
	$\vdash \neg(y\to\neg x) \to y$	by (te0)	
4.	$\vdash \neg(x\to\neg y) \to y$	(te1) by (CR)	$a\cap b\leq b$
	$\vdash x\to y \to \neg y\to\neg x$	by (m1)	
	$\vdash \neg y\to\neg x \to (z\to\neg y)\to(z\to\neg x)$	by (d4)	
	$\vdash (z\to\neg y)\to(z\to\neg x) \to \neg(z\to\neg x)\to\neg(z\to\neg y)$	by (m1)	
5.	$\vdash x\to y \to \neg(z\to\neg x)\to\neg(z\to\neg y)$	by (CR)	if $a\leq b$ then $c\cap a\leq c\cap b$

Thus

$$
\begin{array}{rlll}
z\to x & \vdash \neg(z\to\neg z)\to\neg(z\to\neg x) & \text{by } 5 & \text{if } c\leq a \text{ then } c\cap c\leq c\cap a \\
 & \vdash z \to \neg(z\to\neg z) \to (\neg(z\to\neg z)\to\neg(z\to\neg x))\to(z\to\neg(z\to x)) & \text{by (d2)} & \\
 & \vdash \neg(z\to\neg z)\to\neg(z\to\neg x)) \to (z\to\neg(z\to x)) & \text{by } 2 & \\
z\to x & \vdash z\to\neg(z\to\neg x) & \text{by (CR)} & \text{if } c\leq a \text{ then } c\leq c\cap a \\
z\to x & \vdash z\to\neg(x\to\neg z) & \text{by } 3 & \text{if } c\leq a \text{ then } c\leq a\cap c \\
z\to y & \vdash \neg(x\to\neg z)\to\neg(x\to\neg y) & \text{by } 5 & \text{if } c\leq b \text{ then } a\cap c\leq a\cap b \\
z\to x, z\to y & \vdash z\to\neg(x\to\neg y) & \text{by (CR)} & \text{if } c\leq a,\ c\leq b \text{ then } c\leq a\cap b \\
 & \vdash (z\to x)\to((z\to y)\to(z\to\neg(x\to\neg y)) & \text{(te2)}\ . &
\end{array}
$$

I shall leave it to the reader to find deductions S_{go}, S_{g1}, S_{g2}. This then concludes the proof of Theorem 4. – Another proof could be given making use of the correspondence between deductions and derivations. A cc-deduction S of M $\vdash$ u can be translated first into a LK- and then into an MK-derivation D of M $\Longrightarrow$ u . D will use only formulas from T_n since MK has the subformula property. Retranslating D into a cc-deduction U, also U will only use formulas from T_n and, therefore, only the axioms of cc_{fl}.

In view of the elegance of the axioms of cc_{fl} it may be desirable to have an argument more direct than the reduction to cc in order to see that cc_{fl} coincides with the semantical consequence operation defined by Boolean algebras. As far as cc_{fl}-deductions are concerned, I only need that (d0), (d1) permit to deduce (d2), (d4), that (c4) deduces (m0) from Lemma 6 (D) and that (m0) deduces (m1),(m2),(m3),(m4) from Lemma 4. I also need the deductions given during the proof of the last theorem.

Observe first that for fl-frames the equivalence relation $\approx$ is also a congruence relation on with respect to the operation – as follows from the availability of (m2). Let now $\mathbf{A}_{fl}$ be the class of algebras $A = \langle u(A), \supset, - \rangle$ defined by the equations D_d together with

(a_{fl}) $\qquad (-a\supset -b) \supset (b\supset a) = e$.

THEOREM 1_{fl} The class associated to cc_{fl} is $\mathbf{A}_{fl}$. The pair $T_n, cc_{fl}(0)$ is an fl-frame, and the relation V_{fl} of cc_{fl}-interdeducibility coincides with the congruence relation of this frame. V_{fl} is an invariant congruence relation on T_n and $F_{fl} = T_n/V_{fl}$ is the free algebra in $\mathbf{A}_{fl}$ generated by the image of the set X of variables under the canonical homomorphism p_{fl} from T_n onto F_{fl}. For every finite M and every v the statement $v \in cc_{fl}(M)$ is equivalent to:

for every A in $\mathbf{A}_{fl}$ and for every homomorphism h from T_n into A: if $h(m) = e_A$ for every m in M then $h(v) = e_A$.

The class $\mathbf{A}_{fl}$ consists of the reducts with respect to $\supset$, – of (the expansions with $\supset$ of) Boolean algebras.

The proof of the first part is verbally that of Theorem 1_d. Also, $v \in cc_{fl}(M)$ implies the statement about homomorphisms as follows by induction over a cc_{fl}-deduction S of v from M. Conversely, if not $v \in cc_{fl}(M)$ then I can find an A in $\mathbf{A}_{fl}$ and a homomorphism h into A which maps all of M to e_A but $h(v) \neq e_A$. Because T_n, $cc_{fl}(M)$ is a fl-frame determining as $\approx$ a congruence relation V_M on T_n. Thus $A = T_n/V_M$ is in $\mathbf{A}_{fl}$, and the canonical homomorphism p_M onto A maps all of M to e_A but $h(v) \neq e_A$.

It remains to be shown that every A in $\mathbf{A}_{fl}$ is reduct of a Boolean algebra. Defining the operation $\cap$ by $a \cap b = -(a \supset -b)$, the formulas deduced during the proof of the previous theorem show

$$a \leq a \cap b \ , \qquad a \cap b = b \cap a \ ,$$
$$b \leq a \cap b \ , \qquad \text{if } c \leq a, c \leq b \text{ then } c \leq a \cap b \ .$$

Hence $a \cap b$ is the infimum of a and b. It follows from (c4) via (c1) that the map $-$ is an order reversing involution; hence $-(-a \cap -b)$ must be the supremum of a,b. But $-(-a \cap -b) = -(-(-a \supset --b)) = -a \supset b$ whence I may define $a \cup b = -a \supset b$. Also, the involution $-$ will map the largest element $e = a \supset a$ into a smallest element $-e$. Since $-a \cup a = a \supset a = e$ and $a \cap -a = -(a \supset a) = -e$, I obtain a lattice with complements. Let now B be the expansion of A with the defined operations $\cap$ and $\cup$. In order to see that B is a Boolean algebra, it remains to show distributivity, and this will hold if I see that I have an RPC-lattice, i.e. that

$$a \leq b \supset c \text{ if and only if } a \cap b \leq c \ .$$

Now $a \leq b \supset c$ implies $a \cap b \leq b \supset c$ by whence $(a \cap b) \supset (b \supset c) = e$. It follows by (d1) that $((a \cap b) \supset b \supset) \supset ((a \cap b) \supset c) = e$, but $(a \cap b) \supset b = e$ whence $(a \cap b) \supset c = e$, i.e. $a \cap b \leq c$. Conversely, it follows from $a \cap b \leq c$ that $-(b \supset -a) \leq c$, $-c \leq b \supset -a$. But then $-c \cap b \leq -a$ follows by an application of the already proven part of my equivalence, and thus $a \leq -(-c \cap b) = -(b \cap -c) = --(b \supset --c) = b \supset c$.

Once Theorem 1_{fl} is available, there are several other possibilities to prove Theorem 4: (1) By the deduction theorem, $M \vdash u$ is equivalent to $\vdash M \supset^* u$ where M is viewed as a sequence. So it suffices to show that $cc(0) \cap u(T_n)$ is contained in $cc_{fl}(0)$. $F_{fl} = T_n/V_{fl}$ in $\mathbf{A}_{fl}$ determines an expanded Boolean algebra B_0 by Theorem 1_{fl}. The canonical map p_{fl}, restricted to the variables, determines a homomorphism h from T to B which on T_n coincides with p_{fl}. An element v in cc(0) is, under every homomorphism from T into a Boolean algebra, mapped to the largest element e_B. But $v \in cc(0) \cap u(T_n)$ implies $p_{fl}(v) = h(v)$; hence p_{fl} maps v to the largest element e_0 of B_0. This e_0 is also the largest element of F_{fl} since B_0 expands F_{fl}. By definition of F_{fl} this is $cc_{fl}(0)$, and $p_{fl}(v) = cc_{fl}(0)$ is equivalent to $v \in cc_{fl}(0)$ by definition of the canonical map. (2) Since cc_{fl}-deductions are cc-deductions, every congruence class $p_{fl}(v)$ of T_n/V_{fl} is contained in the congruence class $p_c(v)$ of T/V_c, and there is a homomorphism g from T_n/V_{fl} to (the $\mathbf{A}_{fl}$-reduct of) T/V_c such that $g \cdot p_{fl} = p_c \upharpoonright u(T_n)$. Since $\mathbf{A}_{fl}$ is equationally equivalent to

$\mathbf{A}_c$, g is an isomorphism from the free algebra F_{f1} onto (the reduct of) the free algebra F_c. In particular, g is injective such that, for v,w in T_n, $p_c(v) = p_c(w)$ implies $p_{f1}(v) = p_{f1}(w)$. Hence the congruence classes under V_{f1} are the intersections with $u(T_n)$ of the congruence classes under V_c – in particular $cc_{f1}(0) = u(T_n) \cap cc(0)$. **(3)** The homomorphisms h' from T_n into algebras from $\mathbf{A}_{f1}$ are precisely the restrictions to $u(T_n)$ of homomorphisms h from T into algebras from $\mathbf{A}_c$. Thus if cn' is the semantical consequence operator defined by $\mathbf{A}_{f1}$ on T_n and if cn is the one defined by $\mathbf{A}_c$ on T then $cn'(M) = cn(M) \cap u(T_n)$ for $M \subseteq u(T_n)$. But is was shown that $cn' = cc_{f1}$ and $cn = cc$. Thus for any $M \subseteq u(T_n)$ also $cc_{f1}(M) = cc(M) \cap u(T_n)$.

ŁUKASIEWICZ 29 has proposed another pure MP–calculus cc_{11} on T_n with the axioms

(d2)	$(x \to y) \to ((y \to z) \to (x \to z))$
(i0')	$x \to (\neg x \to y)$
(c0)	$(\neg x \to x) \to x$.

Every cc_{11}–deduction is a cc–deduction. Conversely, starting from a cc–deduction, I may assume that it is already a cc_{f1}–derivation, and thus it will suffice to derive the axioms of cc_{f1} in cc_{11}. The following chain of deduced formulas contains (d0) on line 16 and (c4) on line 17; it ends with (d3) on line 26. This being derived, I apply Lemma 3 and derive (d1) from (d0), (d2) and (d3).

1.	$(\neg x \to y) \to z \ \to\ x \to z$	(d2) applied to (i0')
2.	$(\neg x \to y) \to (\neg y \to y) \ \to\ x \to (\neg y \to y)$	instance of 1
3.	$\neg y \to \neg x \ \to\ (\neg x \to y) \to (\neg y \to y)$	instance of (d2)
4.	$\neg y \to \neg x \ \to\ x \to (\neg y \to y)$	(CR) for 3, 2
5.	$x \to (\neg y \to y) \ \to\ ((\neg y \to y) \to y) \to (x \to y)$	instance of (d2)
6.	$\neg y \to \neg x \ \to\ ((\neg y \to y) \to y) \to (x \to y)$	(CR) for 4, 5
7.	$y \ \to\ \neg y \to \neg x$	instance of (i0')
8.	$y \ \to\ ((\neg y \to y) \to y) \to (x \to y)$	(CR) for 7, 6
9.	$(\neg z \to z) \to z \ \to\ ((\neg u \to u) \to u) \to (x \to ((\neg z \to z) \to z))$	8 for $u = (\neg z \to z) \to z$
10.	$(\neg u \to u) \to u \ \to\ x \to ((\neg z \to z) \to z)$	(MP) for (c0), 9
11.	$x \ \to\ (\neg z \to z) \to z$	(MP) for (c0), 10
12.	$\neg(x \to y) \ \to\ (\neg y \to y) \to y$	instance of 11
13.	$((\neg y \to y) \to y) \to (x \to y) \ \to\ (\neg(x \to y)) \to (x \to y)$	(d2) applied to 12
14.	$(\neg(x \to y)) \to (x \to y) \ \to\ x \to y$	instance of (c0)
15.	$((\neg y \to y) \to y) \to (x \to y) \ \to\ x \to y$	(CR) for 13, 14
16.	$y \ \to\ x \to y$	(CR) for 8, 15
17.	$\neg y \to \neg x \ \to\ x \to y$	(CR) for 6, 15
18.	$\neg x \ \to\ \neg\neg x \to \neg\neg y$	instance of (i0')
19.	$\neg x \ \to\ x \to y$	(CR) for 18, 17 twice
20.	$(x \to y) \to x \ \to\ \neg x \to x$	(d2) applied to 19
21.	$(\neg x \to x) \neg x \ \to\ ((x \to y) \to x) \to x$	(d2) applied to 20
22.	$(x \to y) \to x \ \to\ x$	(MP) for (c0), 21

23. $(x \to (x \to y)) \to (((x \to y) \to x) \to (x \to y))$ (d2) applied to 22
24. $(((x \to y) \to x) \to (x \to y)) \to (x \to y) \to (x \to (x \to y)) \to (x \to y)$ (d2) applied to 23
25. $((x \to y) \to x) \to (x \to y) \to x \to y$ instance of 22
26. $x \to (x \to y) \to x \to y$ (MP) for 25, 24

References

A.Diego: Sobre algebras de Hilbert. Tesis Buenos Aires 1961. Also: Notas di Mat. **12** Bahia Blanca 1965 . French translation: Sur les algebres de Hilbert. Paris 1966

G.Frege: Begriffsschrift, eine der arithmetischen nachgebildete Formelsprache des reinen Denkens. Halle 1879

J.Łukasiewicz Elementy logiki matematyccznej. Warszawa 1929 . 2nd. ed. 1958 . English translation: Elements of Mathematical Logic London 1963

J.Łukasiewicz, A.Tarski: Untersuchungen über den Aussagenkalkül. C.R. Soc. Sci. Lettr. Varsovie 23 (1930) 51–77 . Reprinted in Tarski 56 and Lukasiewicz 70

J.Łukasiewicz: Selected Works. ed. L.Borkowski. 1970 Amsterdam

H.A.Schmidt: Mathematische Gesetze der Logik I . Berlin 1960

A.N.Whitehead, B.Russell: Principia Mathematica, vol. I , Cambridge 1910

A.Tarski: Logic, Semantics, Metamathematics. Papers from 1923 to 1938. Oxford 1956

Historical Notes on Chapters 1 - 7

The first mathematical treatment of logic was that of BOOLE 47 who discovered that propositions satisfy laws which we now call those of Boolean algebra. While Boole investigated laws between Boolean terms and studied their relation to traditional logical situations such as those described in Aristoteles' syllogisms, he did not systematically consider the deduction of laws from other ones and their reduction to a set of axioms. That task was attempted, and solved, by FREGE 79 (with a more detailed presentation in FREGE 93) who wrote in § 13

> Es ist offenbar nicht dasselbe, ob man blos die Gesetze kennt, oder ob man auch weiss, wie die einen durch die andern schon mitgegeben sind. Auf diese Weise gelangt man zu einer kleinen Anzahl von Gesetzen, in welchen, wenn man die in den Regeln enthaltenen hinzunimmt, der Inhalt aller, obschon unentwickelt, eingeschlossen ist. Und auch dies ist ein Nutzen der ableitenden Darstellungsweise, dass sie jenen Kern kennen lehrt. ... Die Zahl der Sätze, die in der folgenden Darstellung den Kern bilden, ist neun. ...

Frege's only deduction rule was the modus ponens, and those propositions forming the *Kern* are what we now call axioms. So there were nine axioms, and as Frege only used the propositional connectives $\rightarrow$ and $\neg$ (though in a different and actually two-dimensional notation), there were six propositional axioms, namely (d0) [Frege's (1.], (d1) [Frege's (2.], (d5) [Frege's (8.], (m1) [Frege's (41.], (m2) [Frege's (28.], (c1) [Frege's (31.]. From these axioms Frege derived all other tautologies he considered.

Frege's two-dimensional notation was linearized by WHITEHEAD-RUSSELL 10 who chose $\vee$ and $\neg$ as primitive connectives and used the axioms mentioned above (they also introduced the symbol $\vdash$ in today's meaning as a remnant of Frege's signs). While propositional (2-valued) truth functions were discussed there, the question of whether such a calculus was *complete,* in that it permitted to deduce all semantically true propositions, was answered independently by BERNAYS in his 1918 Habilitationsschrift (cf. BERNAYS 26) and by POST 21. These proofs essentially consist in the following steps. (1) The calculus makes every formula v interdeducible with a formula v^* in *conjunctive prenormal* form in the sense of Chapter **1.10**; **(2)** v^* is also semantically equivalent to v; **(3)** v^* is a conjunction of disjunctions d_i of variables and negated variables and if v, hence v^*, is true then every of the d_i contains a subformula $x \vee \neg x$ for some x; **(4)** if the calculus deduces (c10), (g0), (g1), (e_{2o}) then it deduces v^* and, therefore, also v.

For the connectives $\rightarrow$ and $\neg$, a calculus with the axioms (d0),(d3),(d4),(d5),(c5) and $x \rightarrow (\neg x \rightarrow y)$ (interdeducible with (i0) by the deduction theorem) was presented in HILBERT 23. Frege's axioms were simplified to those of cc_{f1} by ŁUKASIEWICZ 29 where also cc_{11} was set up. A detailed report on the early work about MP-calculi is ŁUKASIEWICZ-TARSKI 30. A particular interest then was taken in pure MP-calculi employing *one* axiom only, and for the classically deducible formulas in $\rightarrow$ alone, deduced by ccx, a provably shortest single axiom was given by ŁUKASIEWICZ 48, namely $((x \rightarrow y) \rightarrow z) \rightarrow ((z \rightarrow x) \rightarrow (p \rightarrow x))$. For the classically deducible formulas in $\rightarrow$ and $\neg$ the shortest known single axiom

$$(((x \rightarrow y) \rightarrow (\neg z \rightarrow \neg p)) \rightarrow z) \rightarrow q) \quad \rightarrow \quad ((q \rightarrow x) \rightarrow (p \rightarrow x))$$

is due to MEREDITH 53; a proof is indicated in MONK 76, p.132.

The calculus ci of intuitionistic logic was proposed by HEYTING 30, the calculus cm of minimal logic by JOHANSSON 37. Based upon the rule (MP), there is another approach to cm and its fragments which was discussed in HILBERT-BERNAYS 34 and has been carried out in detail in SCHMIDT 60. (1) Consider first formulas v containing only the connective $\rightarrow$. A sequence $<v_1,\ldots,v_n>$ of formulas, ending with a formula v_n which is not an implication (i.e. is a variable), is called a d-*presentation* of v if $v = v^0$ for $v^n = v_n$, $v^{k-1} = v_k \rightarrow v^k$ for $0<k<n$; in that case I write $v = D(v_1,\ldots,v_n)$. Now every formula has a unique d-presentation, namely $x = D(x)$ for variables and $w \rightarrow v = D(w,v_1,\ldots,v_n)$ if $v = D(v_1,\ldots,v_n)$. A formula v is called d-*direct* if for

$v = D(v_1, \ldots, v_n)$ the formula v_n can be obtained by (MP) from (some of) the $v_1, \ldots, v_{n-1}$. A formula is called d-*derivative* if it can be obtained by (MP) from d-direct formulas. For instance, the formulas (d0)–(d5) are d-direct while $(((x \to y) \to x) \to x) \to y$ is d-derivative only. It is easily seen that the d-derivative formulas are precisely those in cd(0). – **(2)** Considering formulas containing $\wedge$ or both $\wedge$ and $\vee$, the e- and g-presentations and the e- and g-direct formulas are defined as above, only that v_n now may be a conjunction or a disjunction. The e- and g-derivative formulas are defined as those deducible by (MP) from direct formulas by (MP) *and* the rules

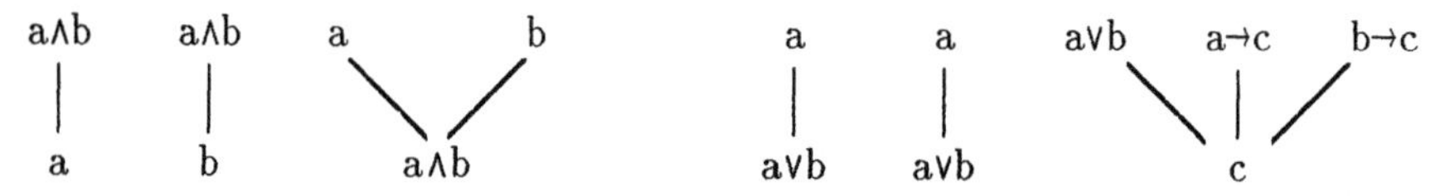

Then the e- and the g-derivative formulas are precisely those in ce(0) and in cg(0). **(3)** Considering formulas v containing $\neg$, a $\neg$- *reduct* of v is defined as obtained by (a) choosing a variable z not occurring in v and (b) replacing all subformulas $\neg u$ of v by $u \to z$. Now if *one* $\neg$-reduct of v is d-, e- or g-direct [derivative] then so is any other, and I define v to be m_d-, m_e- or m-direct [-derivative] if such $\neg$-reduct is d-, e- or g-direct [-derivative]. For instance, the formulas (m0)–(m5) are m_d-direct. It is easily seen that the m-derivative formulas are precisely those in cm(0).

The sequent calculi LJ and MK are the invention of GENTZEN 34 who established their cut elimination process and proved their equivalence with the calculi ci and cc. Much detailed work on sequent calculi has been collected in CURRY 63. Gentzen presented also another type of calculi which he called those of *natural deduction.* Their objects are deduction trees S, with roots e_S, whose nodes x carry *formulas*, together with a function assigning to every x a class $A_S(x)$ of nodes carrying *assumptions* on which [the formula at] x *depends*; if S consists of one node e_S only then $A_S(e_S) = \{e_S\}$. The rules for the logical connectives say that deductions R,S,T can be extended to a deduction U with a new root e_U

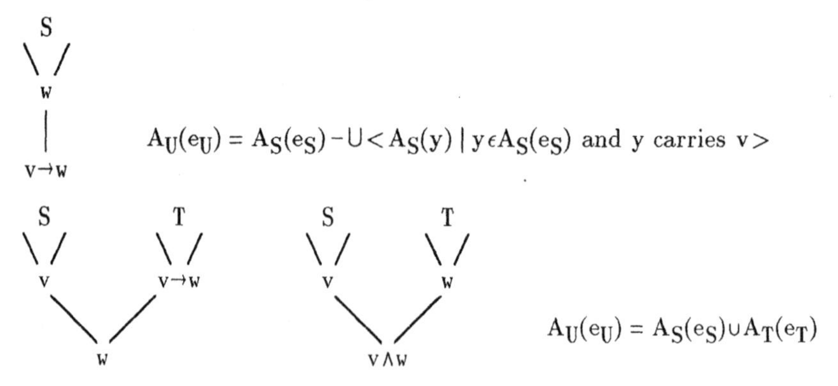

$$A_U(e_U) = A_S(e_S) - \bigcup \langle A_S(y) \mid y \in A_S(e_S) \text{ and } y \text{ carries } v \rangle$$

$$A_U(e_U) = A_S(e_S) \cup A_T(e_T)$$

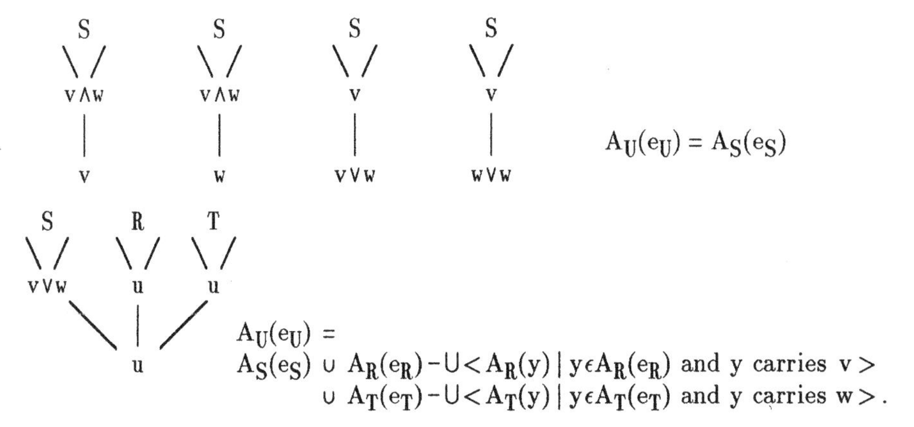

The structural rules are

(a) Two deductions S and T with endformulas v and w can be joined into a new deduction U of v [w] by putting them side by side and then adding a new node e_U below them which carries v, and then $A_U(x) = A_S(x)$ for x in S, $A_U(y) = A_T(y)$ for y in T, $A_U(e_U) = A_S(e_S)$.

(b) Two deductions S and T with endformulas v and w can be merged into a new deduction U of w by implanting S at some of the top nodes y_0 of T which carry v. Then $A_U(x) = A_S(x)$ for x in S, and if y is in T and Y_0 is the set of y_0 with $y \leq y_0$ at which the implantation has been performed then $A_U(y) = (A_T(y) - \bigcup < A_T(y_0) \mid y_0 \in Y_0 >) \cup A_S(e_S)$ if Y_0 is not empty and $A_U(y) = A_T(y)$ otherwise.

Consider now a calculus NKP for K_t-sequents which, starting from the (reflexive) axioms $u \Longrightarrow u$ proceeds by the structural rules (W),(RC),(RP) and (CUT) and by the logical rules

(NI∧) $\dfrac{M \Longrightarrow v \qquad M \Longrightarrow w}{M \Longrightarrow v \wedge w}$ (NE∧l) $\dfrac{M \Longrightarrow v \wedge w}{M \Longrightarrow v}$ (NE∧r) $\dfrac{M \Longrightarrow v \wedge w}{M \Longrightarrow w}$

(NI∨l) $\dfrac{M \Longrightarrow v}{M \Longrightarrow v \vee w}$ (NI∨r) $\dfrac{M \Longrightarrow w}{M \Longrightarrow v \vee w}$

(NE∨) $\dfrac{M \Longrightarrow v \vee w \qquad v, P \Longrightarrow u \qquad w, Q \Longrightarrow u}{M, P, Q \Longrightarrow u}$

(NI→) $\dfrac{v, M \Longrightarrow w}{M \Longrightarrow v \rightarrow w}$ (NE→) $\dfrac{M \Longrightarrow v \qquad M \Longrightarrow v \rightarrow w}{N \Longrightarrow w}$

Then every natural deduction S of u determines an NKP-derivation of a sequent $M \Longrightarrow u$ with $A_S(e_S) \subseteq M$; the structural rule (a) transforms into instances of the weakening rule (W) and the structural rule (b) into instances of (CUT). Conversely, every NKP-derivation determines a natural deduction S (for a different sequential calculus whose derivations correspond to natural deductions cf. ZUCKER 74).

It is easily seen that NKP-derivations can be transformed into KPC-derivations and vice versa.

The extension from positive to minimal and intuitionistic logic is obvious for a language Fm(Δ) where for the latter case only a rule (AINΔ) has to be added. For a language Fm(¬) the sequents M ⟹ ▲ of LM, LJ do not correspond to natural deductions such that I have to expand NKP to NKM, NKJ and NKC by the rules (P_0¬), (P_1¬) of KM_2, (AIN¬) of KJ_2 and (K_0) of KK_2.

While the structural rules (b) and (CUT) are *constituent* for natural deductions and for NKP, there is another type of complications in, say, NKP-derivations which may may be removed by an algorithm of *normalization*. This are the situations in which an introduction of a composite formula (say v→w by (NI→)) is immediately followed by its elimination (say v→w by (NE→)). Such normalization algorithm was described by PRAWITZ 65, and the relationship between cut elimination and normalization has been discussed by ZUCKER 74 and POTTINGER 77 .

References

P.Bernays: Axiomatische Untersuchung des Aussagenkalküls der Principia Mathematica. Math.Z. **25** (1926) 305-320

G.Boole: The mathematical analysis of logic, being an essay towards a calculus of deductive reasoning. London 1847

H.B.Curry: Foundations of Mathematical Logic. New York 1963

G.Frege: Grundgesetze der Arithmetik. Begriffsschriftlich abgeleitet. Vol. I . Jena 1893

G.Gentzen: Untersuchungen über das logische Schliessen I,II. Math.Z. **39** (1934/35), 176-210 and 405-431

A.Heyting: Die formalen Regeln der intuitionistischen Logik. Sitzungsber.Preuss.Akad.Wiss. 1930 , 42-56

D.Hilbert: Die logischen Grundlagen der Mathematik. Math.Ann. **88** (1923) 151-165

D.Hilbert, P.Bernays: Grundlagen der Mathematik I . Berlin 1934

I.Johansson: Der Minimalkalkul, ein reduzierter intuitionistischer Formalismus. Compos. Math. **4** (1937) 119-136

J.Łukasiewicz: The shortest axiom of the implicational calculus of propositions. Proc.Roy. Irish Acad. 52A3 (1948) 25-33 . Reprinted in Lukasiewicz 70

C.A.Meredith: Single axioms for the systems (C,N), (C,O) and (AA,N) of the two-valued propositional calculus. J.of Comp.Systems **1** (1953) 155-164 .

D.Monk: Mathematical Logic . Berlin 1976

E.L.Post: Introduction to a General Theory of Elementary Propositions. Amer.J.Math. **43** (1921) 180

G.Pottinger: Normalization as a homomorphic image of cut elimination. Ann.Math.Logic **12** (1977) 323–357

D.Prawitz: Natural Deduction. Stockholm 1965

J.Zucker: The correspondence between Cut–Elimination and Normalization. Ann.Math.Logic **7** (1974) 1–155

Chapter 8. Sequent Calculi for Quantifier Logic

1. Quantifier Rules for Deductive Situations

I consider a language with quantifiers $\exists$ and $\forall$, as defined in Chapter **1.12**. As I did in Chapter **1**, I begin by considering general deductive situations $M \Longrightarrow v$, and I ask again for fair and reasonable hypotheses about a deduction in order to conclude that there also is a deduction leading to a formula $\exists x\, w$ or to a formula $\forall x\, w$. Let $w(t)$ denote a substitution instance arising from w by substituting the term t for x in w; then in the first case an obvious answer is

($I\exists$) If $M \Longrightarrow w(t)$ is a deduction for the particular case of some t, then it should be possible to conclude that there is a deduction $M \Longrightarrow \exists x\, w$.

In the same manner, I ask for fair and reasonable hypotheses about a deduction in order to conclude that there also is a deduction leading to a formula u from a formula $\exists x w$, or from a formula $\forall x w$, together with further assumptions M. This time, in the second case an obvious answer is

($E\forall$) If there is a deduction $w(t), M \Longrightarrow u$ for the particular case of some t, then it should be possible to conclude that there is a deduction $\forall x\, w, M \Longrightarrow u$.

But how am I to establish a deductive situation $M \Longrightarrow \forall x\, w$? Am I to look at *all* particular cases of substitution instances and, enumerating the terms by $t_0, t_1, \ldots$, should I propose

($\Omega\forall$) Given deductions $M \Longrightarrow w(t_0)$, $M \Longrightarrow w(t_1), \ldots,$ $M \Longrightarrow w(t_n)$, ... for every particular case of the term t_n, then it should be possible to conclude that there is a deduction $M \Longrightarrow \forall x\, w$?

To do so would mean to leave the realm of finite proof trees and require the study of trees with infinite ramifications. [It would be impracticable also for the model theorist who, in his imagination, means to interpret the quantifier language in structures which he considers to be uncountably large such that the $t_0, t_1, \ldots$ still would not capture all their possibilities.] And since mortal humans cannot generate *infinitely* many, but otherwise *unspecified* deductions as required here as premisses, there would arise the question of how to put such a rule to any use at all.

What *may* be generated, however, are single specified deduction *schemata* in which a (fixed) free variable takes the place of the t_i. And so a fair and reasonable premiss to deduce $\forall x\, w$ is

($I\forall$) If $M \Longrightarrow w(y)$ is a deduction with a free variable y, then it should be possible to conclude that there is a deduction $M \Longrightarrow \forall x\, w$.

That the variable y be *free* here shall mean that nothing else depends on it, i.e. that it is *not* subjected to further hypotheses and restrictions by occurring in the assumptions M from which w(y) is deduced. So the *one* deducibility $M \Longrightarrow w(y)$ has taken the place of the infinitely many ones assumed in $(\Omega\forall)$: it assumes a *uniform* situation $M \Longrightarrow w(y)$ instead of the infinitely many special ones. Uniform computations with free variables are well known from elementary school algebra, e.g. from formulas such as $a^2 - b^2 = (a+b)(a-b)$, and, beginning with the computations with elements in Boolean algebras in Chapter **1.1**, the reader will have met such computations throughout all of this book as well. It is plausible to expect that a uniform deduction $M \Longrightarrow w(y)$ can be specialized to a deduction $M \Longrightarrow w(t_i)$ replacing y everywhere by t_i, and this will be proven in the following sections in the Replacement Theorem and the Substitution Theorem. It should, however, be emphasized that $(I\forall)$, by its insistence on uniformity, has a much stronger hypothesis than $(\Omega\forall)$ and that, therefore, $(I\forall)$ is a much weaker tool than would be $(\Omega\forall)$ - only that applying *this* tool would require more strength than I can muster.

[The rule $(I\forall)$ was first stated by FREGE 79, p.21, in the paragraph beginning with the phrase "*Auch ist einleuchtend, dass man aus ...* "; Frege did not include it among his *Kern* propositions. In FREGE 93 it is stated at the start of p.32 in symbolical notation ("*Wir schreiben einen solchen Uebergang so ...* ") and in verbalized form at the end of p.33 ("*Wir fassen dies in folgende Regel ...* "). RUSSELL 08 stated $(I\forall)$ as rule (8) of section VI, p.246, and in WHITEHEAD-RUSSELL 10 it is principle **9.13** (whose uses can be found in the proofs of **9.21** ff.).]

As for the remaining case of deductions from a formula $\exists x w$, a fair and reasonable premiss to deduce $\forall x w$ is

$(E\exists)$ If $w(y), M \Longrightarrow u$ is a deduction with a free variable y, then it should be possible to conclude that there is a deduction $\exists x w, M \Longrightarrow u$.

This describes the argumentation familiar among mathematicians when, in order to prove a statement u from the assumption that a certain set W be not empty, the statement u is derived under a hypothesis expressed as "let y be an arbitrary element of W".

The rules $(I\forall)$ and $(E\exists)$, together with the arithmetical induction principle, form the *seam* connecting (a) the naive mathematical argumentation about 'mathematical objects' (such as numbers or sets) and (b) the explicit use of language in argumentations. In so far, they lie *at the heart* of mathematical logic.

Reference

B.Russell: Mathematical Logic as based on the Theory of Types. American.J.of Math. **30** (1908) 222-262 .

and the references listed at the end of the last Chapter.

2. Sequent Calculi with Q-rules

Having stated sufficient conditions on the forms of reasonable rules for quantifiers, I now turn the tables in the same manner as I did in the propositional case: I use these conditions as *defining rules* for the use of quantifiers in a sequential calculus.

So I continue with a language L as defined in Chapter **1.12**, and I shall now make use of the maps rep and sub studied in Chapter **1.13**. For any of the propositional sequent calculi developed in Chapters **1-4**, I define *two* corresponding quantifier calculi. For both of them I

> consider sequents consisting of formulas from the quantifier languages of Chapter **1.12**,
>
> use *atomic formulas* instead of *propositional variables* in the axioms of the calculus.

For the first type of calculi I add as new logical rules the four Q-*rules with replacement*

$$(\mathrm{I}\forall)\ \frac{M \Longrightarrow \mathrm{rep}(x,y\,|\,w)}{M \Longrightarrow \forall x\ w}\ (\mathrm{EV}) \qquad (\mathrm{E}\forall)\ \frac{\mathrm{rep}(x,t\,|\,w),\ M \Longrightarrow u}{\forall x\ w,\ M \Longrightarrow u}$$

$$(\mathrm{I}\exists)\ \frac{M \Longrightarrow \mathrm{rep}(x,t\,|\,w)}{M \Longrightarrow \exists x\ w} \qquad (\mathrm{E}\exists)\ \frac{\mathrm{rep}(x,y\,|\,w),\ M \Longrightarrow u}{\exists x\ w,\ M \Longrightarrow u}\ (\mathrm{EV})$$

where $\zeta(x,y)$ and $\zeta(x,t)$ are *free* for w, and with the additional condition for (I∀) and (E∃)

(EV) y does not occur free in any formula of the conclusion.

For the second corresponding quantifier calculus I add as new logical rules the four Q-*rules with substitution*

$$(\mathrm{I}\forall)\ \frac{M \Longrightarrow \mathrm{sub}(x,y\,|\,w)}{M \Longrightarrow \forall x\ w}\ (\mathrm{EV}) \qquad (\mathrm{E}\forall)\ \frac{\mathrm{sub}(x,t\,|\,w),\ M \Longrightarrow u}{\forall x\ w,\ M \Longrightarrow u}$$

$$(\mathrm{I}\exists)\ \frac{M \Longrightarrow \mathrm{sub}(x,t\,|\,w)}{M \Longrightarrow \exists x\ w} \qquad (\mathrm{E}\exists)\ \frac{\mathrm{sub}(x,y\,|\,w),\ M \Longrightarrow u}{\exists x\ w,\ M \Longrightarrow u}\ (\mathrm{EV})$$

again with the additional condition (EV) for (I∀) and (E∃).

The variable y in (I∀) and (E∃) is called the *eigenvariable* of that particular instance of the rule, and (EV) is called the eigenvariable condition. The rules (I∀) and (E∃) are called *critical* because (EV) has to be observed in their applications. For the extremal calculi K_uP,... the rules (E∀) and (E∃) have to be rephrased as

$$(\mathrm{E}\forall)\ \frac{\forall x\, w,\ \mathrm{rep}(x,y \mid w),\ M \Longrightarrow u}{\forall x\ \ w,\ M \Longrightarrow u} \qquad (\mathrm{E}\exists)\ \frac{\exists x\, w,\ \mathrm{rep}(x,y \mid w),\ M \Longrightarrow u}{\exists x\ \ w,\ M \Longrightarrow u}\ (\mathrm{E}\vee)$$

and analogously in the case of sub. The quantified formulas indicated in the conclusions are again the *principal* formulas, and the w(y) and w(t) are the *side* formulas of the instance of the respective Q–rule; all the other occurrences of formulas are parametric. In (I∃) and (E∀) the term t is the *replacement* or *substitution term.*

There are certain fundamental differences between the calculi using Q–rules with rep and the calculi using Q–rules with sub; they will be discussed in a moment.

I shall *not* introduce new names for the quantifier calculi arising in this manner – say as $\mathrm{KPQ}_{\mathrm{rep}}$ and $\mathrm{KPQ}_{\mathrm{sub}}$ from KP – but shall, from now on, always use (a) the names for the propositional calculi also for their expansions to quantifier calculi and (b) distinguish the use of replacement and substitution explicitly in the context. For the time being, there will be no reason to consider a particular form of the underlying propositional calculi. Most examples will be written with $\mathrm{K_t}$–sequents in order to write fewer formulas.

[A reader, interested in a semantical interpretation of the *classical* quantifier calculus MK only, may at this point proceed immediately to Chapter **10**.9.]

Derivations are finite 2–ary trees T together with functions d, r as in the propositional case, but now the Q–rules are admitted as one premiss rules. By the *skeleton* of a derivation D I understand the underlying tree T together with the function r assigning the rules (and should I wish to, I could also associate the sequents formed from the skeletons of formulas in the sequents of D). The *length* of a derivation is defined as in the propositional case with instances of Q–rules being counted.

Corresponding occurrences and *predecessor* of occurrences are defined as in the propositional case. There will be occasion to use the more general notion of *precursors* of an occurrence i in the endsequent of a derivation D which is defined as follows. Every predecessor of i is also a precursor of i, and every predecessor of a precursor is a precursor. Consider now a sequent s in D which does not lay on a maximal node and thus is conclusion of an instance of a rule. If the principal occurrence in s (with respect to that instance) is a precursor of i then also the side occurrences in the premiss(es) are precursors of i. It follows that, in contrast to the predecessors, the precursors of i do not necessarily carry the same formula. It also follows that all occurrences in sequents of D are precursors of occurrences in the endsequent. This situation, however, does not continue to hold in derivations employing also a cut rule.

Again, every reflexivity axiom v ⟹ v (and every generalized reflexivity axiom) has a derivation, because it follows from $v(x) = \mathrm{rep}(x,x \mid v) = v$ and $v(x) = \mathrm{sub}(x,x \mid v) = v$ that I can prolong such derivations by

$$\begin{array}{l} v(x) \Longrightarrow v(x) \\ \forall x\ v \Longrightarrow v(x) \\ \forall x\ v \Longrightarrow \forall x\ w \end{array} \quad (\mathrm{EV}) \qquad \begin{array}{l} v(x) \Longrightarrow v(x) \\ v(x) \Longrightarrow \exists x\ v \\ \exists x\ w \Longrightarrow \exists x\ v \end{array} \quad (\mathrm{EV})$$

Lemma 1.1 remains in effect since the transformations considered there only change multiplicities of formulas in sequents but not the sets of formulas themselves; hence eigenvariable conditions remain unaffected.

Also in effect remains the proof in Lemma 1.2 that (RC) is admissible for the extremal calculi (K_u, L_u, ...), as does the transformation of extremal derivations into non extremal ones. Conversely, however, a non-extremal derivation can, in general, not be transformed into an extremal one: that construction employed the weakening of entire branches of derivations (with principal formulas) and so eigenvariable conditions may become violated. For instance, if p(y) is a 1-ary atomic formula then the derivation

$$\begin{array}{r} p(y) \Longrightarrow p(y) \\ p(y) \Longrightarrow \exists x\ p(x) \\ \exists x\, p(x) \Longrightarrow \exists x\ p(x) \\ q(y) \wedge \exists x\, p(x) \Longrightarrow \exists x\ p(x) \end{array}$$

cannot be transformed by adding the principal formula of the (E∧)-instance to the upper sequents, since this would destroy (EV) for (E∃).

In view of this situation, and of related ones to be met soon, I introduce a further name: a derivation D shall be called *eigen* if every eigenvariable, belonging to an instance of a critical Q-rule at a node e of D, *occurs only* in sequents at nodes *above* e.

For non-extremal derivations which are eigen the transformation from Lemma 1.2 remains in effect. The above derivation can be transformed into one which is eigen, replacing y in the first two sequents by x. I shall show in the following Chapters that *every* derivation can be transformed into one which is eigen, has the same endsequent, the same skeleton and, therefore, the same length.

For the positive (intuitionistic) inversion rule (JI∨) the definition of Harrop formulas is extended by defining ∀xv and ∃xv as Harrop for any v. Then the propositional inversion rules (JI→), (JI∧), (JI∨), (JE→L), (JE∧), (JE∨), (JI¬) remain in effect because an eigenvariable condition cannot be violated by introducing a new occurrence of a subformula of the (at least) previosly present principal formula. – Also, the intuitionistic inversion operators **JS∨L**, **JS∨R**, **JS→¬**, **JS¬∧**, **JS¬∨**, **JS¬¬** from Lemma **4**.4 and the classical operators **JI∨**, **JE→R**, **JE¬** from Chapter **5** continue to operate as before. – The inversion of quantifier rules requires additional tools and will be taken up in the last section of this Chapter.

The propositional calculi, unless they employed (CUT) or a classical rule such as (K_0), had the *propositional* subformula property. I call *immediate replacement* [*substitution*] *part* of a formula v every $\mathrm{rep}(x,t \mid v_0)$ with $\zeta(x,t)$

free for v [every sub(x,t | v_0)] where v_0 is a subformula of v. I call *replacement* [*substitution*] *part* of v the formula v itself as well as all the replacement [substitution] parts of immediate replacement [substitution] parts of v. The quantifier calculi, which arise from propositional calculi obeying the propositional subformula principle, then have the *quantificational*

> *Subformula Property*: In any derivation of an endsequent M $\Longrightarrow$ u there occur, in its sequents, only such formulas which are replacement [substitution] parts of formulas from the endsequent.

The replacement parts w of a formula v can be characterized by :

(1) there exists a subformula u of v and there exist two sequences $<x_0, \dots, x_{k-1}>$, $<t_0, \dots, t_{k-1}>$ such that $w = rep(x_0, t_0 \mid rep(x_1, t_1 \mid \dots rep(x_{k-1}, t_{k-1} \mid u) \dots))$.

This is a consequence of

(2) If v_0 is subformula of v then for every subformula w_0 of $rep(x,t \mid v_0)$ there exists a subformula u of v and a term t' such that $w_0 = rep(x,t' \mid u)$

because this implies that an immediate replacement part $rep(y,s \mid w_0)$ of an immediate replacement part $rep(x,t \mid v_1)$ of v is of the form $rep(y,s \mid rep(x,t' \mid u))$ for a subformula u of v; I then continue by induction. - As for the statement (2), it is trivial if $w_0 = rep(x,t \mid v_0)$ since then $u = v_0$ and $t' = t$. In particular, this includes the case that v_0 is atomic. I now proceed by induction on the degree of the subformula v_0 and assume the statement to be proven for all proper subformulas of v_0. Also, I can assume that w_0 is a proper subformula of $rep(x,t \mid v_0)$. If $v_0 = v_{00} \rightarrow v_{01}$ then $rep(x,t \mid v_0) = rep(x,t \mid v_{00}) \rightarrow rep(x,t \mid v_{01})$ such that w_0 is a subformula of $rep(x,t \mid v_{0i})$, $i = 0,1$. By inductive hypothesis, there exists a subformula u of v such that $w_0 = rep(x,t' \mid u)$. The case of the other propositional connectives is analogous. If $v_0 = Qy\, v_{00}$ then $rep(x,t \mid v_0) = Qy\, rep(x,t' \mid v_{00})$ with $t' = t$ for $x \neq y$ and with $t' = x$ for $x = y$. Now w_0 is a subformula of $rep(x,t' \mid v_{00})$, and by inductive hypothesis there exists a subformula u of v such that $w_0 = rep(x,t'' \mid u)$.

Unfortunately, there is no simple characterization of substitution parts analogous to (1). As an example, consider

$$\forall x_0 \forall x_1 \forall x_2\, p(x_0,x_1,x_2)$$
$$sub(x_0,x_1 \mid \forall x_1 \forall x_2\, p(x_0,x_1,x_2)) = \forall x_3 \forall x_2\, p(x_1,x_3,x_2)$$
$$sub(x_3,x_2 \mid \forall x_2\, p(x_1,x_3,x_2)) = \forall x_0\, p(x_1,x_2,x_0) .$$

As in the propositional case, the subformula principle for a calculus implies the property of

> *Directness*: In any derivation of an endsequent M $\Longrightarrow$ u only such logical rules are used which refer to connectives and quantifiers occurring in the endsequent.

Every derivation in one of the propositional calculi becomes a derivation of the corresponding quantifier calculus if the propositional variables are replaced by formulas of a quantifier language. I observe as an application of directness

> Let $M \Longrightarrow v$ be propositional sequent of which it can be shown that it has no derivation in some propositional sequent calculus which has the subformula property. Let $M^q \Longrightarrow u^q$ be the sequent, obtained by replacing the propositional variables in $M \Longrightarrow u$ by quantifier free formulas. Then $M^q \Longrightarrow u^q$ has no derivation in the corresponding quantifier calculus.

Because $M^q \Longrightarrow u^q$ contains no quantifiers, and hence would have to have a derivation D not employing Q-rules. But then D could be immediately rewritten as a derivation of $M \Longrightarrow v$.

As for the relationship between calculi with Q-rules using rep and those with Q-rules using sub, it is clear that a derivation with rules using rep can be viewed as one with rules using sub because $\mathrm{rep}(x,t \mid w) = \mathrm{sub}(x,t \mid w)$ if $\zeta(x,t)$ is free for w. Next I observe

(3) For the calculi using Q rules with rep, there is, for every formula v and for every map every η, a derivation of $v \Longrightarrow \mathrm{tot}(Y_0,Y_1,\eta \mid v)$ and a derivation of $\mathrm{tot}(Y_0,Y_1,\eta \mid v) \Longrightarrow v$.

This follows by induction on v, and only the step from v to Qx v requires consideration. The case that $\mathrm{tot}(Y_0,Y_1,\eta \mid Qx\, v) = Qx\, \mathrm{tot}(Y_0,Y_1,\eta_x \mid v)$ is trivial. Otherwise, I have

$$\mathrm{tot}(Y_0,Y_1,\eta \mid Qx\, v) = Qy\, \mathrm{rep}(x,y \mid \mathrm{tot}(Y_0,Y_1,\eta_x \mid v))$$

with y not in $\mathrm{fr}(\mathrm{tot}(Y_0,Y_1,\eta_x \mid v))$ and $\zeta(x,y)$ free for $\mathrm{tot}(Y_0,Y_1,\eta_x \mid v))$. Since $\mathrm{rep}(y,x \mid \mathrm{rep}(x,y \mid \mathrm{tot}(Y_0,Y_1 \mid \eta_x \mid v))) = \mathrm{tot}(Y_0,Y_1 \mid \eta_x \mid v)$ holds by Lemma **1.13**.9.(iii), it follows from the inductive hypothesis that, say,

$$\begin{aligned} v &\Longrightarrow \mathrm{rep}(y,x \mid \mathrm{rep}(x,y \mid \mathrm{tot}(Y_0,Y_1,\eta_x \mid v))) \\ \forall x\, v &\Longrightarrow \mathrm{rep}(y,x \mid \mathrm{rep}(x,y \mid \mathrm{tot}(Y_0,Y_1,\eta_x \mid v))) \\ \forall x\, v &\Longrightarrow \forall y\, \mathrm{rep}(x,y \mid \mathrm{tot}(Y_0,Y_1,\eta_x \mid v)) \\ \forall x\, v &\Longrightarrow \mathrm{tot}(Y_0,Y_1,\eta_x \mid v) \end{aligned}$$

$$\begin{aligned} \mathrm{rep}(y,x \mid \mathrm{rep}(x,y \mid \mathrm{tot}(Y_0,Y_1,\eta_x \mid v))) &\Longrightarrow v \\ \forall y\, \mathrm{rep}(x,y \mid \mathrm{tot}(Y_0,Y_1,\eta_x \mid v)) &\Longrightarrow v \\ \forall y\, \mathrm{rep}(x,y \mid \mathrm{tot}(Y_0,Y_1,\eta_x \mid v)) &\Longrightarrow \forall x\, v \\ \mathrm{tot}(Y_0,Y_1,\eta \mid \forall x\, v) &\Longrightarrow \forall x\, v \ . \end{aligned}$$

(4) A derivation employing Q-rules with sub can embedded into a derivation employing Q-rules with rep *and* using cuts.

Consider a premiss, say of ($I\forall$) with sub, from which I want to derive the conclusion. Since $\mathrm{sub}(x,y \mid w) = \mathrm{rep}(x,y \mid \mathrm{tot}(x,y \mid w))$ and $\zeta(x,y)$ is free for $\mathrm{tot}(x,y \mid w)$, I find the derivation using Q-rules with rep and using a cut:

$$\begin{array}{ccc} & & \mathrm{tot}(x,t\,|\,w) \Longrightarrow w \\ M \Longrightarrow \mathrm{sub}(x,y\,|\,w) & & \forall x\ \mathrm{tot}(x,t\,|\,w) \Longrightarrow w \\ M \Longrightarrow \forall x\ \mathrm{tot}(x,y\,|\,w) & & \forall x\ \mathrm{tot}(x,t\,|\,w) \Longrightarrow \forall x\ w \\ & \diagdown \quad \diagup & \\ & M \Longrightarrow \forall x\ w & \end{array}$$

In order to apply (I$\forall$) on the left, I have used that tot(x,y | w) and w have the same free variables, hence also $\mathrm{fr}(\forall x\, \mathrm{tot}(x,y \mid w) = \mathrm{fr}(\forall x\, w)$, such that y is not free in the conclusion. The case of the other rules is analogous.

It is a consequence of the above that the calculus employing Q-rules with sub derives the same sequents as derives the calculus employing Q-rules with rep *and* using the cut rule. Let now p and q be 1-ary predicate symbols. The following *Example 1*

$$\begin{array}{ccc} p(y) \Longrightarrow p(y) & & \\ q(x)\wedge p(y) \Longrightarrow p(y) & & \\ \forall x\ (q(x)\wedge p(y)) \Longrightarrow p(y) & & \\ \forall y\ \forall x\ (q(x)\wedge p(y)) \Longrightarrow p(y) & & p(x) \Longrightarrow p(x) \\ \forall y\ \forall x\ (q(x)\wedge p(y)) \Longrightarrow \forall y\ p(y) & & \forall y\ p(y) \Longrightarrow p(x) \\ & \diagdown \quad \diagup & \\ & \forall y\ \forall x\ (q(x)\wedge p(y)) \longrightarrow p(x) & \end{array}$$

shows that there are sequents derivable without cut by Q-rules with sub and *not* derivable without cut by Q-rules with rep. Because this is a derivation using Q-rules with rep and ending with a cut. Without a cut, the only premiss to obtain the endsequent would be of the form

$$\forall x\ (q(x)\wedge p(t)) \Longrightarrow p(x) \qquad \text{with x not in occ(t)} ,$$

and the only premiss to obtain this would be of the form

$$q(s) \wedge p(t) \Longrightarrow p(x) .$$

But this could only be derived if t were x. – On the other hand, in the calculi with sub my sequent can easily be derived without cut. Let me write x_0 for x and x_1 for y, such that again

$$\begin{array}{r} q(x_2) \wedge p(x_0) \Longrightarrow p(x_0) \\ \forall x_2\, (q(x_2) \wedge p(x_1)) \Longrightarrow p(x_0) \end{array}$$

Now $\forall x_2\,(q(x_2) \wedge p(x_0))$ is $\mathrm{sub}(x_1,x_0 \mid \forall x_0\,(q(x_0) \wedge p(x_1))$ since x_2 is the first variable not free in $q(x_0) \wedge p(x_1)$ and not in $\mathrm{occ}(\zeta(x_1,x_0)(x_1))$. The last line being rewritten, I thus continue by

$$\begin{array}{r} \mathrm{sub}(x_1, x_0 \mid \forall x_0\, (q(x_0) \wedge p(x_1)) \Longrightarrow p(x_0) \\ \forall x_1\ \forall x_0\, (q(x_0) \wedge p(x_0)) \Longrightarrow p(x_0) \end{array} .$$

Let me add the observation that a simple renaming of bound variables will suffice to change the formula $v = \forall x_1 \forall x_0\,(q(x_0) \wedge p(x_1))$ into one for which I find a cut free derivation of an analogous endsequent, for instance with $Y_0 = Y_1 = \{x_0\}$

$$\mathrm{tot}(Y_0, Y_1 \mid \forall x_1 \forall x_0 (q(x_0) \wedge p(x_1)) = \forall x_1 \forall x_2 (q(x_2) \wedge p(x_1)) .$$

the derivation

$$\begin{array}{rcl} p(x_0) & \Longrightarrow & p(x_0) \\ q(x_2) \wedge p(x_0) & \Longrightarrow & p(x_0) \\ \forall x_2 \, (q(x_2) \wedge p(x_0)) & \Longrightarrow & p(x_0) \\ \forall x_1 \forall x_2 \, (q(x_2) \wedge p(x_1)) & \Longrightarrow & p(x_0) \ . \end{array}$$

And I know from (3) that v and its renamed copy $\mathrm{tot}(Y_0, Y_1 \mid v)$ are interdeducible, i.e. $\forall x_1 \forall x_0 (q(x_0) \wedge p(x_1)) \Longrightarrow \forall x_1 \forall x_2 (q(x_2) \wedge p(x_1))$ has a derivation. But again a cut will be required to obtain from these two derivations the endsequent $\forall x_1 \forall x_0 (q(x_0) \wedge p(x_1)) \Longrightarrow p(x_0)$.

It follows from the above that the cut rule cannot be admissible for the calculi using Q–rules with rep – at least not without additional restrictions. Still, suitable restrictions under which the cut elimination procedures work will be discussed in the next section. The rest of this Chapter then will be devoted to show that cut elimination algorithms work *without* restrictions for the calculi using Q–rules with sub.

As a matter of fact, the mechanics of the elimination algorithms developed for propositional calculi in Chapter **2** will work without any change once a certain auxiliary fact will have been established. Consider Theorem **2**.1 whose proof will remain in effect once the exchange operator $\mathbf{E}_1$ has been extended to also accept the Q–rules. Making use of the concepts developed in this proof, I first consider the

Case E1

(R) = E$\forall$

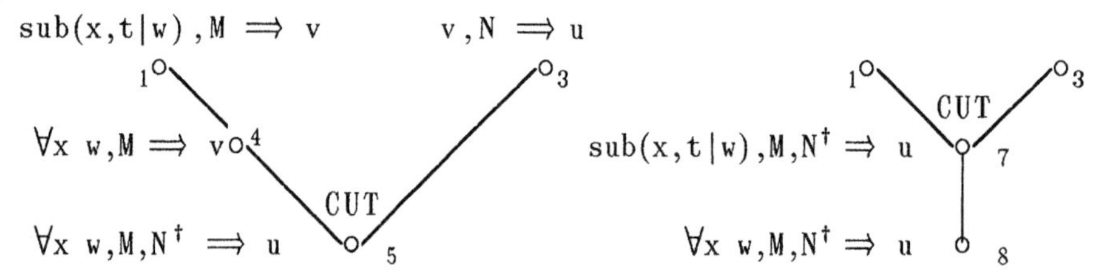

The case of rep in place of sub is analogous.

(R) = E$\exists$

Here I have the same diagram (with $\exists$ in place of $\forall$) as above, but now t is an eigenvariable. In order to go from node 7 to 8 by (E$\forall$) I have to know that t does not occur free in $N^\dagger$. This is no problem for the symmetric rule (CUT_0); for other cut rules it suffices to assume that the given derivation is eigen whence the variable t does not occur in formulas at node 3 or above.

Case E2 . This is analogous with (R) as I∃ and I∀ .

Case E3

$v = \forall x\, w$

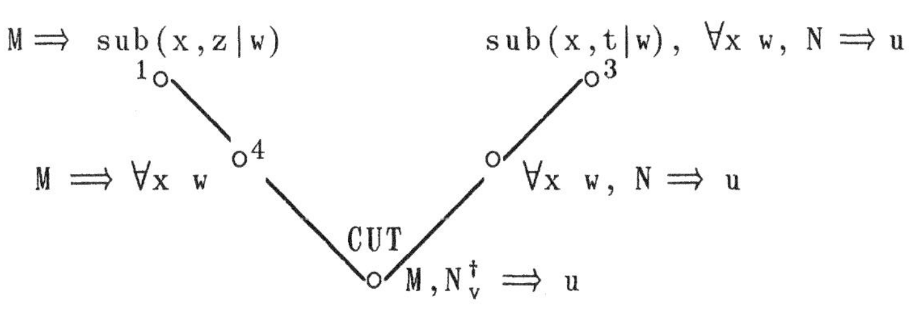

Here z is an eigenvariable and t is a term. All I need to continue as in the propositional case is a *Replacement Theorem*, respectively a *Substitution Theorem*, saying that the derivation D leading to node 1 can be transformed into a derivation $D^{\#}$ with the same length and with cut formulas of unchanged degrees, but ending with the sequent $M \Longrightarrow \text{rep}(x,t\,|\,w)$, respectively with $M \Longrightarrow \text{sub}(x,t\,|\,w)$. I shall prove such theorems in the next sections, first for calculi employing Q-rules with rep and under additional hypotheses, and then for calculi employing Q-rules with sub. Assuming such theorems to be available, I replace D by $D^{\#}$ and proceed as I did in the propositional case from

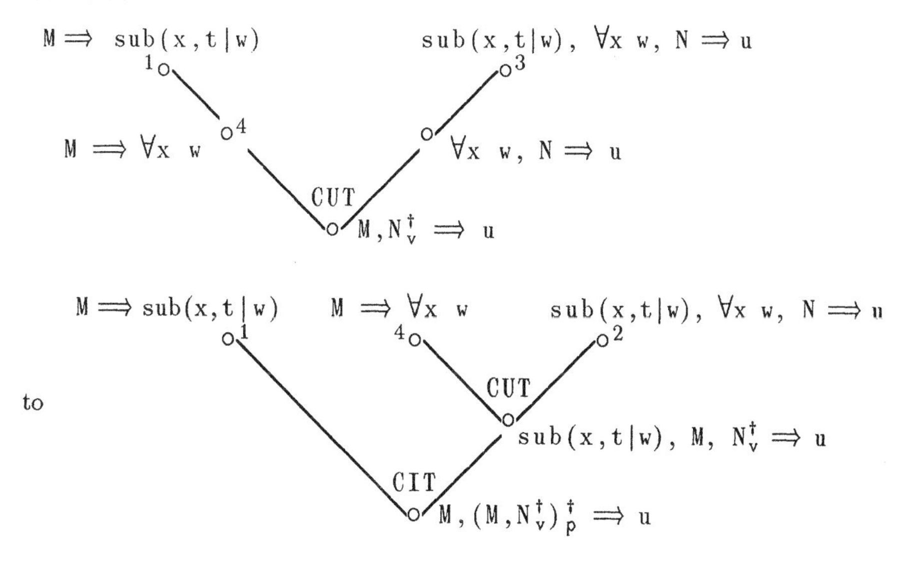

where (CIT) is a cut with the cut formula $\text{sub}(x,t\,|\,w)$. – The case of $v = \exists x\, w$ is analogous. In this manner, then, the exchange operator $\mathbf{E}_1$ has been defined and the proof of Theorem **2**.1 remains in effect with the same bounds on lengths.

3. The Replacement Theorem and Cut Elimination for Calculi with rep

The first aim of this section is to prove a Replacement Theorem which will say that, under additional hypotheses still to be elucidated,

> for a given term s, a derivation H of an endsequent $M \Longrightarrow rep(x,z \mid v)$ or $rep(x,z \mid v), M \Longrightarrow u$ can be transformed into a derivation $H^{\#}$ of the endsequent $M \Longrightarrow rep(x,s \mid v)$, respectively $rep(x,s \mid v), M \Longrightarrow u$, which has the same skeleton.

The idea of this transformation is quite simple: apply rep(z,s | –) to *every* formula in *every* sequent of H – and hope that the result is the desired $H^{\#}$. In order to leave the other formulas of the endsequent unchanged, a first requirement evidently will be that z is not free in the formulas of M (nor, in the second case, in u). Further, I will need that rep(z,s | rep(x,z) | v)) becomes rep(x,s | v). And I will also need that instances of Q–rules within H remain such instances:

> If Qyw is principal formula with a side formula rep(y,t | w), y≠z, thenthese formulas are transformed into Qy rep(z,s | w) and rep(z,s | rep(y,t | w)). But the latter should be usable as a side formula again, and so it should be of the form $rep(y, h_{\varsigma(z,s)}(t) \mid rep(z,s \mid w))$.

Clearly, such *commutativity relations* between replacements will not hold without any assumptions, and in Lemma **1.13**.6 to Lemma **1.13**.10 I have assembled their detailed investigation.

Consider a sequent calculus with Q–rules employing rep and which may contain a cut rule. Making use of Lemma **1.13**.9, it is easy to prove the *Replacement Lemma* :

LEMMA 1 Let D be a derivation and let s be a term such that occ(s) does not contain

(a) variables bound in principal formulas of instances of Q-rules,
(b) eigenvariables of D.

Let x be a variable and assume that there holds *one* of

(c0) x is not an eigenvariable of D, or
(c1) s is a variable and is not free in the conclusions of instances of critical Q–rules.

Then D is transformed into a derivation rep(x,s | D) if every formula v in the sequents of D is replaced by rep(x,s | v). Instances of rules and their principal, side and cut formulas are preserved; substitution terms and eigenvariables are transformed by $h_{\varsigma(x,s)}$.

It is obvious that axioms are transformed into axioms and that propositional rules, weakenings and cuts are preserved. Consider now the principal formula Qz w of an instance of Q–rule. If $z \neq x$ then it is transformed into Qz rep(x,s | w). By Lemma **1.13**.9.(i) and (ii), the side formula rep(z,t | w) is transformed

$$\text{for } x \neq z \text{ into } \operatorname{rep}(x,s \mid \operatorname{rep}(z,t \mid w)) = \operatorname{rep}(z, h_{\varsigma(x,s)}(t) \mid \operatorname{rep}(x,s \mid w)) ,$$
$$\text{for } x = z \text{ into } \operatorname{rep}(z,s \mid \operatorname{rep}(z,t \mid w)) = \operatorname{rep}(z, h_{\varsigma(x,s)}(t) \mid w) .$$

Here the assumptions of Lemma **1.13**.9.(i),(ii) are satisfied, because $\zeta(z,t)$ is free for w and occ(s) does not contain the bound variable of Qz w. Further, also bd(w) is disjoint to occ(s), and so the map $\zeta(z, h_{\varsigma(x,s)}(t))$ is free for rep(x,s | w) respectively for w. Thus instances of uncritical Q–rules are preserved. In the critical case, t is an eigenvariable, whence in case $x \neq t$ also $h_{\varsigma(x,s)}(t) = t$. In this case, therefore, t remains eigenvariable in the transformed situation, since (b) implies that t is not free in the transformed conclusion also. In case x = t this variable is mapped to s, and s is neither free in the original conclusion by (c1), nor can it arrive in the transformed one since x = t was not free there in the first place. This concludes the proof of Lemma 1 .

But this Lemma does require its hypotheses. Recall now the situation in which I want to apply it: there will be a derivation D ending with a cut with a cut formula Qy w, and both the subderivations D_l and D_r leading to its premisses end with instances of Q–rules for which the cut formula is principal, e.g. if Q is ∃ then

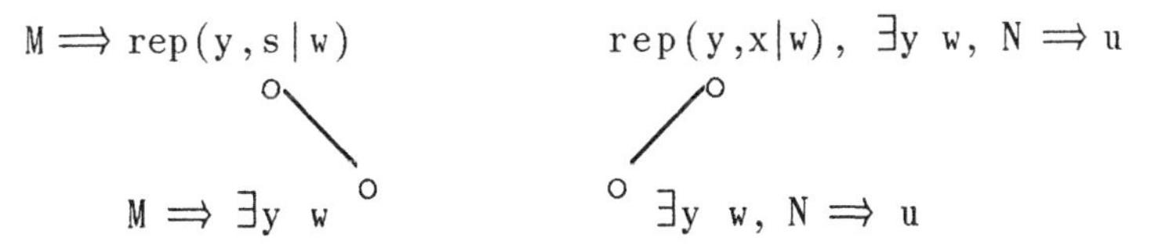

where x is an eigenvariable and s is a term. I then want to apply the Lemma to the subderivation D_{ro} leading to the premiss of the critical principal formula. So my term s must contain neither variables bound in principal formulas of D_{ro}, nor eigenvariables of D_{ro}, and x must not occur as an eigenvariable in D_{ro}. For this to be the case, sufficient conditions to hold for the derivation D would be

(f) No variable from a replacement term of a Q–rule occurs bound in any formula of D ,

(e) D is eigen, i.e. every eigenvariable, belonging to an instance of a critical Q–rule at a node e of D, occurs *only* in sequents at nodes *above* e .

Because (e) first would imply that the eigenvariable x does not occur in D_l at all, and thus not in s either. And it secondly would imply that the eigenvariables used in D_{ro} do not occur below that usage, hence not in the endsequent of D_r, and so x will be distinct from each of them.

A condition even stronger than (f) would be

(g) No variable, occurring free in some formula in some sequent of D, occurs bound in some (in particular: the same) formula in some sequent of D .

It suffices, however, to assume (g) only for the endsequent S of D:

(h) No variable, occurring free in some formula in S, occurs bound in some (in particular: the same) formula in S

because a derivation satisfying (h) can be transformed, or *prepared,* into a derivation satisfying (g); this will be the content of the *Preparation Theorem* to be shown below.

Sequents satisfying (h) I call *pure.* In general, of course, this situation will not prevail. And in general, as the Example 1 with an impure endsequent shows, cut elimination cannot be expected to be possible.

In now first state the *Eigenvariable Lemma* :

LEMMA 2 Let D be a derivation, and let X_1 be a set of variables with infinite complement $X-X_1$. There is an operator transforming D into a derivation D^* such that

(i) D^* and D have the same endsequent and the same skeleton,

(ii) D^* is eigen and has no eigenvariables in X_1.

The construction of D^* proceeds by recursion on the length of D. It is trivial if D consists of an axiom only; assume now that D has positive length and that the Lemma has been proven for all shorter derivations and for all admissible choices of X_1. Let var(S) be the set of variables occurring in the endsequent S of D. Let D_l, D_r be the subderivations leading to the premisses of the last rule (R) of D (and set $D_l = D_r$ if there is only one premiss). Consider first the case that (R) is structural (including cut) or propositional. By inductive hypothesis, D_l can be transformed into D_l^* which is eigen and without eigenvariables in $X_1 \cup \text{var}(S)$, and if $D_l \neq D_r$ then D_r can be transformed into D_r^* which is eigen and has no eigenvariables in $X_1 \cup$ var(S) nor in the sequents of D_l^*. Also, D_l^*, D_r^* end with the endsequents of D_l, D_r respectively. Proceeding by (R) from them to S, I obtain D^*. Consider next the case that (R) is a Q-rule. Again, the inductive hypothesis gives D_l^* without eigenvariables in $X_1 \cup \text{var}(S)$, and if (R) is not critical I proceed with (R), obtaining D^*. If R is critical then the endsequent S_l of D_l^* contains the side formula $\text{rep}(x,z \mid w)$, and z does not occur free in any other formula of S_1. I apply Lemma 1 to D_1^*, choosing the variable s outside of $X_1 \cup var(S)$ and distinct from all variables occurring in D_1^*. The end-sequent $S^{\ddagger}$ of resulting derivation $rep(z,s \mid D_1^*)$ has $rep(z,s \mid rep(x,z \mid w))$ in place of of $rep(x,z \mid w)$, and all its other formulas remain unchanged. But $rep(z,s \mid rep(x,z \mid w)) = rep(x,s \mid w)$ by Lemma **1.13**.9.(iii), and so I obtain D^* by continuing $rep(z,s \mid D_1^*)$ with (R) and the eigenvariable s.

The main work for the proof of Preparation Theorem will be performed in the *Purification Lemma,* and this, unfortunately, may require the renaming of certain bound variables. I thus turn my attention first to the *Renaming Lemma* :

LEMMA 3 Let D be a derivation, let Y_0 be an arbitrary set of variables, and let Y_1 be a set of variables, with infinite complement in X, which contains

(a) every variable quantified in an instance of a Q–rule,
(b) the variables of the substitution terms and the eigenvariables.

Then D is transformed into a derivation $tot(Y_0,Y_1 \mid D)$ if every formula v in sequents of D is replaced by $tot(Y_0,Y_1 \mid v)$. Instances of rules and their principal, side and cut formulas are preserved.

Once more, only the transformation of instances of Q–rules is not trivial. Consider first a principal formula Qz w for which z is not in Y_0. It is replaced by $Qz\, tot(Y_0,Y_1 \mid w)$ while the side formula $rep(z,t \mid w)$ is transformed into

$$tot(Y_0,Y_1 \mid rep(z,t \mid w)) = rep(z,t \mid tot(Y_0,Y_1 \mid w))$$

and $\zeta(z,t)$ is free for $tot(Y_0,Y_1 \mid w)$: this follows from Lemma **1.13**.10 in view of (a) and (b).Thus inferences of uncritical Q–rules are preserved in this case. Now, for every formula u, there holds $fr(u) = fr(tot(Y_0,Y_1 \mid u))$. Hence an eigenvariable cannot occur in the transformed conclusion either. Consider now the case that z is in Y_0. Then Qz w will be transformed into $Qz_1 rep(z,z_1 \mid tot(Y_0,Y_1 \mid w))$ where z_1 is not free in $tot(Y_0,Y_1 \mid w)$ and $\zeta(z,z_1)$ is free for $tot(Y_0,Y_1 \mid w)$. Now Lemma **1.13**.9.(iii) implies

$$rep(z,t \mid tot(Y_0,Y_1 \mid w)) = rep(z_1,t \mid rep(z,z_1 \mid tot(Y_0,Y_1 \mid w)))$$

and $\zeta(z_1,t)$ is free for $rep(z,z_1 \mid tot(Y_0,Y_1 \mid w))$. Hence also in this case instances of Q–rules remain preserved.

Rephrasing the terminology used above, a sequence or a set M of formulas is called *pure* if $fr(M) \cap bd(M) = 0$, i.e. if there is no variable which occurs free in some formula of M and occurs bound in some (possibly different) formula of M. I now state the *Purification Lemma* :

LEMMA 4 Let D be a derivation of pure endsequent. Let X_0, X_1 be sets of variables such that the complement $X-(X_0 \cup X_1)$ is infinite and such that X_0 does not contain variables free in the endsequent. Then there is an operator transforming D into a derivation $D^\#$ such that:

(i) $D^\#$ and D have the same endsequent and the same skeleton,

(ii) $D^\#$ is eigen and all sequents of $D^\#$ are pure,

(iii) X_0 does not contain variables free in sequents of $D^\#$, nor does it contain variables from substitution terms of $D^\#$,

(iv) $X_0 \cup X_1$ does not contain eigenvariables of $D^\#$.

The construction of $D^\#$ proceeds again by recursion on the length of D. It is trivial if D consists of an axiom only; assume now that D has positive length and that the Lemma has been proven for all shorter derivations *and* for all admissible choices of X_0, X_1. I denote as D_1, D_r the subderivations leading to the premisses of the last rule (R) of D (and set $D_r = D_1$ in case there is only one premiss), and I write M, M_1, M_r for the sets of formulas in the endsequents of D, D_1, D_r. If (R) is propositional or structural, but no cut, then $fr(M_1) \subseteq fr(M)$, $bd(M_1) \subseteq bd(M)$ and analogously for M_r. Thus also the endsequents of D_1, D_r are pure and without free variables in X_0. By inductive hypothesis, D_1 can be transformed into $D_1^\#$ with X_1 replaced by $X_1 \cup var(M)$, and then D_r can be transformed into $D_r^\#$ with X_1 replaced by $X_1 \cup var(M)$ enlarged by the variables occurring in sequents of $D_1^\#$. Continuing with an instance of (R), I obtain the derivation $D^\#$ from $D_1^\#$ and $D_r^\#$.

Consider now the case that (R) is a Q-rule with principal formula Qz w in M and side formula rep(z,t | w) in M_1. If z is *not* free in w then rep(z,t | w)= w and so $fr(M_1) = fr(M)$. I then apply the inductive hypothesis to D_1, X_0, X_1 , and the resulting $D_1^\#$ I continue to $D^\#$ with help of (R). Here the role of t as substitution term or eigenvariable is purely formal, and so in place of t any other, sufficiently new variable may be written. Consider now the case that z *is* free in w, hence $bd(M_1) \subseteq bd(M)$ and $fr(M) \subseteq fr(M_1) \subseteq fr(M) \cup occ(t)$. In order to apply the inductive hypothesis to D_1 with X_0, I would have to know that the set

$$N(D) = (fr(M_1) \cap bd(M_1)) \cup (fr(M_1) \cap X_0)$$

is empty. But this, in general, will not be the case, and so I shall first transform, by repeated applications of Lemma 1, the derivation D into a derivation $D^\S$ for which the corresponding set $N(D^\S)$ indeed is empty.

Observe first that $x \in N(D)$ implies $x \in bd(M) \cup X_0$, and since M was pure and fr(M) was disjoint to X_0, there follows that not $x \in fr(M)$. Consequently, $x \in fr(M_1)$ implies $x \in occ(t)$. I now choose, in Lemma 1, the term s to be a variable y satisfying (c1) and not in $bd(M_1) \cup X_0$. It then follows from *not* $x \epsilon fr(M)$ that $rep(x,y \mid -)$ does not affect the endsequent of D, and it follows from

$$bd(rep(x,y \mid M_1)) = bd(M_1) \quad \text{and} \quad fr(rep(x,y \mid M_1)) = (fr(M_1) - \{x\}) \cup \{y\}$$

that $N(rep(x,y \mid D))$ contains one element less than N(D). Moreover, if (R) was critical and $t = x$ its eigenvariable, then I choose y also outside of X_1. After a finite number of steps, I thus obtain $D^§$ from D. Observe that, as mentioned during the proof of Lemma 1, in $D^§$ the formula $rep(z,t \mid w)$ is replaced by $rep(z, t^§ \mid w)$ for a suitable $t^§$ which, in case of a critical (R), is an eigenvariable $y^§$ outside of X_1.

Having obtained $D^§$, let $D_1{}^§$ be its subderivation ending with the premiss of the last rule (R). Since $N(D^§)$ is empty, I can apply the inductive hypothesis to $D_1{}^§$ and the sets X_0 and $X_1{}^§ = X_1 \cup var(M)$, arriving at a derivation $D_1{}^{§\#}$ ending with the endsequent of $D_1{}^§$. Thus it contains $rep(z,t^§ \mid w)$, and if (R) is critical, then the eigenvariable $y^§ = t^§$ is not in $X_1{}^§$ and thus not in var(M). Continuing $D_1{}^{§\#}$ with an application of (R), I obtain the desired $D^\#$.

It remains to consider the case that (R) is a cut. Let v be the cut formula such that $M_l \subseteq M \cup \{v\}$ and $M_r \subseteq M \cup \{v\}$. If M_l, or M_r, are not pure or have free variables in X_0, then this can be so only because

$$fr(v) \cap bd(M) \neq 0 \qquad \text{or} \qquad fr(v) \cap X_0 \neq 0 \qquad \text{or} \qquad bd(v) \cap fr(M) \neq 0 \ .$$

If fr(v) contains variables from bd(M) or X_0, then they will not be in fr(M). Applying Lemma 1 a finite number of times, I can replace them by variables outside of $var(M) \cup X_0 \cup X_1$ without changing the endsequent. As for bd(v), I can replace, in the same manner, the variables in fr(v)−fr(M), belonging to bd(v), by others outside of $bd(v) \cup var(M) \cup X_0 \cup X_1$. I now assume that such replacements have been performed already and keep my previous notations. Then

$$(fr(M_l) \cup fr(M_r)) \cap X_0 = 0 \ , \quad fr(v) \cap bd(M) = 0 \ , \quad (fr(v) - fr(M)) \cap bd(v) = 0 \ .$$

Consequently, if M_l or M_r still are not pure, then this can only be if there are variables both in fr(v) and in fr(M) which belong to bd(v), and so the set $Y(D) = bd(v) \cap fr(M)$ will be non-empty.

I now want to apply Lemma 3 with a map $tot(Y_0,Y_1 \mid -)$. I choose $Y_0 = Y(D)$ and Y_1 as a set containing $fr(M) \cup fr(v)$ and satisfying the hypotheses of the Lemma. Consider now the derivation $tot(Y_0,Y_1 \mid D)$. The bound variables from Y(D) in bd(v) have been moved outside of Y_1. The bound variables in M have not been changed since $Y_0 \subseteq fr(M)$ implies $Y_0 \cap bd(M) = 0$; hence the endsequent has not been changed. But Y_1 contains both the sets $fr(M_l) = fr(tot(Y_0,Y_1 \mid M_l))$ and $fr(M_r) = fr(tot(Y_0,Y_1 \mid M_r))$; hence both $tot(Y_0,Y_1 \mid M_l)$ and $tot(Y_0,Y_1 \mid M_r)$ are pure. Thus the inductive hypothesis can be applied to the subderivations $tot(Y_0,Y_1 \mid D_l)$, $tot(Y_0,Y_1 \mid D_r)$, producing purified derivations $D_l^{\#}$, $D_r^{\#}$ of their endsequents. These I continue to $D^{\#}$ by a cut with $\mathrm{tot}(Y_0,Y_1 | v)$; the validity of the condition on eigenvariables I ensure as I did in the case of a propositional rule (R). This concludes the proof of Lemma 4.

A *derivation* D is called *pure* if it is eigen and if all of its sequents are pure. Two disjoint sets X_f and X_b of variables are said to be *separating* for a derivation D – or, more generally, for a set of derivations – if X_f contains all the variables free in formulas of D or occurring there in substitution terms or as eigenvariables, and if X_b contains all variables bound in formulas of D. A pure derivation D – or a set of pure derivations – is called *separated* if there are two separating sets for D. There then holds the *Preparation Theorem*:

THEOREM 1 There is an operator **P** transforming a derivation D with a pure endsequent into a separated derivation **P**(D) with the same endsequent and the same skeleton.

I first apply the Purification Lemma 4 to D where I choose X_0 to be bd(M) for the set M of formulas of the endsequent. For the resulting pure derivation $D^{\#}$, it follows from the parts (v), (vi) of this Lemma that bd(M) is disjoint to the set X_f of all variables free in formulas of $D^{\#}$ or occurring there in substitution terms or as eigenvariables.

I next apply the Renaming Lemma 3 to $D^{\#}$. As I do not want to change the variables in bd(M), I choose Y_0 as the complement X−bd(M). All other bound variables shall be renamed such as to be not in X_f, such that Y_1 should contain X_f. But Lemma 3.(b) also requires Y_1 to contain the variables quantified by Q-rules; hence I choose Y_1 to be the set of all variables occurring in formulas of $D^{\#}$. The derivation $\mathbf{P}(D) = \mathrm{tot}(Y_0,Y_1 | D^{\#})$ then has the same free variables, the same variables in substitution terms and the same eigenvariables as has $D^{\#}$. All variables which are bound in formulas of **P**(D) are either unchanged variables from bd(M) or result from the renaming and thus are not in Y_1, hence not in X_f. Hence I may set $X_b = \mathrm{bd}(M) \cup (X - Y_1)$.

Finally, there holds the *Replacement Theorem* :

THEOREM 2 Let D be a separated derivation, the last rule of which has two premisses to which there lead subderivations D_l, D_r.

Assume that the last rule of D_l is an uncritical Q-rule with the substitution term s.

Assume that the last rule of D_r is a critical Q-rule with the eigenvariable x and that D_{ro} is the subderivation of D_r leading to its premiss.

Then $rep(x,s \mid D_{ro})$ is a pure and separated derivation with the same eigenvariables as D_r, and the separating sets X_f, X_b of D can be chosen as those for D_{ro}.

I wish to apply Lemma 1 to D_{ro}, and to this end I have to verify its hypotheses. The hypothesis (a) holds since D is separated. The hypothesis (b) holds since it follows from the condition (vi) of Lemma 4 that the eigenvariable x of D_r does not occur in any sequents of D_1. The hypothesis (c0) holds since condition (vi) implies that x is distinct from all previous eigenvariables in D_{ro}. Thus $rep(x,s \mid D_{ro})$ is derivation. Its bound variables remain those of D_{ro}, while it free variables may, in addition to those of D_{ro}, also contain those in occ(s). But occ(s) still was contained in the separating set X_f for D.

From the remarks made at the end of the previous section, it now follows that the exchange operators and the cut elimination algorithm **A** from Theorem 2.1 remain in effect for separated derivations in quantifier calculi with rep. The preparation of derivation with a pure endsequent leaves unchanged both its length and its cut degree. Hence there holds the

THEOREM 3 Let D be a derivation of a pure endsequent in a calculus employing Q-rules with rep and possibly employing cuts. Then **A**(**P**(D)) is a derivation without cuts of the same endsequent and of length at most $e_3(k,n)$.

4. The Substitution Theorem and Cut Elimination for Calculi with sub

I consider a sequent calculus employing Q-rules with sub. For the sake of convenience, I assume the calculus to be extremal (i.e. K_u-, L_u- ...). My first aim, the *First Substitution Theorem,* I state right away:

THEOREM 4 There is an operator, acting on

derivations D, and variables y_o and terms t_o, such that D has no eigenvariables in $occ(t_o) \cup \{y_o\}$

and transforming D into a derivation $D^{\#}$ as follows.

Let u_o, u_1,... be all those formulas in the endsequent of D in which y_o is free; let $x_0, x_1,...$ be variables, and let v_o, v_1,... be formulas such that $u_i = sub(x_i, y_o \mid v_i)$ and such that y_o is not free in v_i, $i = 0, 1, ...$.

> Then D and $D^{\#}$ have the same skeleton, and the instances of critical Q-rules employ the same eigenvariables in D and in $D^{\#}$. The endsequent of $D^{\#}$ arises from that of D by replacing u_i by $sub(x_i, t_0 \mid v_i)$; all other sequents of $D^{\#}$ arise by appropriate replacements from those of D.

The construction of $D^{\#}$ from D is easy to describe. I define, by recursion from below, for every sequent S of D the sequent $S^{\#}$ of $D^{\#}$ by assigning to every formula r in S a formula $r^{\#}$; this will happen in such a way that r and $r^{\#}$ have the same skeleton (in the sense of Chapter **1.13**).

I begin with the endsequent of D: for the formulas u_i in which y_0 is free I set $u_i^{\#} = sub(x_i, t_0 \mid v_i)$, and I set $r^{\#} = r$ for all other formulas in the endsequent. I now proceed upwards and assume that the assignment has already been defined for the formulas in the conclusion T of an instance of a rule (R). For the formulas r in a premiss S there then are the following possibilities:

(a1) (R) is a cut with cut formula r. Then I set $r^{\#} = sub(y_0, t_0 \mid r)$.

(a2) r occurs as its immediate predecessor r_0 in the conclusion T. Then $r_0^{\#}$ was defined in $T^{\#}$, and I define $r^{\#}$ as $r_0^{\#}$ at the appropriate position in $S^{\#}$.

(a3) r is the side formula of a propositional rule (R). Then the principal formula q occurs in the T and $q^{\#}$ is defined. Since $q^{\#}$ has the same skeleton as q, there corresponds to r a unique propositional subformula of $q^{\#}$ which I choose as $r^{\#}$.

(a4) r is the side formula of a Q-rule (R) with principal formula $Qx\,w$, hence $r = sub(x,s \mid w)$ with some term or some eigenvariable s. Since $(Qx\,w)^{\#}$ is defined and has the same skeleton as $Qx\,w$, there holds $(Qx\,w)^{\#} = Qx'\,w'$. With $\eta = \zeta(y_0, t_0)$ I set $r^{\#} = sub(x', h_\eta(s) \mid w')$.

This concludes the definition of $D^{\#}$. I thus have defined the algorithm producing $D^{\#}$, and it remains to be seen that it is correct, i.e. that $D^{\#}$ is again a derivation.

To this end, I have to show that, under the transformation from D to $D^{\#}$, the axioms remain axioms and the instances of rules (R) remain instances of the same rules. This follows for cuts and propositional rules immediately from (a2), (a1) and (a3), and for uncritical Q-rules it follows from (a4). In the case of a critical Q-rule, all that remains to be seen is the condition on eigenvariables. It follows from the hypothesis on D that in (a4) the eigenvariable s is different from y_0 whence $h_\eta(s) = s$. I thus have to prove that s is not free in the transformed formulas from the conclusion, and this will hold, if I can show that

(5) $fr(r^{\#}) \subseteq fr(r) \cup occ(t_0)$

holds for every r, because s was not in fr(r) by the original eigenvariable condition, and s is not in $occ(t_0)$ by the hypothesis in D. As for the preservation of axioms, it will suffice to show that, for quantifier free r, there holds

(6) $$r^{\#} = rep(y_0, t_0 \mid r)$$

because then an atomic formula r, occurring on the different sides of an axiom, will be transformed into the same $r^{\#}$ on both these sides.

Unfortunately, the proofs of these apparently simple statements require a detailed study of the way in which the formulas $r^{\#}$ arise under the repeated applications of the substitution map sub. Assume, for instance, that a formula r_0 of the endsequent (not yet containing y_0) gives rise, further up, to a side formula r_1 which itself, still further up, gives rise to another side formula r_2. If $r_0 = Qx_0\,Qx_1\,Qx_2\,w_0$ then

$$r_0 = Qx_0\,Qx_1\,Qx_2\,w_0$$
$$r_1 = sub(x_0, s_0 \mid Qx_1\,Qx_2\,w_0) = Qy_1\,Qy_2\,w_1$$
$$r_2 = sub(y_1, s_1 \mid Qy_2\,w_1) = Qz_2\,w_2$$

$$r_0^{\#} = Qx_0\,Qx_1\,Qx_2\,w_0$$
$$r_1^{\#} = sub(x_0, h_\eta(s_0) \mid Qx_1\,Qx_2\,w_0) = Qy_1'\,Qy_2'\,w_1'$$
$$r_2^{\#} = sub(y_1', h_\eta(s_1) \mid Qy_2'\,w_1') = Qz_2'\,w_2' \,.$$

Clearly, z_2 is a renaming of y_2 which is a renaming of x_2, and z_2' is a renaming of y_2' which also is a renaming of x_2 – but all these renamings may differ. Still, for the proofs of (5) and (6), it will be important to know that, for instance, z_2 is free in w_2 if, and only if, z_2' is free in w_2'.

Thus I am forced to interrupt the proof of Theorem 4 in order to develop as an auxiliary tool

5. The Sets SUB

For the replacement map rep, I know that the identity

$$rep(y,t \mid rep(x,y \mid v)) = rep(x,t \mid v)$$

will hold under suitable assumptions (cf. Lemma **1.13**.9.(iii)). An analogous identity

$$sub(y,t \mid sub(x,y \mid v)) = sub(x,t \mid v) \,,$$

however, *cannot* be expected to hold, because the renaming of bound variables performed by sub(x,y | –) alone may lead to quite different bound variables in the left and in the right expression.

I shall now associate, to every map η and every formula v, a set $SUB(\eta \mid v)$, consisting of *variants* of $sub(\eta \mid v)$, differing only in the (names of) bound variables. The definition of the sets $SUB(\eta \mid v)$ must be chosen sufficiently *general* in order to ensure that

$$sub(y,t \mid sub(x,y \mid v)) \text{ is still in } SUB(\zeta(y,t) \mid v)$$

and it must be sufficiently *restrictive* in order to ensure that formulas $r^{\#}$, constructed during the algorithm of the Substitution Theorem, can still be recognized as belonging to $SUB(\zeta(y_0,t_0) \mid v)$. Throughout the following, I shall relay on concepts and terminology explained in Chapter **1.13** .

Before formulating the precise definition of the sets SUB, let me introduce one more notation for branches: if $\alpha = \langle \alpha(j) \mid j<k \rangle$ is a branch of a formula $\alpha(0)$, then I denote, for $i<k$, by $\alpha \restriction i$ the branch $\langle \alpha(j) \mid i \leq j<k \rangle$ ascending from the formula $\alpha(i)$.

Let v and u be two formulas having the same skeleton, and let α, α' be naturally corresponding branches of v and u. Assume that there holds

(B) For every $i<def(\alpha)-1$: if $\alpha(i) = Qx_i w$ and $\alpha'(i) = Qy_i w'$ then $x_i \epsilon fr(\alpha \restriction i+1)$ if, and only if, $y_i \epsilon fr(\alpha' \restriction i+1)$.

In that case, I obtain a bijection λ from $bd(\alpha)$ onto $bd(\alpha')$ if, for every $x \epsilon bd(\alpha)$, I choose the largest i such that $\alpha(i) = Qx\,w$ (i.e. the index of x) and set $\lambda(x) = y$ for the y such that $\alpha'(i) = Qy\,w'$. I shall say that λ is the *natural* (index preserving) bijection from $bd(\alpha)$ onto $bd(\alpha')$.

For any map η and any formula v, I define the set $SUB(\eta \mid v)$ to consist of all formulas u which have the following properties:

u and v have the same skeleton.

Let α be a branch of v and let α' be the branch of u naturally corresponding to α.

(A) If $x \epsilon fr(\alpha)$ then no variable in $occ(\eta(x))$ has a binding in α' .

(B) For every $i<def(\alpha)-1$: if $\alpha(i) = Qx_i w$ and $\alpha'(i) = Qy_i w'$ then $x_i \epsilon fr(\alpha \restriction i+1)$ if, and only if, $y_i \epsilon fr(\alpha' \restriction i+1)$.

For the natural bijection λ from $bd(\alpha)$ onto $bd(\alpha')$ there holds

(C) If $\alpha(k-1)$, $\alpha'(k-1)$ are the last formulas of α, α' then $\alpha'(k-1) = rep(\eta_\lambda \mid \alpha(k-1))$ where η_λ coincides with η on $fr(\alpha)$ and with λ on $bd(\alpha)$.

From this definition there follows:

(7) If $v = \neg v_0$ or $v = v_0 \% v_1$ with a propositional connective %, then u in $SUB(\eta \mid v)$ implies $u = \neg u_0$, respectively $u = u_0 \% u_1$, with formulas u_i in $SUB(\eta \mid v_i)$.

Also, if v contains no quantifiers then $SUB(\eta \mid v)$ consists of the single formula $rep(\eta \mid v)$. The following is the *Key Lemma* about SUB :

LEMMA 5 Let i be the identity and let η be arbitrary. Then

(i) $\mathrm{tot}(\eta, Y_0, Y_1 \mid v) \in \mathrm{SUB}(i \mid v)$,

(ii) $\mathrm{sub}(\eta, Y_0, Y_1 \mid v) \in \mathrm{SUB}(\eta \mid v)$.

During the proof, I shall not write the sets Y_0 and Y_1 in order to simplify the notation.

Let me begin with the special case of (ii): if η is free for v then $\mathrm{rep}(\eta \mid v)$ is in $\mathrm{SUB}(\eta \mid v)$. Here (A) follows from the freeness of η. Now let α be a branch of v and let $\alpha' = \mathrm{rep}(\eta \mid \alpha)$ be the naturally corresponding branch of $\mathrm{rep}(\eta \mid v)$.

If $\alpha(i) = Qx\,w$ then $\alpha'(i) = \mathrm{rep}(\eta \mid \alpha)(i) = \mathrm{rep}(\eta_Z \mid \alpha(i))$ where Z is the set of those z such that $\alpha(j) = Qz\,\alpha(j+1)$ for some $j<i$. Hence $\alpha'(i) = Qx\,w'$ with $w' = \mathrm{rep}(\eta_Y \mid w)$ where Y is $Z \cup \{x\}$, and thus also $\alpha' \restriction i+1 = \mathrm{rep}(\eta_Y \mid \alpha \restriction i+1)$ with $x \in Y$. Consequently, $x \in \mathrm{fr}(\alpha \restriction i+1)$ if and only if $x \in \mathrm{fr}(\alpha' \restriction i+1)$, and this establishes (B).

The variables which have a binding in α are the same as those which have a binding in α' because $\mathrm{rep}(\eta \mid -)$ does not change quantified variables . Thus the natural bijection λ is the identity. (C) holds because $\alpha'(k-1) = \mathrm{rep}(\eta_X \mid \alpha(k-1))$ where X is the set of all variables which have a binding in α. This concludes the proof the special case.

I now prove (i) by induction on v; the statement holds if v is atomic, and it is preserved under propositional composition of formulas. Consider now the case that $v = Qx\,w$, and let y be the variable in

$$\mathrm{tot}(\eta \mid Qx\,w) = Qy \;\; \mathrm{rep}(x,y \mid \mathrm{tot}(\eta_x \mid w)) \; .$$

I shall write ζ for $\zeta(x,y)$. I shall use that, by definition of tot, if $y \neq x$ then y is not free in w and $\zeta(x,y)$ is free for $\mathrm{tot}(\eta_x \mid w)$.

Let α be a branch of v and let $\alpha' = \mathrm{tot}(\eta \mid \alpha)$ be the naturally corresponding branch of $\mathrm{tot}(\eta \mid v)$. Removing the first members of α, α', I obtain β, β' such that

β, β' are corresponding branches of w and $\mathrm{rep}(x,y \mid \mathrm{tot}(\eta_x \mid w))$;

further, choose $\beta'' = \mathrm{tot}(\eta_x \mid \beta)$ such that

(8) β, β'' are corresponding branches of w and $\mathrm{tot}(\eta_x \mid w)$.

Observe that now

(9) β'', $\beta' = \mathrm{rep}(x,y \mid \beta)$ are corresponding branches of $\mathrm{tot}(\eta_x \mid w)$ and $\mathrm{rep}(x,y \mid \mathrm{tot}(\eta_x \mid w))$.

Applied to (8), the inductive hypothesis for (i) presents me with

(10_1) (B) holds for β, β'' ,
(10_2) $\beta''(k-2) = \mathrm{rep}(i_\mu | \beta(k-2))$ for the natural bijection μ from $\mathrm{bd}(\beta)$ onto $\mathrm{bd}(\beta'')$.

As $\zeta(x,y)$ is free for $\mathrm{tot}(\eta_x | w)$, the special case, applied to (9), presents me with

(11_1) (B) holds for β'', β'
(11_2) $\beta'(k-2) = \mathrm{rep}(\zeta_j | \beta''(k-2))$ for the identity j from $\mathrm{bd}(\beta'')$ onto $\mathrm{bd}(\beta')$.

Now the first two statements in

(12_1) (B) holds for β, β'
(12_2) μ is the natural bijection from $\mathrm{bd}(\beta)$ onto $\mathrm{bd}(\beta') = \mathrm{bd}(\beta'')$,
(12_3) $\beta'(k-2) = \mathrm{rep}(\zeta_\mu | \beta(k-2))$

follow immediately. As for (12_3), it follows from (11_2) and (10_2) that

$$\beta'(k-2) = \mathrm{rep}(\zeta_j | \mathrm{rep}(i_\mu | \beta(k-2))) .$$

As these are atomic formulas, I only need to show that, for the homomorphism h extending ζ_j, the map $h \cdot i_\mu$ coincides with ζ_μ on $\mathrm{var}(\beta(k-2)) = \mathrm{fr}(\beta) \cup \mathrm{bd}(\beta)$. But $h \cdot i_\mu \restriction \mathrm{bd}(\beta)$ maps $\mathrm{bd}(\beta)$ by μ onto $\mathrm{bd}(\beta'')$ and then continues as the identity j. And $\mathrm{fr}(\beta'') = \mathrm{fr}(\beta)$ by Lemma **1.13**.4.(i), whence $h \cdot i_\mu \restriction \mathrm{fr}(\beta)$ maps $\mathrm{fr}(\beta)$ by the identity i onto $\mathrm{fr}(\beta'')$ and then continues as ζ.

Let me observe that it follows from (11_2), say, that ζ_j maps $\mathrm{var}(\beta''(k-2)$ onto $\mathrm{var}(\beta'(k-2))$ and maps $\mathrm{bd}(\beta'')$ onto $\mathrm{bd}(\beta')$. Hence the elements in $\mathrm{fr}(\beta')$ must be images already under ζ acting on $\mathrm{fr}(\beta'')$: thus ζ maps $\mathrm{fr}(\beta'')$ *onto* $\mathrm{fr}(\beta')$. This I shall use in showing that

(13) $x \in \mathrm{fr}(\beta)$ if, and only if, $y \in \mathrm{fr}(\beta')$.

If $x \in \mathrm{fr}(\beta)$ then ζ by (12_3) maps x into $\mathrm{fr}(\beta')$. Conversely, the implication to the left is trivial in case $x = y$. If $x \neq y$ then y is not in fr(w), hence not in $\mathrm{fr}(\mathrm{tot}(\eta_x | w))$. For the branch β'' of $\mathrm{tot}(\eta_x | w)$, therefore, y cannot be in $\mathrm{fr}(\beta'')$. But ζ maps $\mathrm{fr}(\beta'')$ onto $\mathrm{fr}(\beta')$, and so y must be an image under this map. As it cannot arrive as $\zeta(y) = y$, it must arrive as $\zeta(x) = y$, and so x will be in $\mathrm{fr}(\beta'')$. But once more $\mathrm{fr}(\beta'') = \mathrm{fr}(\beta)$ by Lemma **1.13**.4.(i).

It follows from (13) and (12_1) that (B) holds for α, α'.

It follows from (B) that the natural bijection λ from $\mathrm{bd}(\alpha)$ onto $\mathrm{bd}(\alpha')$ is μ unless one (and then both) of the statements $x \in \mathrm{fr}(\beta)$ and $y \in \mathrm{fr}(\beta')$ is true; in the latter case, λ extends μ by $\lambda(x) = y$.

Since $\mathrm{fr}(\beta') = (\mathrm{fr}(\beta) - \{x\}) \cup \{y\}$ by (12_3), $x \in \mathrm{fr}(\beta)$ implies $\mathrm{fr}(\alpha) = \mathrm{fr}(\beta) - \{x\} = \mathrm{fr}(\beta') - \{y\} = \mathrm{fr}(\alpha')$, and otherwise I have $\mathrm{fr}(\alpha) = \mathrm{fr}(\beta) = \mathrm{fr}(\beta') = \mathrm{fr}(\alpha')$. In the both cases, therefore, the map ζ_μ from (12_3) is the identity on $\mathrm{fr}(\alpha)$ and coincides with λ on $\mathrm{bd}(\alpha)$. Thus (12_3) gives

$$\alpha'(k-1) = rep(i_\lambda \mid \alpha(k-1)) ,$$

and this proves (C), and thereby ends the proof of statement (i).

Statement (ii) now follows from (i) and the special case, since $sub(\eta \mid v) = rep(\eta \mid tot(\eta \mid v))$ and η is free for $tot(\eta \mid v)$. If α is a branch of v then $\alpha' = sub(\eta \mid \alpha)$ may be written as $rep(\eta \mid \gamma)$ for $\gamma = tot(\eta \mid \alpha)$. That (A) holds follows from the freeness of η for $tot(\eta \mid v)$ and from the fact that variables having bindings in α' are the same as those having bindings in γ. As (B) holds for α, γ and for γ, α', it also holds for α, α'. As $bd(\alpha') = bd(\gamma)$ by the special case, the natural bijection ν with respect to α, γ is also the natural bijection λ with respect to α, α'. Finally

$$\alpha'(k-1) = rep(\eta_X \mid \gamma(k-1)) \quad \text{and} \quad \gamma(k-1) = rep(i_\nu \mid \alpha(k-1))$$

where X is the set of all variables which have a binding in γ. If h is the homomorphism extending η_X then

$$\alpha'(k-1) = rep(h \cdot i_\nu \mid \alpha(k-1)) .$$

But $h \cdot i_\nu$ maps $bd(\alpha)$ by ν onto $bd(\gamma)$ and then continues as the identity since $bd(\gamma) \subseteq X$. And it maps $fr(\alpha)$ by the identity onto $fr(\gamma)$ and then continues by η since $fr(\gamma) \cap X = 0$. Thus $h \cdot i_\nu$ acts as η_λ. This establishes (C), concludes the proof of (ii) and hence that of Lemma 5.

LEMMA 6 Assume that $u_0 = sub(x,y \mid v)$, $u_1 = sub(x,t \mid v)$, and let y be not free in v. Then u_1 is in $SUB(\zeta(y,t) \mid u_0)$.

The formulas v, u_0 and u_1 have the same skeleton. Let α be a branch of v and let $\alpha' = sub(x,y \mid \alpha)$, $\alpha'' = sub(x,t \mid \alpha)$ be corresponding branches of u_0, u_1. It follows from the Lemma 5 that the pairs α, α' and α, α'' each satisfy (A)–(C). Then (B) follows by transitivity for α', α'', and if λ, μ are the natural bijections from $bd(\alpha)$ onto $bd(\alpha')$ and onto $bd(\alpha'')$, $\lambda^{-1} = \nu$, then $\kappa = \mu \cdot \nu$ is the natural bijection from $bd(\alpha')$ onto $bd(\alpha'')$.

In order to verify (A), I have to show that if $y \in fr(\alpha')$ then no variable in $occ(t)$ has a binding in α''. It follows from (C) for α, α' that $\alpha'(k-1) = rep(\zeta(x,y)_\lambda \mid \alpha(k-1))$. As y was not free in v, hence not free in α, it can be free in α' only by arriving as image under $\zeta(x,y)$ of x; hence x will be free in α. But (A) holds for α, α'', and this implies the claim about $occ(t)$.

In order to verify (C), I have to show that $\alpha''(k-1) = rep(\zeta(x,t)_\kappa \mid \alpha'(k-1))$. Now

$$\alpha'(k-1) = rep(\zeta(x,y)_\lambda \mid \alpha(k-1)) \quad \text{implies} \quad \alpha(k-1) = rep(\zeta(y,x)_\nu \mid \alpha'(k-1)) .$$

Because on $bd(\alpha')$ the map ν reverses λ. And on $fr(\alpha')$ the map $\zeta(y,x)$ affects y if, and only if, it is free in α'. But this is the case if, and only if, x is in $fr(\alpha)$ and *was* affected by $\zeta(x,y)$.

Now it follows from (C) for α, α'' that $\alpha''(k-1) = rep(\zeta(x,t)_\mu \mid \alpha(k-1))$. So

$$\alpha''(k-1) = \text{rep}(\zeta(x,t)_\mu \cdot \zeta(y,x)_\nu \mid \alpha'(k-1)) \ .$$

But $\zeta(x,t)_\mu \cdot \zeta(y,x)_\nu$ is κ on $\text{bd}(\alpha')$ and $\zeta(y,t)$ on $\text{fr}(\alpha')$, since $x \epsilon \text{fr}(\alpha)$ implies $y \epsilon \text{fr}(\alpha')$. [Observe that, in this proof, I have used that the natural bijection actually *is* bijective.]

LEMMA 7 If u is in $\text{SUB}(\eta \mid v)$ then $\text{fr}(u) = \bigcup \langle \text{occ}(\eta(x)) \mid x \epsilon \text{fr}(v) \rangle$.

Observe that fr(u) is the union of all sets $\text{fr}(\alpha')$ for branches α' of u. If α is the branch of v corresponding to α' then $\alpha'(k-1) = \text{rep}(\eta_\lambda \mid \alpha(k-1))$ by (C) whence $\text{fr}(\alpha') = \bigcup \langle \text{occ}(\eta(x)) \mid x \epsilon \text{fr}(\alpha) \rangle$.

After this technical intermission, I now continue with

6. The Substitution Theorem Resumed

The proof of Theorem 4 will be completed once the statements (5) and (6) can be shown for the formulas $r^\#$ assigned to formulas r in the algorithm constructing $D^\#$ from D. To this end, I shall show for each of these r and $r^\#$ that

(14) $r^\# \epsilon \text{SUB}(\eta \mid r)$ for $\eta = \zeta(y_0, t_0)$.

Then (5) will follow immediately from Lemma 6, and (6) follows from the fact that for a quantifier free r the set $\text{SUB}(\eta \mid r)$ consists of $\text{rep}(\eta \mid r)$ only.

That (14) holds for the formulas u_i of the endsequent follows from Lemma 6, and for the remaining formulas r of the endsequent, in which y_0 is not free, $r^\# = r$ is trivially in $\text{SUB}(\eta \mid r)$. I now use induction on the construction of $r^\#$. If $r^\#$ is defined by (a1) then (14) follows from Lemma 5.(ii). If $r^\#$ is defined by (a2) then (14) follows from the inductive hypothesis; if $r^\#$ is defined by (a3) then (14) follows from the inductive hypothesis and the observation (7).

So it remains to consider the case that $r^\# = \text{sub}(x', h_\eta(s) \mid w')$ is defined by (a4) from $r = \text{sub}(x,s \mid w)$, and I shall prove (14) by verifying the conditions (A)–(C) of the definition of SUB for two naturally corresponding branches α and α' of r and $r^\#$. But now not only r and $r^\#$ have the same skeleton, but all the four of r, w, w', $r^\#$ do. Thus to α there naturally corresponds a branch β of w and a branch β' of w'. Also, the branch β extends to a branch γ of Qx w, and since Qx w and Qx' w' have the same skeleton, γ naturally corresponds to a branch γ' of Qx' w' which extends β' :

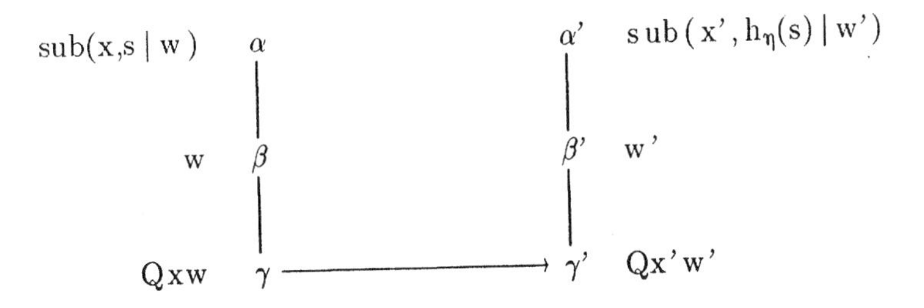

Observe that

$$\mathrm{sub}(x,s \mid w) \in \mathrm{SUB}(x,s \mid w) \quad ,$$

establishing a correspondence between β, α , and that

$$\mathrm{sub}(x', h_\eta(s) \mid w') \in \mathrm{SUB}(x', h_\eta(s) \mid w') \quad ,$$

establishing a correspondence between β', α' , and that by inductive hypothesis

$$Qx'\,w' \in \mathrm{SUB}(y_0, t_0 \mid Qx\,w) \quad ,$$

establishing a correspondence between γ, γ' .

I begin by verifying the condition (A). I have to show that $z \epsilon \mathrm{fr}(\alpha)$ implies $\mathrm{occ}(\eta(z)) \cap \mathrm{bd}(\alpha') = 0$. Observe that $z \epsilon \mathrm{fr}(\alpha)$ implies that either ($z \epsilon \mathrm{occ}(s)$ and $x \epsilon \mathrm{fr}(\beta)$) or ($z \epsilon \mathrm{fr}(\beta)$ and $z \neq x$).

I first consider the case $z \epsilon \mathrm{occ}(s)$ and $x \epsilon \mathrm{fr}(\beta)$. It follows from (B), applied to γ, γ', that $x \epsilon \mathrm{fr}(\beta)$ implies $x' \epsilon \mathrm{fr}(\beta')$. Hence x' *is* changed in w' under the map $\mathrm{sub}(x', h_\eta(s) \mid -)$, and thus none of the variables in $h_\eta(s)$ will have a binding in α'. But if $z \neq y_0$ then $\eta(z) = z$, and $z \epsilon \mathrm{occ}(s)$ implies $z \epsilon \mathrm{occ}(h_\eta(s))$. And if $z = y_0$ then $z \epsilon \mathrm{occ}(s)$ also implies $\mathrm{occ}(\eta(z)) \subseteq \mathrm{occ}(h_\eta(s))$.

Let me now consider the case $z \epsilon \mathrm{fr}(\beta)$ and $z \neq x$. Then also $z \epsilon \mathrm{fr}(\gamma)$, and (A), applied to γ, γ', shows that none of the variables in $\eta(z)$ has a binding in γ'; in particular, none of them equals x'. Further, (C) shows that $\mathrm{occ}(\eta(z))$ is in $\mathrm{fr}(\gamma)$ and, therefore, in $\mathrm{fr}(\beta)$. Being distinct from x', none the variables in $\mathrm{occ}(\eta(z))$ then will obtain a binding in α'. This concludes the verification of (A).

The condition (B) for α, α' is immediately inherited from α, β, γ, γ', β', α'. Further, there are the natural bijections

λ^1 from $\mathrm{bd}(\beta)$ onto $\mathrm{bd}(\alpha)$,
λ^2 from $\mathrm{bd}(\beta')$ onto $\mathrm{bd}(\alpha')$,
λ^0 from $\mathrm{bd}(\gamma)$ onto $\mathrm{bd}(\gamma')$.

It follows from (B), applied to γ, γ', that $\mathrm{bd}(\beta)$ differs from $\mathrm{bd}(\gamma)$ by x if, and only if, $\mathrm{bd}(\beta')$ differs from $\mathrm{bd}(\gamma')$ by x'. Thus I obtain an index preserving bijection λ^* from $\mathrm{bd}(\beta)$ onto $\mathrm{bd}(\beta')$ if I set $\lambda^* = \lambda - \{<x,x'>\}$ if $x \epsilon \mathrm{bd}(\beta)$, and $\lambda^* = \lambda$ otherwise. Consequently,

$$\lambda = \lambda^2 \cdot \lambda^* \cdot (\lambda^1)^{-1}$$

is the natural bijection from $bd(\alpha)$ onto $bd(\alpha')$.

In order to prove (C), I have to show that

(15) $\alpha'(k-1) = rep(\eta_\lambda \mid \alpha(k-1))$ where η_λ coincides with η on $fr(\alpha)$ and with λ on $bd(\alpha)$.

Observe that

(15_1) $\alpha(k-1) = rep(\zeta(x,s)_1 \mid \beta(k-1))$,
(15_2) $\alpha'(k-1) = rep(\zeta(x', h_\eta(s))_2 \mid \beta'(k-1))$,
(15_3) $\gamma'(k) = rep(\eta_o \mid \gamma(k))$,

where

$\zeta(x,s)_1$ coincides with $\zeta(x,s)$ on $fr(\beta)$ and with λ^1 on $bd(\beta)$,
$\zeta(x', h_\eta(s))_2$ coincides with $\zeta(x', h_\eta(s))$ on $fr(\beta')$ and with λ^2 on $bd(\beta')$,
η_o coincides with η on $fr(\gamma)$ and with λ^o on $bd(\gamma)$.

It follows from (15_3) and the definition of λ^* that

(15_4) $\beta'(k-1) = rep(\eta^* \mid \beta(k-1))$

where η^* coincides with η of $fr(\beta)-\{x\}$ and coincides with λ^* on $bd(\beta)$ and where $x \epsilon fr(\beta)$ implies $\eta^*(x) = x'$. Thus

$\alpha'(k-1) = rep(\zeta(x', h_\eta(s))_2 \mid \beta'(k-1)) = rep(\zeta(x', h_\eta(s))_2 \mid rep(\eta^* \mid \beta(k-1)))$,
$rep(\eta_\lambda \mid (\alpha(k-1)) = rep(\eta_\lambda \mid rep(\zeta(x,s)_1 \mid \beta(k-1)))$,

and (15) will follow if the two right expressions are the same, i.e. if

(16) the maps $h_2 \cdot \eta^*$ and $h \cdot \zeta(x,s)_1$ coincide on $var(\beta(k-1)$,

where h_2 is the homomorphic extension of $\zeta(x', h_\eta(s))_2$, and h is the homomorphic extension of η_λ .

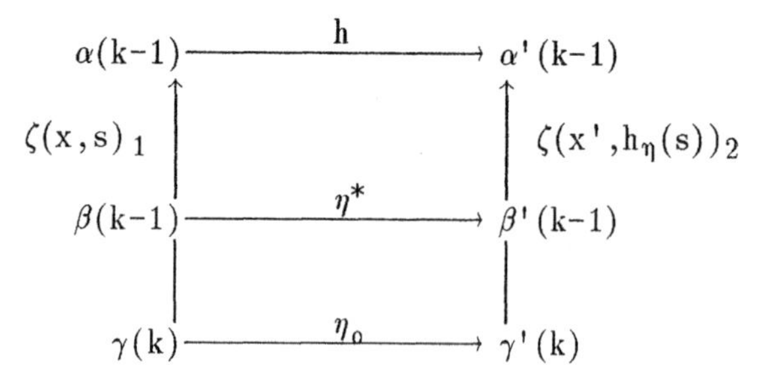

On $bd(\beta)$, $h_2 \cdot \eta^*$ acts as $\lambda^2 \cdot \lambda^*$, and $h \cdot \zeta(x,s)_1$ acts as $\lambda \cdot \lambda^1$. By definition of λ^*, both these maps are the same. It thus remains to be shown that the maps coincide on $fr(\beta)$. Recall that η is $\zeta(y_o, t_o)$.

If z in $fr(\beta)$ is neither x nor y_o then it remains fixed under $\eta^* = \eta$. In order to know that it then remained fixed under h_2 as well, I have to know that

$z \neq x'$. But $z \neq x$ implies that z is also free in γ. Hence (15_3) implies that η_0 (which keeps z fixed) maps z into $fr(\gamma')$, whence $z \neq x'$. Consequently, z remains fixed under the first map in (16). Concerning the second map, z first remains fixed under $\zeta(x,s)_1$, and being in $fr(\alpha)$ it then remains fixed under η_λ, i.e. under h.

If y_0 is in $fr(\beta)$ and is not x, then it is mapped to t_0 by η^*. It follows from (A), applied to γ, γ', that x' is not in $occ(t_0)$; thus y_0 is mapped to t_0 by the first map in (16). On the other hand, $y_0 \neq x$ implies that y_0 remains fixed under $\zeta(x,s)_1$, and being so in $fr(\alpha)$ it then is mapped to t_0 by η_λ.

If x is in $fr(\beta)$ then it is mapped to x' by η^*, and x' then is mapped to $h_\eta(s)$ by h^2. On the other hand, x is mapped to s by $\zeta(x,s)_1$. But no variable in occ(s) has a binding in the branch α of $sub(x,s \mid w)$, hence $occ(s) \subseteq fr(\alpha)$, and so s is mapped to $h_\eta(s)$ by the extension h of η_λ.

This concludes the proof of (16), hence that of (15), hence that of (14), and thereby it concludes the proof of Theorem 4.

As I did already earlier, I shall call a derivation D *eigen* if, for every node e carrying an instance of a critical Q-rule, the eigenvariable at e occurs as variable *only* in formulas of sequents *above* e. I then state the *Eigenvariable Lemma* :

LEMMA 8 There is an operator, acting on derivations H and finite sets Y of variables, which transforms H into a derivation H^* with the same skeleton which is eigen and has no eigenvariables in Y.

I argue by induction on the length of H. The statement being trivial if H consists only of an axiom, assume that H ends with some instance of a rule (R). Consider first the case that (R) is a critical Q-rule, and let Qx w be its principal and $sub(x,y \mid w)$ be its side formula; let H_0 be the subderivation of H leading to its premiss. By inductive assumption, I transform H_0 into H_0^* such that no eigenvariables of H_0^* are in $Y \cup \{y\}$. Next, I chose a variable z different from the eigenvariables of H_0^*, not in the endsequent of H and outside of Y. Then the hypotheses of Theorem 4 hold for H_0^*, y, z in place of D, y_0, t_0, and Theorem 4 produces a derivation $H_1 = (H_0^*)^\#$. The endsequent of H_1 differs from that of H_0 and H_0^* only in having $sub(x,y \mid w)$ replaced by $sub(x,z \mid w)$. Since z was not in the endsequent of H, I may continue H_1 to H^* by applying (R) with the eigenvariable z. But H_1 and H_0^* have the same eigenvariables in the same places, and as they were all distinct from z, H^* has the property requested.

If (R) is an uncritical rule with one premiss, then it suffices to apply the inductive hypothesis to the subderivation H_0 of H leading to that premiss (and if (R) is permitted to be a weakening, then Y must be enlarged such as to omit the variables of the enlarged endsequent), continuing afterwards with the application of (R). If (R) is a rule with two premisses, let H_0 and H_1 be the subderivations leading to them. I then apply the inductive hypothesis to H_0 with Y enlarged by the variables in H_1, and to H_1 with Y enlarged by the variables in H_0, and afterwards I again continue with the application of (R).

I now state the *Second Substitution Theorem* :

THEOREM 5 There is an operator, acting on derivations D, variables y_0 and terms t_0, and transforming D into a derivation $D^\ddagger$ as follows.

Let u_0, u_1,... be all those formulas in the endsequent of D in which y_0 is free; let x_0, x_1,... be variables, and let v_0, v_1,... be formulas such that $u_i = \text{sub}(x_i, y_0 \mid v_i)$ and such that y_0 is not free in v_i, $i = 0,1,\ldots$.

Then D and $D^\ddagger$ have the same skeleton, and $D^\ddagger$ is eigen. The endsequent of $D^\ddagger$ arises from that of D by replacing u_i by $\text{sub}(x_i, t_0 \mid v_i)$; all other sequents of $D^\ddagger$ arise by appropriate replacements from those of D.

I first apply Lemma 8 to D, obtaining a derivation D^* with the same endsequent as D and satisfying the hypothesis of Theorem 4, i.e. such that none of its eigenvariables is in $\text{occ}(t_0) \cup \{y_0\}$. Applying Theorem 4 to D^*, I obtain the desired $D^\ddagger$ as $(D^*)^\#$.

In concluding, I wish to observe that in Lemma 8 the derivation H^* may also be defined from H by a recursion from below. To do so, I argue as at the start of the proof of Theorem 4 and define, for every node e of H carrying a sequent S, the sequent S^* of H^* together with a map σ_e from variables to variables; again S^* is obtained by assigning to each r in S a formula r^* with the same skeleton.

If e is minimal and S is the endsequent of H then I set $r^* = r$ for every formula r in S, i.e. $S^* = S$, and take for σ_e the identity. I now proceed upwards and assume that the assignment has already been defined for all nodes of lesser height than e, and in particular for the lower neighbour h of e carrying the conclusion T of an instance of a rule (R).

If (R) is not a critical Q–rule then I set $\sigma_e = \sigma_h$. If (R) *is* a critical Q–rule with the principal formula Qx w and the eigenvariable z at e, then I choose as z' the first variable *not* occurring in H or in any of the sequents of H^* at nodes below e, and I define σ_e by

$$\sigma_e(y) = \sigma_h(y) \quad \text{for} \quad y \neq z \quad , \quad \sigma_e(z) = z' \quad .$$

Observe that σ_e and σ_h coincide on all variables free in formulas of the conclusion T. – For the formulas r in the premiss S of (R) there are the following possibilities:

(a1) (R) is a cut with cut formula r. Then I set $r^* = \mathrm{sub}(\sigma_e \mid r)$.

(a2) r is parametric (in particular: r is principal). Then r is in T, and I take r^* as it was defined for T^*.

(a3) r is the side formula of a propositional rule (R). Then the principal formula q occurs in the T and q^* is defined. Since q^* has the same skeleton as q, there corresponds to r a unique propositional subformula of q^* which I choose as r^*.

(a4) r is side formula of a Q-rule (R) with principal formula Qx w, hence $r = \mathrm{sub}(x,s \mid w)$ with some term or some eigenvariable s. Since $(Qx\,w)^*$ is defined and has the same skeleton as Qx w, there holds $(Qx\,w)^* = Qx'\,w'$. I then set $r^* = \mathrm{sub}(x', h(s) \mid w')$ with the homomorphic extension h of σ_e.

The proof then proceeds in analogy to that of Theorem 4.

Observe, however, that this construction still requires that sufficiently many unused new variables be available, in order to choose the new eigenvariable $\sigma_e(z)$. They *are* available since my derivations are finite and so will use only finitely many variables. A generalization of the theorem to the case of infinite derivations H will, therefore, be possible only under additional hypotheses, ensuring the availability of unused variables.

7. Cut Elimination Resumed

From the remarks made in the introductory section it now follows that the exchange operators and the cut elimination algorithm **A** from Theorem 2.1 remain in effect for derivations in a sequent calculus employing Q-rules which use the substitution map sub. Hence there holds the

THEOREM 6 Let D be a derivation of an endsequent in a calculus employing Q-rules with sub and possibly employing cuts. In the case of symmetric cuts, set $D^* = D$; in the case of non-symmetric cuts let D^* arise from D by the transformation in the Eigenvariable Lemma. Then $\mathbf{A}(D^*)$ is a derivation without cuts of the same endsequent and of length at most $e_3(k,n)$.

As an illustration, let me discuss once more the earlier *Example 1*:

$$
\begin{array}{rcl}
p(x_1) & \Rightarrow & p(x_1) \\
q(x_0)\wedge p(x_1) & \Rightarrow & p(x_1) \\
\forall x_0\,(q(x_0)\wedge p(x_1)) & \Rightarrow & p(x_1) \\
\forall x_1\,\forall x_0\,(q(x_0)\wedge p(x_1)) & \Rightarrow & p(x_1) \\
\forall x_1\,\forall x_0\,(q(x_0)\wedge p(x_1)) & \Rightarrow & \forall y_1\, p(x_1)
\end{array}
\qquad
\begin{array}{rcl}
p(x_0) & \Rightarrow & p(x_0) \\
\forall x_1\, p(x_1) & \Rightarrow & p(x_0)
\end{array}
$$

$$\forall x_1\,\forall x_0\,(q(x_0)\wedge p(x_1)) \Rightarrow p(x_0) \quad .$$

While viewed originally for the calculus with rep, it may also be seen as a derivation with sub such that the subderivation of the premiss of the left cut formula becomes

$$
\begin{array}{lrcl}
1. & p(x_1) & \Rightarrow & p(x_1) \\
2. & \mathrm{sub}(x_0, x_0 \mid q(x_0)\wedge p(x_1)) & \Rightarrow & p(x_1) \\
3. & \mathrm{sub}(x_1, x_1 \mid \forall x_0\,(q(x_0)\wedge p(x_1))) & \Rightarrow & p(x_1) \\
4. & \forall x_1\,\forall x_0\,(q(x_0)\wedge p(x_1)) & \Rightarrow & p(x_1)\ .
\end{array}
$$

Applying the algorithm of the Substitution Theorem with $\eta = \zeta(x_1, x_0)$, line 4 becomes

$$4^{\#}. \qquad \forall x_1 \forall x_0 (q(x_0)\wedge p(x_1)) \Rightarrow p(x_0)\ .$$

Thus line 3 becomes

$$\mathrm{sub}(x_1,\ \eta(x_1) \mid \forall x_0(q(x_0)\wedge p(x_1))) \Rightarrow p(x_0)$$

or

$$\mathrm{sub}(x_1, x_0 \mid \forall x_0(q(x_0)\wedge p(x_1))) \Rightarrow p(x_0)$$

i.e.

$$3^{\#}. \quad \forall x_2(q(x_2)\wedge p(x_0)) \Rightarrow p(x_0)\ .$$

Consequently, line 2 becomes

$$\mathrm{sub}(x_0,\ \eta(x_0) \mid q(x_2)\wedge p(x_0)) \Rightarrow p(x_0)$$

i.e.

$$2^{\#}. \quad q(x_2)\wedge p(x_0) \Rightarrow p(x_0)\ .$$

Hence the transformed subderivation is

$$
\begin{array}{lrcl}
1^{\#}. & p(x_0) & \Rightarrow & p(x_0) \\
2^{\#}. & q(x_2)\wedge p(x_0) & \Rightarrow & p(x_0) \\
3^{\#}. & \forall x_2(q(x_2)\wedge p(x_0)) & \Rightarrow & p(x_0) \\
4^{\#}. & \forall x_1 \forall x_0(q(x_0)\wedge p(x_1)) & \Rightarrow & p(x_0)\ .
\end{array}
$$

In this particular situation, therefore, no further cut with the premiss $x_0 \Rightarrow x_0$ of the right cut formula is required. The reader will notice that the same derivation was mentioned (as to *be found easily*) in the earlier discussion of Example 1 .

The cut elimination algorithm discussed so far was based on the use of exchange operators as in Theorem **2**.1. The method of *explicit retracing* developed in Lemma **3**.1–2 extends to the case of Q–rules as well, both for those working with rep and those working with sub. For Lemma **3**.1 the derivations H and J considered there must be such that the derivation D, obtained by prolonging H and J by a cut with v, is eigen. In that case the various weakenings of copies (of subderivations) of H in the proof can be performed without destroying the eigenvariable property. What needs to be considered in addition is the case that J(e) arises from $J(e_0)$ under a Q-rule (R) for which the omitted occurrence of v is principal. Keeping the notations used in the proof in Chapter **3**, if (R) = E∀ and v = ∀xw then H ends with I∀ such that

$$\begin{array}{llll} J(e_0): & sub(x,t \mid w),v,A \Longrightarrow a & \qquad J^*(e_0): & sub(x,t \mid w),A \Longrightarrow a \\ J(e): & v,A \Longrightarrow a & \qquad J^*(e): & A \Longrightarrow a \end{array}$$

while $H(k_0)$: M ⟹ sub(x,y | w) and H(k): M ⟹ v . Applying Theorem 5 to the subderivation H_{ko}, I obtain the derivation $H^{\ddagger}$ of M ⟹ sub(x,t | w) which, by proper use of Lemma 8, still will not have eigenvariables occurring in J. Weakening the right sides of all antecendents with A/M then will not disturb the eigenvariable conditions in $H^{\ddagger}$, and concluding with a permutation, $H^{\ddagger}$ prolongs to the derivation H^*_{eko} of A ⟹ sub(x,t | w). From here on I can argue as in the earlier proof. If (R) = E∃ and v = ∃xw then I proceed analogously, applying Theorem 5 this time to J^*_{eo}.

The continuous cut elimination operators $\mathbf{R}_0$ and $\mathbf{R}_1$ extend to the case of Q-rules working with sub, but can be defined only for derivations which from the outset are assumed to be eigen or, in the case of H and J, are subderivations of a derivation which is eigen. The definition of $\mathbf{R}_0$ then has to be combined with that of the, equally ascending, algorithm producing $D^{\#}$ from D in Theorem 4, in order to stepwise construct the sequents resulting from the substitution of t for y which come from the transformation of H_{ko} into $H^{\ddagger}$ in the situation just discussed for elementary retracing. In the case of v = ∃xw several such substitutions will have to be protocolled simultaneously if there are repeated principal occurrences of v in J.

8. Inversion Rules

I consider a sequent calculus which shall be extremal (i.e. K_u- ,L_u- ...).

(JI∀) Let D be a derivation which is eigen and without cuts, and with the endsequent M ⟹ ∀x v . Then I can find a suitable z and can transform D into a derivation D' of M ⟹ rep(x,z | v), respectively of M ⟹ sub(x,z | v), and D' has at most the length of D.

I consider first the case of Q-rules with rep. The formula ∀xw, at the maximal nodes e of its predecessor tree P, is introduced from premisses M_e ⟹ rep(x,z_e | w) with eigenvariables z_e; let D_e be the subderivations leading to these premisses. Let z be a new variable not occuring anywhere in D. By

Lemma 1, D_e can be transformed into a derivation $D_e' = \text{rep}(z_e,z \mid D_e)$. Under this transformation, $\text{rep}(x,z_e \mid w)$ will be replaced by the formula $\text{rep}(z_e,z \mid \text{rep}(x,z_e \mid w)) = \text{rep}(x,z \mid w)$ where I have made use of Lemma **1.13**.9(iii). As D is eigen, the z_e do not occur in the sequents of the predecessor tree P and, therefore, not in M_e or N_e. Hence D_e' derives $M_e \Longrightarrow \text{rep}(x,z \mid w)$, and implanting these sequents already at the maximal nodes e of P (instead of at their upper neighbours outside of P), I can continue the D_e' by the sequents and rules from P, none of which contained any z_e (nor the new z). – In the case of Q–rules with sub, I proceed analogously, but now form the derivations D_e' by applying the Substitution Theorem 5 and replace $\text{sub}(x,z_e \mid w)$ by $\text{sub}(x,z \mid w)$. – In perfect analogy, I find

(JE∃) Let D be a derivation which is eigen and without cuts, and with the endsequent $\exists x\, w,\ M \Longrightarrow u$. Then I can find a suitable z and can transform D into a derivation D' of $\text{rep}(x,z \mid v),\ M \Longrightarrow u$, respectively of $\text{sub}(x,z \mid v),\ M \Longrightarrow u$, and D' has at most the length of D.

The inversion of non-critical Q–rules requires additional restrictions. In the case of $M \Longrightarrow \exists x\, v$, for instance, the premisses above the maximal nodes e of the predecessor tree P of $\exists x\, v$ will be of the form $N_e \Longrightarrow \text{rep}(x,t_e \mid w)$ with terms t_e, and if there are several of them, I cannot, in general, expect to find a unifying term t which might replace the various t_e. A sufficient condition to prevent this situation will be that P is not ramified, and this will be the case if M contains only propositional Harrop formulas or, more generally, if no instance of (E∨) occurs below one of (I∃). In this situation there will be only one term t_e, but now a second problem arises: replacing $\exists x\, w$ by $\text{rep}(x,t_e \mid w)$, or $\text{sub}(x,t_e \mid w)$, in the sequents of P will bring in the possibly new variables in t_e, and these now may disturb the eigenvariable conditions of critical Q–rules used within P. Such instances of critical Q–rules, occurring of the left side of the sequents in P, can only be instances of (E∃). As a sufficient condition to exclude this possibility is that M consists of ∃-*Harrop* formulas in the following sense: every atomic formula is ∃-Harrop, if v and w are ∃-Harrop then so are $v \wedge w$, $\neg v$, $\forall x\, v$, if w is ∃-Harrop then $v \to w$ is ∃-Harrop. I thus find

(JI∃) Let D be a derivation of the endsequent $M \Longrightarrow \exists x\, w$. Assume that M consists of ∃-Harrop formulas. Then I can find a term t and can transform D into a derivation D' of $M \Longrightarrow \text{rep}(x,t \mid w)$ with $\zeta(x,t)$ free for w, respectively of $M \Longrightarrow \text{sub}(x,t \mid w)$, and D' has at most length of D.

(JE∀) Let D be a derivation of the endsequent $\forall x\, w,\ M \Longrightarrow u$. Assume that M consists of ∃-Harrop formulas and that u does not contain ∧ or ∀ . Then I can find a term t and can transform D into a derivation D' of $\text{rep}(x,t \mid w),\ M \Longrightarrow u$ with $\zeta(x,t)$ free for w , respectively of $\text{sub}(x,t \mid w),\ M \Longrightarrow u$, and D' has at most length of D.

Observe that (JI$\forall$) and (JE$\exists$) hold also for the multiple sequents of the classical calculi, whereas (JI$\exists$) and (JE$\forall$) in their stated formulations hold only for singular sequents, e.g. for intuitionistic logic.

Chapter 9. Semantical Consequence Operations and Modus Ponens Calculi

1. The Calculi cxqt and cxqs

The (sequent) calculi, established for propositional logic in Chapter **1** and for quantifier logic in Chapter **8**, were based on an analysis of the linguistical use of logical connectives and quantifiers. There arises the question of their relationship to the semantical consequence operations studied in Book 1.

In the propositional case, the logical calculi were associated with classes of propositional algebras in Chapter **6**, and it was an easy consequence of this association, stated as Corollary **6**.3, that classical derivability coincides with classical consequence in the sense of Chapter **1.10**.

On the other side, I have described finitary algorithms generating the consequence operations ct and cs in Chapter **1.15**. As it was just the existence of such algorithms, but not their particular form, which played a rôle for the questions treated in Book 1, I did not care there to present them in more detail. It now turns out, however, that these algorithms, arising partly from general algebraic notions about invariant congruence relations, and partly from properties of suprema and infima in lattices, permit an almost immediate translation into modus ponens calculi which will be given on the next few pages. The connection between these modus ponens calculi and the sequential calculi from the previous Chapter then will be an easy extension of the same connection established for the propositional case in Theorem **7**.2, and it will be carried out at the beginning of Chapter **10**.

The modus ponens calculi, arising immediately from the algorithms of Chapter **1.15**, are slight variants of the calculi studied since FREGE 79 in the literature. In the later sections of this Chapter, I shall translate between several such variants of quantificational modus ponens calculi, and I shall also use the occasion to make precise the different possibilities of their positive, minimal and classical forms.

I have shown in Corollary **1.15**.1 that the set ct(C) of model theoretical consequences of a set C of formulas is the congruence class of an arbitrary propositional tautology ∇ under $[C\times\{\nabla\}]^{\mathrm{inv}}$, and by Theorem **1.15**.2 this congruence relation is the *smallest* relation R having certain properties (r0)-(r5). The first three of them, (r0–2), say that R is a congruence relation containing $C\times\{\nabla\}$ and *Boolean* in the sense that it contains all instances of Boolean equations. Every such congruence relation R is uniquely determined by the congruence class $R(\nabla)$ of ∇ under R since

(BR) $\quad <u,w> \epsilon R$ if, and only if, both $u \rightarrow w$ and $w \rightarrow u$ are in $R(\nabla)$, and thus for two Boolean congruence relations R, R' the inclusion $R \subseteq R'$ is equivalent to $R(\nabla) \subseteq R'(\nabla)$. Hence the congruence class of ∇ under the smallest relation with (r0)–(r5) may be characterized also as

> the smallest subset D of Fm(L) containing C, which is congruence class $R(\nabla)$ of a Boolean congruence relation R satisfying (r3)–(r5).

This I shall use in order to translate the characterization of $[C\times\{\nabla\}]^{inv}$ into one of the congruence class $[C\times\{\nabla\}]^{inv}(\nabla)$.

A subset D of Fm(L) is a congruence classes of ∇ under a Boolean congruence relation of Fmp(L) if, and only if, it is closed under the propositional consequence operation on the propositional algebra Fmp(L). Choosing cc to be any of the various MP–calculi for classical propositional logic discussed in Chapter **7**, closed sets D of this type are precisely the sets of the form cc(E) for subsets E of Fmp(L). But $D = cc(E)$ implies that $C \subseteq D$ is equivalent to $cc(C) \subseteq D$:

> $[C\times\{\nabla\}]^{inv}(\nabla)$ is the smallest subset D of Fm(L) such that $cc(C) \subseteq D$ and which is of the form $R(\nabla)$ for a Boolean congruence relation R satisfying (r3)–(r5).

The conditions (r3) and (r4) refer to the properties (lf0)–(lf3) in Lemma **1.15**.2 which themselves already were conditions on $D = R(\nabla)$. In (lf2), (lf3) there occurs a property J(u,w,t) which, by Lemma **1.15**.3, I may choose as $J_f(u,w,t)$, saying that t is a variable neither free in u nor in w; as observed after Theorem **1.15**.1, I may also choose J as $J_v(u,w,t)$, saying that t is a variable occurring neither in u nor in w. – As for the last condition, (r5), it refers to

(i1) for every η: $<u,w> \in R$ implies $<sub(\eta \mid u), sub(\eta \mid w)> \in R$.

Since ∇ was a propositional tautology, also $sub(\eta \mid \nabla)$ will be so, i.e. there holds

for every η: if $u \in D$ then $sub(\eta \mid u) \in D$,

and by (BR) this in turn implies (i1). I thus arrive at the following characterization

> $[C\times\{\nabla\}]^{inv}(\nabla)$ is the smallest subset D of Fm(L) such that $cc(C) \subseteq D$
> $\forall x\, w \to sub(x,t \mid w) \in D$
> $sub(x,t \mid w) \to \exists x\, w \in D$
> if $u \to sub(x,y \mid w) \in D$ then $u \to \forall x w \in D$ provided y not in $var(u) \cup var(w)$
> if $sub(x,y \mid w) \to u \in D$ then $\exists x w \to u \in D$ provided y not in $var(u) \cup var(w)$
> for every η: if $u \in D$ then $sub(\eta \mid u) \in D$.

Every MP–calculus cc for classical propositional logic thus gives rise to a calculus ccqt if I enlarge his set of axioms by all formulas

(ao0) $\forall x\, w \to \mathrm{sub}(x,t \mid w)$
(ao1) $\mathrm{sub}(x,t \mid w) \to \exists x\, w$

and enlarge his set of rules by the 1–premiss quantifier rules

(Ro0) $u \to \mathrm{sub}(x,y \mid w) \vdash u \to \forall x\, w$ provided y not in var(u)∪var(w),
(Ro1) $\mathrm{sub}(x,y \mid w) \to u \vdash \exists x\, w \to u$ provided y not in var(u)∪var(w),
(Rσ) for every η : $u \vdash \mathrm{sub}(\eta \mid u)$.

Thus the

THEOREM 1_c The calculus ccqt is is both correct and complete for ct: $v \in \mathrm{ct}(C)$ if, and only if, $v \in \mathrm{ccqt}(C)$.

is but a reformulation of Corollary **1.15**.1. The completeness statement, i.e. that $v \in \mathrm{ct}(C)$ implies $v \in \mathrm{ccqt}(C)$, depends via that Corollary on the prime ideal axiom for Boolean algebras, and its proof, depending on that of Theorem **1.15**.3, is only indirect: it shows with help of a countermodel that if *not* $v \in \mathrm{ccqt}(C)$ then *not* $v \in \mathrm{ct}(C)$. Of course, the Theorem also holds for the Boolean valued consequence $\mathrm{ct}^{\mathbf{B}}$ in place of ct, and in that case it does not depend on the prime ideal axiom.

I observed in Chapter **1.15**, at the end of the section on Boolean valued quantifier logic, that the developments leading to Theorem **1.15**.2 remain in effect without any change if, instead of the Boolean valued consequence $\mathrm{ct}^{\mathbf{B}}$, I consider an **X**–valued consequence $\mathrm{ct}^{\mathbf{X}}$ defined with respect to structures for which **X** is an equational class of algebras with the properties

there is an equationally definable 0–ary operation with a value called 1^{X},

there is an equationally definable 2–ary operation $\supset$ such that, on every algebra X in **X**, the relation $\leq$, defined on u(X) by $a \leq b$ if $a \supset b = 1^{\mathrm{X}}$, is an order.

By now I know the classes $\mathbf{A}_d$, $\mathbf{A}_e$, $\mathbf{A}_g$, $\mathbf{A}_m$, $\mathbf{A}_i$ discussed in Chapter **6**, and each of them can serve as such a class **X**. I know from Chapter **7** the modus ponens calculi cd, cd, cg, cm, ci for the propositional semantical consequence operations defined by **X**. Enlarging these calculi cx by the above axioms and rules, I obtain calculi cxqt for which the same argument as above shows

THEOREM 1_x The calculus cxqt is is both correct and complete for $\mathrm{ct}^{\mathbf{X}}$: $v \in \mathrm{ct}^{\mathbf{X}}(C)$ if, and only if, $v \in \mathrm{cxqt}(C)$.

The notion of a calculus *defined* by a set of *rules* $\mathcal{R}$ (the ones of arity 0 being called *axioms*) I introduced in Chapter **7**. Looking at matters more closely, it will be clear that it is *not* actual formulas which occur in the axioms and rules above; rather it is objects which I might call *metaformulas* and which, in an instantiation, are replaced by actual formulas. The precise explanation of how to perform such an instantiation will be but a pedantic description of the use of language, as it was employed above (and which the reader no doubt will suppose to have understood without such effort). The algebra MF of metaformulas is absolutely freely generated by a set V of (meta) variables u,v,w,... under (a) the propositional operations, (b) 1-ary operations $\forall$p and $\exists$p, indexed by natural numbers p, and (c) 1-operations *sub*(p,t | -) and *sub*(η | -), depending on *term parameters* t and *mapping parameters* η. An instantiation g of MF in Fm(L) starts with three maps: (1) g_0 from V into fm(L), (2) g_1 from the numbers p into the set X of variables of T(L), and (3) g_2 from term variables and mapping variables into terms and mappings. Then g is computed by

homomorphically extending g_0 for propositional operations ,
setting $g(Qp\, w) = Qg_1(p)\, g(w)$ for quantifiers ,
setting $g(sub(p,t \mid w)) = sub(g_1(p), g_2(t) \mid g(w))$,
$g(sub(\eta \mid w)) = sub(g_2(\eta) \mid g(w))$.

Additional stipulations, such as those in (Ro0), (Ro1), always refer to the situation of the *result* of the instantiation; they thus *restrict* the possible instantiations admitted.

Turning to the consequence cs for satisfaction, I have shown in Corollary **1.15**.2 that cs(C) is the congruence class of ∇ under $[C\times\{\nabla\}]^{sat}$, the smallest Boolean congruence relation containing $C\times\{\nabla\}$ and $[0]^{inv}$. Again, the set $[C\times\{\nabla\}]^{sat}(\nabla)$ is

the smallest subset D of Fm(L) containing C and $[0]^{inv}(\nabla) = ct(0)$, which is of the form $R(\nabla)$ for a Boolean congruence relation R.

Since such congruence classes are generated by the propositional calculi cc, I define the calculus

ccqs to be the pure MP-calculus, obtained from cc by enlarging its set of axioms by all formulas in ct(0)

and obtain the

THEOREM 2_c The calculus ccqs is is both correct and complete for cs: $v \in cs(C)$ if, and only if, $v \in ccqs(C)$.

A node e of a deduction is said to *depend* on the maximal nodes above it and on the formulas carried by these nodes. In particular, e does *not depend on* C if the maximal nodes above e all carry instantiations of axioms. Given a (pure MP-) deduction of ccqs, I can implant the ccqt-deductions producing members of ct(0) at the maximal nodes carrying such. The resulting object may be characterized as a ccqt-deduction such that

(PS) The premisses of instances of quantifier rules do not depend on C.

Let ccqs' be the calculus obtained from cc by adding the axioms (ao0), (ao1) and the rules (Ro0), (Ro1) with the additional stipulation (PS) [but without the substitution rule (Rσ)]. Then also ccqs' is correct and complete for cs: $v \epsilon$cs(C) if, and only if, $v \epsilon$ccqs'(C). Because the empty set is invariant, and so it follows from Lemma **1.15**.4 that $[0]^{\text{inv}} = [0]^{\log}$ where $[-]^{\log}$ is generated without closure under (li). Hence the consequences ct(0) of the empty set can already be obtained with the calculus ccqt', obtained from ccqt by omitting (Rσ), which then in turn gives rise to ccqs'.

Making use of ct and cs, I thus conclude that $v \epsilon$ccqt(0) implies $v \epsilon$ccqt'(0) and that $v \epsilon$ccqs(C) implies $v \epsilon$ccqs'(C). Observe, however, that this is a semantical and inconstructive argument: I have *not* provided an algorithm transforming a ccqt-deduction from the empty set, or a ccqs-deduction from C, into one which avoids the use of (Rσ). The difficulties such algorithm would have to conquer were mentioned in the discussion following Lemma **1.15**.4.

As for equality logic and its consequence operations $ct_=$ and $cs_=$, I have shown in Theorem **1.12**.1 and its Corollary that $ct_=(C) = ct(C \cup EA)$ and $cs_=(C) = cs(C \cup EA_u)$ for suitable systems EA and EA_u of equality axioms. Consequently, I obtain correct and complete calculi $ccqt_=$ and $ccqs_=$ from ccqt and ccqs (and analogously from any of their variants to be discussed below) by defining $ccqt_=(C) = ccqt(C \cup EA)$ and $ccqs_=(C) = ccqs(C \cup EA_u)$. Observe that for $ccqs_=$ all the substitution instances in the system EA_0 of equality axioms are derivable from EA_u, making use of the availability of the logical axioms (ao0) in ct(0).

2. The Variant $cxqt_0$ of cxqt

Starting from a propositional MP-calculus cx, I define the calculus $cxqt_0$ by adding to the axioms of cx

(a00) $\forall x\, w \rightarrow rep(x,t \mid w)$ where $\zeta(x,t)$ is free for w
(a01) $rep(x,t \mid w) \rightarrow \exists x\, w$ where $\zeta(x,t)$ is free for w

and adding the quantifier rules

(R00) $u \rightarrow w \vdash u \rightarrow \forall x\, w$ provided x not in fr(u)
(R01) $w \rightarrow u \vdash \exists x\, w \rightarrow u$ provided x not in fr(u).

I shall show that cxqt and $cxqt_0$ are *deductively equivalent:* there are algorithms transforming cxqt-deductions into $cxqt_0$-deductions and vice versa.

Transforming a $cxqt_0$-deduction is easy. First, the axioms (a0i) are special cases of the (aoi). And an instantiation of (R00) can be replaced by

$u \to w$
$u \to sub(x,y \mid w)$ by $(R\sigma)$ if x is not in fr(u)
$u \to \forall x\, w$ by (R0) if y was chosen not in $var(u) \cup var(w)$.

Transforming a cxtq–deduction requires some preparatory work. Let $\simeq$ denote the relation of interdeducibility under $cxqt_0$: $u \simeq w$ if both $u \to w$ and $w \to u$ are in $cxqt_0(0)$. As cxqt contains a propositional calculus, $\simeq$ will be a propositional congruence relation. I now show the

LEMMA 1 (i) $\simeq$ is a congruence relation on Fm(L): if $u \simeq w$ then $\forall x\, u \simeq \forall x\, w$ and $\exists x\, u \simeq \exists x\, w$.

(ii) If y is not in fr(w) and $\zeta(x,y)$ is free for w then $\forall x w \simeq \forall y\, rep(x,y \mid w)$ and $\exists x w \simeq \exists y\, rep(x,y \mid w)$.

(iii) $v \simeq tot(Y_0, Y_1, \eta \mid v)$ for every η and every v .

Here (i) follows from

$\forall x\, u \to u \quad u \to w$ by (a00)
$\forall x u \to w$ by transitivity
$\forall x u \to \forall x w$ by (R00) .

(ii) follows from (i) if $x = y$. If $x \neq y$ then it follows from

$\forall x\, w \to rep(x,y \mid w)$ by (a00)
$\forall x\, w \to \forall y\, rep(x,y \mid w)$ by (R00)

$\forall y\, rep(x,y \mid w) \to rep(y,x \mid rep(x,y \mid w))$ by a(00)
$\forall y\, rep(x,y \mid w) \to w$
$\forall y\, rep(x,y \mid w) \to \forall y\, w$ by (R00)

where I use Lemma **1.13**.9.(iii) saying that, under the hypotheses stated, $\zeta(y,x)$ is free for $rep(x,y \mid w)$ and $w = rep(y,x \mid rep(x,y \mid w))$.

The proof of (iii) is by induction on v; I omit to write the sets Y_0, Y_1. The statement is trivial if v is atomic, and it remains preserved under propositional compositions since $\simeq$ is a congruence relation. Consider the case $v = Qx\, w$; if $tot(\eta \mid v) = Qx\, tot(\eta_x \mid w)$ then $w \simeq tot(\eta_x \mid w)$ by inductive hypothesis such that (i) can be applied. If $tot(\eta \mid v) = Qy\, rep(x,y \mid tot(\eta_x \mid w))$ then y is not in fr(w), hence not free in $tot(\eta_x \mid w)$, and $\zeta(x,y)$ is free for $tot(\eta_x \mid w)$. Thus

$w \simeq tot(\eta_x \mid w)$ by inductive hypothesis
$Qx w \simeq Qx\, tot(\eta_x \mid w)$ by (i)
$Qx\, tot(\eta_x \mid w) \simeq Qy\, rep(x,y \mid tot(\eta_x \mid w))$ by (ii)
$v \simeq Qx\, tot(\eta_x \mid w)$
$Qx\, tot(\eta_x \mid w) \simeq tot(\eta \mid v)$ both by definition
$v \simeq tot(\eta \mid v)$ by transitivity

and this shows how to build the required $cxqt_0$–deductions.

With Lemma 1 being available, I can deduce the axioms (aoi) in cxqt_0. They coincide with the (a0i) if $\zeta(x,t)$ is free for w; otherwise $\mathrm{sub}(x,t \mid w) = \mathrm{rep}(x,t \mid \mathrm{tot}(\zeta(x,t) \mid w))$ where $\zeta(x,t)$ is free for $\mathrm{tot}(\zeta(x,t) \mid w)$. Thus

$$\begin{array}{ll} w \to \mathrm{tot}(\zeta(x,t) \mid w) & \text{by (iii)} \\ \forall x w \to \forall x\, \mathrm{tot}(\zeta(x,t) \mid w) & \text{by (i)} \\ \forall x\, \mathrm{tot}(\zeta(x,t) \mid w) \to \mathrm{sub}(x,t \mid w) & \text{by (a00)} \\ \forall x w \to \mathrm{sub}(x,t \mid w) & \text{by transitivity .} \end{array}$$

In the same way, I can see that the rules (Ro0), (Ro1) are admissible in cxqt_0. Because if not $y \in \mathrm{var}(w)$ then $\mathrm{sub}(x,y \mid w) = \mathrm{rep}(x,y \mid w)$; thus if also not $y \in \mathrm{var}(u)$ then

$$\begin{array}{ll} u \to \mathrm{rep}(x,y \mid w) & \\ u \to \forall y\, \mathrm{rep}(x,y \mid w) & \text{by (R00)} \\ \forall y\, \mathrm{rep}(x,y \mid w) \to \forall x\, w & \text{by (ii)} \\ u \to \forall w & \text{by transitivity .} \end{array}$$

Before attacking the the admissibility of $(\mathrm{R}\sigma)$, I first observe the admissibility of the *generalization rule*

(R1) $\qquad w \vdash \forall x\, w$,

because if ∇ is a propositional tautology not containing x then

$$\begin{array}{ll} w & \\ \nabla \to w & \text{propositionally} \\ \nabla \to \forall x\, w & \text{by (R00)} \\ \forall x\, w & \text{propositionally .} \end{array}$$

Now the admissibility of $(\mathrm{R}\sigma)$ in ccqt_0 will follow semantically from the fact that $\mathrm{ct}(C) = \mathrm{ccqt}_0(C)$, and since I know that $\mathrm{ccqt}_0(C) \subseteq \mathrm{ccqt}(C) = \mathrm{ct}(C)$ it remains to be shown that $\mathrm{ccqt}(C) \subseteq \mathrm{ccqt}_0(C)$. Observe first that

$$\mathrm{ct}(C) = \mathrm{cs}((\forall)\, C)$$

follows from the definitions, and so it will suffice to show that $\mathrm{cs}((\forall)\, C) \subseteq \mathrm{ccqt}_0(C)$ or, in view of Theorem 2_c,

$$\mathrm{ccqs}((\forall)\, C) \subseteq \mathrm{ccqt}_0(C) \ .$$

Since $\mathrm{ccqs}((\forall)\, C) = \mathrm{ccqs}'((\forall)\, C)$ by the semantical argument from the previous section, this will follow from

$$\mathrm{ccqs}'((\forall)\, C) \subseteq \mathrm{ccqt}_0(C) \ .$$

Given the set C, successive applications of (R1) produce ccqt_0-deductions from C for every universal closure $(\forall)$ v of a formula v in C; hence $(\forall)\, C \subseteq \mathrm{ccqt}_0(C)$. But ccqs' does not employ $(\mathrm{R}\sigma)$, and so ccqt_0 admits all rules of ccqs'.

For the general case, a syntactical proof of the admissibility of $(\mathrm{R}\sigma)$ is the following. First, it will suffice to show the admissibility of

$$(\mathrm{Rr}\sigma) \qquad \mathrm{v} \vdash \mathrm{rep}(\eta \mid \mathrm{v}) \qquad \text{where } \eta \text{ is free for v}$$

since $\mathrm{sub}(\eta \mid \mathrm{v}) = \mathrm{rep}(\eta \mid \mathrm{tot}(\eta \mid \mathrm{v}))$ with η free for $\mathrm{tot}(\eta \mid \mathrm{v})$ and since the rule $\mathrm{v} \vdash \mathrm{tot}(\eta \mid \mathrm{v})$ is admissible by (iii). Further, the special case

$$(\mathrm{Rrs}\sigma) \qquad \mathrm{v} \vdash \mathrm{rep}(\mathrm{x,t} \mid \mathrm{v}) \qquad \text{where } \zeta(\mathrm{x,t}) \text{ is free for v}$$

is admissible by (R1), (a00) and modus ponens. It thus remains to reduce $(\mathrm{Rr}\sigma)$ to $(\mathrm{Rrs}\sigma)$ by showing that every *simultaneous* replacement $\mathrm{rep}(\eta \mid \mathrm{v})$ can be obtained by successive simple replacements $\mathrm{rep}(\mathrm{z,s} \mid -)$. That this is the case was shown in Lemma **1.13**.14 .

The modus ponens calculus ccqt_0 and its variants are often said to be of *Hilbert type*, and it seems that ccqt_0 made its first appearance in HILBERT 31. In HILBERT-ACKERMANN 49 it is stated explicitly that the formulation of ccqt_0 is due to Bernays.

The completeness Theorem 1_c (for ccqt_0 in place of ccqt) was proved first by GÖDEL 30 for countable and by HENKIN 49 for arbitrary languages; a highly readable presentation of these proofs is contained in CHURCH 56. Some references to other approaches to completeness theorems were mentioned in the conclusion of Chapter **1.15**.

References

A. Church: Introduction to Mathematical Logic. I . Princeton 1956

K. Gödel: Die Vollständigkeit der Axiome des logischen Funktionenkalkuls. Monatsheft.Math. Phys. **37** (1930) 349–360

L. Henkin: The completeness of the first order functional calculus. J.Symb.Logic **14** (1949) 159–166

D. Hilbert: Die Grundlegung der elementaren Zahlentheorie. Math.Ann. **104** (1931) 485–494

D. Hilbert and W. Ackermann: Grundzüge der Theoretischen Logik. 3te Auflage. Berlin 1949

3. The Variants cxqt_1 and cxqt_2 of cxqt

From now on I assume that cx is at least ce. I define the calculus cxqt_1 by enlarging the axioms of cxqt_0 to become

$$
\begin{array}{lll}
(\mathrm{a00}) & \forall \mathrm{x\, w} \to \mathrm{rep}(\mathrm{x,t} \mid \mathrm{w}) & \text{where } \zeta(\mathrm{x,t}) \text{ is free for w} \\
(\mathrm{a01}) & \mathrm{rep}(\mathrm{x,t} \mid \mathrm{w}) \to \exists \mathrm{x\, w} & \text{where } \zeta(\mathrm{x,t}) \text{ is free for w} \\
(\mathrm{a12}) & \forall \mathrm{x}\,(\mathrm{u} \to \mathrm{w}) \to (\mathrm{u} \to \forall \mathrm{x\, w}) & \text{if x is not in fr(u)} \\
(\mathrm{a13}) & \forall \mathrm{x}\,(\mathrm{w} \to \mathrm{u}) \to (\exists \mathrm{x\, w} \to \mathrm{u}) & \text{if x is not in fr(u)}
\end{array}
$$

and employing as rule *only* the generalization rule (R1) (and MP).

I shall show that cxqt_0 and cxqt_1 are deductively equivalent.

For the one direction, I know already that (R1) is admissible in $ccqt_0$. And (a12), (a13) have proofs there:

$\forall x\,(u \to w) \to (u \to w)$	by (a00)
$(\forall x\,(u \to w) \wedge u) \to w$	propositionally , cf. (a_3) in Th. **6**.3
$(\forall x\,(u \to w) \wedge u) \to \forall x\, w$	by (R00)
$\forall x\,(u \to w) \to (u \to \forall x\, w)$	propositionally as above
$\forall x\,(w \to u) \to (w \to u)$	by (a00)
$w \to (\forall x\,(w \to u) \to u)$	propositionally
$\exists x\, w \to (\forall x\,(w \to u) \to u)$	by (R01)
$\forall x\,(w \to u) \to (\exists x\, w \to u)$	propositionally .

As for the other direction, (R00) and (R01) are admissible in $cxqt_1$.Because if I generalize their hypotheses by (R1), I obtain the left parts of the implications (a12), (a13), from which MP then will deduce their conclusions.

I define the calculus $cxqt_2$ by enlarging the the axioms of $cxqt_0$ to become

(a00)	$\forall x\, w \to rep(x,t \mid w)$	where $\zeta(x,t)$ is free for w
(a01)	$rep(x,t \mid w) \to \exists x\, w$	where $\zeta(x,t)$ is free for w
(a14)	$\forall x\,(v \to w) \to (\forall x\, v \to \forall x\, w)$	
(a15)	$\forall x\,(v \to w) \to (\exists x\, v \to \exists x\, w)$	
(a16)	$u \to \forall x\, u$	if x is not in fr(u)
(a17)	$\exists x\, u \to u$	if x is not in fr(u)

and employing as rule *only* the generalization rule (R1) (and MP).

I shall show that $cxqt_1$ and $cxqt_2$ are deductively equivalent.

In $cxqt_1$, I deduce (a16) from (a12) and the deducible formulas $u \to u$ and $\forall x\,(u \to u)$; in the same way, I deduce (17) from (a13). The formulas (a14), (a15) have the $cxqt_1$-deductions

$\forall x\,(v \to w) \to (v \to w)$	by (a00)
$(v \to w) \to (\forall x\, v \to w)$	propositionally from (a00)
$\forall x\,(v \to w) \to (\forall x\, v \to w)$	by transitivity from the above
$\forall x\,(v \to w) \to \forall x\,(\forall x\, v \to w)$	by (R00)
$\forall x\,(\forall x\, v \to w) \to (\forall x\, v \to \forall x\, w)$	by (a12)
$\forall x\,(v \to w) \to (\forall x\, v \to \forall x\, w)$	by transitivity from the above
$\forall x\,(v \to w) \to (v \to w)$	by (a00)
$(v \to w) \to (v \to \exists x\, w)$	propositionally from (a01)
$\forall x\,(v \to w) \to (v \to \exists x\, w)$	by transitivity from the above
$v \to (\forall x\,(v \to w) \to \exists x\, w)$	propositionally
$\exists x\, v \to (\forall x\,(v \to w) \to \exists x\, w)$	by (R01)
$\forall x\,(v \to w) \to (\exists x\, v \to \exists x\, w)$	propositionally .

On the other hand, I can deduce (a12) and (a13) in $cxqt_2$:

$u \to \forall x\, u$, x not in fr(u)	by (a16)
$(\forall x\, u \to \forall x\, w) \to (u \to \forall x\, w)$	propositionally from the above

$\forall x\,(u \to w) \to (\forall x\,u \to \forall x\,w)$	by (a14)
$\forall x\,(u \to w) \to (u \to \forall x\,w)$	by transitivity
$\exists x\,u \to u$, x not in fr(u)	by (a17)
$(\exists x\,w \to \exists x\,u) \to (\exists x\,w \to u)$	propositionally from the above
$\forall x\,(w \to u) \to (\exists x\,w \to \exists x\,u)$	by (a15)
$\forall x\,(w \to u) \to (\exists x\,w \to u)$	by transitivity .

4. The Calculi $cxqs_i$

Each of the calculi $cxqt_i$ gives rise to a variant $cxqs_i$ of cxqs, and all these variants are deductively equivalent. The calculi $cxqs_i$ may be described by saying that either

$cxqs_i$ has as axioms the formulas $cxqt_i$-deducible from the empty set, and has MP as its *only* rule , or

$cxqs_i$ has the same axioms and rules as has $cxqt_i$, but in a deduction from C the quantifier rules may be applied only to premisses which do not depend on C.

For non classical cx, however, the calculi $cxqs_i$ are, up to this moment, devoid of any semantical meaning. Of course, the definition of the consequence operation cs for 2-valued structures immediately generalizes to that of a consequence operation $cs^{\mathbf{X}}$ for **X**-valued structures, but the characterizations of cs in Theorem **1.15**.4 and its Corollaries do not work even for $cs^{\mathbf{B}}$ [the reason being that a prestructure defined from Fmp(L)/R will be a structure only if R is logical, and already in Lemma **1.15**.12 the relation S(C) may not be logical at all]. Semantically, however, the connection $ct(C) = cs((\forall)\,C)$ immediately generalizes to

$$ct^{\mathbf{X}}(C) = cs^{\mathbf{X}}((\forall)\,C)$$

for arbitrary classes **X**. If cx is at least ce then I can show the

LEMMA 2 If C consists of sentences then $cxqt_i(C) = cxqs_i(C)$.

Here $B = cxqs_i(C)$ is trivially contained in $cxqt_i(C)$. Conversely, let D be a $cxqt_1$-deduction of a formula u from C; I shall prove by induction on the length of D that B contains u *and* all formulas, arising under iterated applications of (R1) from u. If u is a sentence then $u \to \forall x\,u$ is in $cxqt_i(0)$ by (R00); so my claim is trivial if D has length 0 since in that case u is in C or on $csqt_i(0)$. Assume next that $u \mathrel{\dot=} \forall x\,w$ arises in D from w under (R1); since w has a shorter deduction than u, it follows from the inductive hypothesis that also $\forall x_0 \forall x_1 \ldots \forall x_n u = \forall x_0 \forall x_1 \ldots \forall x_n \forall x\,w$ is in B. Consider now the case that u arises from v and $v \to u$ under (MP); by inductive hypothesis v and $v \to u$ are in B and so will be u. Now (a14) is the case $n = 0$ of

$$(a14_n) \quad \forall x_0 \forall x_1 \ldots \forall x_{n-1}(v \to w) \to (\forall x_0 \forall x_1 \ldots \forall x_{n-1} v \to \forall x_0 \forall x_1 \ldots \forall x_{n-1} w),$$

and this implies $(a14_{n+1})$ since

$$\forall x_0 \forall x_1 \ldots \forall x_{n-1} \forall x_n (v \to w) \to \forall x_0 \forall x_1 \ldots \forall x_{n-1}(\forall x_n v \to \forall x_n w)$$
$$\forall x_0 \forall x_1 \ldots \forall x_{n-1}(\forall x_n v \to \forall x_n w) \to (\forall x_0 \forall x_1 \ldots \forall x_n v \to \forall x_0 \forall x_1 \ldots \forall x_n w)$$

by (a14) and $(a14_n)$. By inductive hypothesis, both $\forall x_0 \forall x_1 \ldots \forall x_n v$ and $\forall x_0 \forall x_1 \ldots \forall x_n (v \to u)$ are in B, and so also $\forall x_0 \forall x_1 \ldots \forall x_n u$ will be in B by $(a14_n)$.

In this way then, the calculi $cxqs_i$ *do* acquire a semantical meaning if cx is at least ce : $cxqs_i((\forall) C) = cs^{\mathbf{X}}((\forall) C)$.

The calculi $cxqs_i$, being describable as pure MP calculi, are deductive in the sense of Chapter 7. I now shall show that they admit certain metarules for quantifiers. I denote by $\vdash$ the deducibility relation for the $cxqt_i$ and by $\vdash_s$ that for the $cxqs_i$.

LEMMA 3 There is a linear operator, transforming a $cxqs_i$-deduction D of $C \vdash_s rep(x,z \mid w)$ into one of $C \vdash_s \forall x\, w$, provided that z is not free in $C \cup \{w\}$ and $\zeta(x,z)$ is free for w.

There is a linear operator, transforming a $cxqs_i$-deduction D of $C, rep(x,z \mid w) \vdash_s u$ into one of $C, \exists x\, w \vdash_s u$, provided that z is not free in $C \cup \{w,u\}$ and that $\zeta(x,z)$ is free for w.

Since D can involve only finitely many members of C, I may assume that C is finite. As the $cxqs_i$ are pure MP-calculi, the deductivity operator from Lemma 7.2 is available. For sequences of formulas I use the notation

$$<\gamma(i) \mid i<1> \to^* v = \gamma(0) \to v ,$$
$$<\gamma(i) \mid i<n+1> \to^* v = \gamma(0) \to (<\gamma(i) \mid 0<i<n> \to^* v)$$

employed for frames already in Chapter 6. Thus deductivity leads from $\{\gamma(i) \mid i<n\} \vdash_s v$ to $\vdash <\gamma(i) \mid i<n> \to^* v$. Further, induction shows that, if x is not free in any formula of $<\gamma(i) \mid i<n>$, then (a12) generalizes to

$$(a12^*) \quad \forall x (<\gamma(i) \mid i<n> \to^* v) \to <\gamma(i) \mid i<n> \to^* \forall x\, v .$$

I also need two deductions generalizing (R00), (R01):

$$(R00^*) \quad <\gamma(i) \mid i<n> \to^* rep(x,z \mid w) \vdash <\gamma(i) \mid i<n> \to^* \forall x\, w$$
$$(R01^*) \quad <\gamma(i) \mid i<n> \to^* (rep(x,z \mid w) \to u) \vdash <\gamma(i) \mid i<n> \to^* (\exists x\, w \to u)$$

if z is not free in w or in any formula of $<\gamma(i) \mid i<n>$ and if $\zeta(x,z)$ is free for w. Because in the first case I deduce

$\forall z\,(<\gamma(i)\mid i<n> \dashv^* \text{rep}(x,z\mid w))$ by (R1)
$<\gamma(i)\mid i<n> \dashv^* \forall z\,\text{rep}(x,z\mid w))$ by (a12*)
$\forall z\,\text{rep}(x,z\mid w) \dashv \forall x\,w$ by Lemma 1.(ii)
$<\gamma(i)\mid i<n> \dashv^* \forall z\,\text{rep}(x,z\mid w)) \dashv <\gamma(i)\mid i<n> \dashv^* \forall x\,w$ propositionally
$<\gamma(i)\mid i<n> \dashv^* \forall x\,w$ by transitivity

and in the second

$\forall z\,(<\gamma(i)\mid i<n> \dashv^* (\text{rep}(x,z\mid w) \dashv u))$ by (R1)
$<\gamma(i)\mid i<n> \dashv^* \forall z\,(\text{rep}(x,z\mid w) \dashv u)$ by (a12*)
$\forall z\,(\text{rep}(x,z\mid w) \dashv u) \dashv (\exists z\,\text{rep}(x,z\mid w) \dashv u)$ by (a13)
$\exists x\,w \dashv \exists z\,\text{rep}(x,z\mid w)$ by Lemma 1.(ii)
$(\exists z\,\text{rep}(x,z\mid w) \dashv u) \dashv (\exists x\,w \dashv u)$ propositionally
$\forall z\,(\text{rep}(x,z\mid w) \dashv u) \dashv (\exists x\,w \dashv u)$ by transitivity
$<\gamma(i)\mid i<n> \dashv^* (\exists x\,w \dashv u)$ by transitivity **6**.(f4*).

Thus if $C = \{\gamma(i)\mid i<n\}$ then $C \vdash_s \text{rep}(x,z\mid w)$ gives $\vdash <\gamma(i)\mid i<n> \dashv^* \text{rep}(x,z\mid w)$ by deductivity and $\vdash <\gamma(i)\mid i<n> \dashv^* \forall x\,w$ by (R00*) whence $C \vdash_s \forall x\,w$ by MP. And $C, \text{rep}(x,z\mid w) \vdash_s u$ gives $\vdash <\gamma(i)\mid i<n> \dashv^* (\text{rep}(x,z\mid w) \dashv u)$ by deductivity and $\vdash <\gamma(i)\mid i<n> \dashv^* (\exists x\,w \dashv u)$ by (R01*) whence $C, \exists x\,w \vdash u$ by MP.

These operators are linear in the sense that the length of the transformed derivation is bounded by a linear function of *length*(D) with the number w(C) of elements in C as a multiplicative constant. In order to see this, recall the proof of Theorem **7**.2_d and consider first the calculus obtained from $cxqs_i$ by adding the deduction rule. Then the length of D is enlarged by the lengths of the derivations (R00*), (R01*) which depend on w(C), and removing the deduction rule then will at most double this bound.

The Lemma remains in effect if, in the hypotheses, I replace the variable z by a constant not occurring in $C\cup\{w\}$ or $C\cup\{w,u\}$ respectively. In order to see this, suffices to observe that a $cxqs_i$–deduction S of $C \vdash_s \text{rep}(x,c\mid w)$, say, becomes a deduction $\text{rep}(c,z\mid S)$ of $C \vdash_s \text{rep}(x,z\mid w)$ if all of its formulas are transformed by $\text{rep}(c,z\mid -)$ where z is chosen sufficiently new. But the transformation $\text{rep}(c,z\mid -)$ preserves propositional axioms as well as instantiations of MP. The axioms (a12)–(a17) as well as the instances of (R00), (R01) and (R1) are preserved if z is distinct from the variable x quantified in them. The axioms (a00), (a01) are preserved if z is not in w, since then $\text{rep}(c,z\mid \text{rep}(x,t\mid w)) = \text{rep}(x,h_{\varsigma(c,z)}(t)\mid \text{rep}(c,z\mid w))$ by Lemma **1**.**13**.4.(i) where $h_{\varsigma(c,z)}(t)$ is free for $\text{rep}(c,z\mid w)$. Finally, the same Lemma shows that $\text{rep}(c,z\mid \text{rep}(x,c\mid w)) = \text{rep}(x,z\mid w)$ with $\zeta(x,z)$ free for w.

There are two more, obvious metarules for the calculi $cxqs_i$, namely

There is an operator, transforming a deduction of $C \vdash_s \text{rep}(x,t\mid w)$ into one of $C \vdash_s \exists x\,w$, provided that $\zeta(x,t)$ is free for w.

There is an operator, transforming a deduction of C, rep(x,t | w) $\vdash_s$ u into one of C, $\forall$x w $\vdash_s$ u , provided that ζ(x,t) is free for w.

In the first case I only need to proceed with the axiom (a01), and in the second case I use the axiom (a00) in order to deduce rep(x,t | v) from $\forall$x w .

5. The Deduction Theorem and Other Metarules for The Calculi $ccqt_i$

Being pure MP-calculi, the $cxqs_i$ were deductive. Making use of semantical arguments, the calculi $ccqt_i$ *cannot* be deductive, because they are correct and complete for ct, and the consequence operation ct is *not* deductive: there are sets C of formulas and formulas v,w such that

w ϵ ct(C$\cup$\{v\}) and *not* v $\rightarrow$ w ϵ ct(C) .

For instance, let C be an axiom system for rings with a neutral element 1, denoted by a constant *1*. Let v be $\exists$y (x = y+y) and let w be $\exists$y (*1* = y+y) . Clearly, if in a ring every a is of the form 2b then so is 1. But in the ring of integers the formula v$\rightarrow$w is not satisfied if x is evaluated by an even integer.

Under additional hypotheses, though, a deduction property can be shown to hold as a metarule for the $cxqt_i$.

In an instantiation of (R00), (R01), (R1) I call *eigenvariable* the variable which is quantified in that instantiation (named x in the above formulations of these rules). If S is a $cxqt_i$-deduction then E(S) shall be the set of all eigenvariables of instantiations performed in S.

Let S be deduction *from* a set C, let c be in C and x be in E(S). I shall say that x is *relevant for* c if x is eigenvariable of an instantiation with a premiss depending on c.

A deduction S from a set C$\cup$\{v\} is called *deductive for* v if no variable free in v is relevant for v. There holds the *deduction principle* for $cxqt_i$:

(QD) There is an operator, transforming deductions S : C, v $\vdash$ w , deductive for v, into deductions S' : C $\vdash$ v $\rightarrow$ w such that E(S) = E(S') .

I denote, for every node e of S, by S(e) the formula carried by e. As I did in the propositional case in Chapter 7, I construct, by recursion in the tree of S, for every node e a deduction S_e such that

the subtree of S ending at e is embedded into the tree of S_e, and if h_0 denotes the image of a node h with h $\geq$ e from S in S_e then e_0 is the smallest node of S_e, the maximal nodes of S_e carry either axioms or elements of C,

if h is a node of S with $h \geq e$ then $S_e(h_o) = v \to S(h)$.

Then S_m will be a deduction establishing $C \vdash v \to w$.

At the maximal nodes of S and at instances of propositional rules I proceed as I did in the propositional case. It remains to consider the instances of (R00), (R01), (R1). Let S(e) be the conclusion of such a rule and let f be the upper neighbour of e. If S(e) (and hence S(f)) does not depend on v then I take the subdeduction of S leading to S(e), put the axiom $S(c) \to (v \to S(e))$ at its side and obtain S_e with the result $v \to S(e)$ by MP. If S(e) depends on v then I define S_e by continuing the S_f which I do have by the hypothesis of my recursion:

S(f): $u \to r$	$v \to (u \to r)$	endsequent of S_f
S(e): $u \to \forall x\, r$	$v \to \forall x\, (u \to r)$	by (R00) since x not in fr(v)
	$\forall x\, (u \to r) \to (u \to \forall x\, r)$	by (a12) since x not in fr(u)
	$v \to (u \to \forall x\, r)$	by transitivity

S(f): $u \to r$	$v \to (u \to r)$	endsequent of S_f
S(e): $\exists x\, u \to r$	$u \to (v \to r)$	propositionally
	$\exists x\, u \to (v \to r)$	by (R01) as x is not in fr(v)$\cup$fr(r)
	$v \to (\exists x\, u \to r)$	propositionally

S(f): r	$v \to r$	endsequent of S_f
S(e): $\forall x\, r$	$v \to \forall x\, r$	by (R00) since x not in fr(r) .

It follows from these definitions of S_e that, indeed, the set of eigenvariables is not changed (since also in the last case the replacement of (R00) by (R1) (i.e. its admissibility for $cxqt_1$) uses the same eigenvariable x).

Once (QD) has become available, various more metarules can be stated with its help. For instance,

> If S: $C, u \vdash w$ and S': $C, u' \vdash w$ are deductive for u and u' respectively, then a deduction S": $C, u \vee u' \vdash w$ can be found such that E(S") = E(S)$\cup$E(S')

follows by (QD) from the propositional tautology $(u \to w) \to ((u' \to w) \to (u \vee u' \to w))$. Observe, though, that if $C,c,v \vdash w$ is deductive for v then $C,c \vdash v \to w$ is not necessarily deductive for c; hence repeated applications of (QD), as they were used in the proof of Lemma 3, may not be possible. Still, there holds

> If S: $C, rep(x,z \mid v) \vdash w$ is deductive for $rep(x,z \mid v)$, if z is not in fr(w) or fr(v), and if $\zeta(x,z)$ is free for v, then a deduction S': $C, \exists x\, v \vdash w$ can be found such that E(S') = E(S)$\cup$\{x\} .

Because (QD) gives a deduction of $C \vdash rep(x,z \mid v) \to w$ which by (R01) extends to one of

$$C \vdash \exists z\, rep(x,z \mid v) \to w$$

and this extends by Lemma 1.(ii) to a deduction of $C \vdash \exists x\, v \to w$.

As another illustration I deduce, once more, in $cxqt_1$ the formulas (a14), (a15), but this time making use of (QD):

$\forall x\,(v\to w), \forall x\, v \vdash v\to w, v$	by (a00)
$\forall x\,(v\to w), \forall x\, v \vdash w$	by MP [relevant for $\forall x\, v$
$\forall x\,(v\to w), \forall x\, v \vdash \forall x\, w$	by (R1), x eigenvariable,
$\forall x\,(v\to w) \vdash \forall x\, v \to \forall x\, w$	by (QD), x relevant
$\vdash \forall x\,(v\to w) \to (\forall x\, v \to \forall x\, w)$	by (QD) [for $\forall x\,(v\to w)$

$\forall x\,(v\to w), v \vdash v\to w, v$	by (a00) and MP
$\forall x\,(v\to w), v \vdash w$	by MP
$\forall x\,(v\to w), v \vdash \exists x\, w$	(a01) and MP, no eigen-
$\forall x\,(v\to w) \vdash v \to \exists x\, w$	by (QD) [variables
$\forall x\,(v\to w) \vdash \exists x\, v \to \exists x\, w$	by (R01), x eigenvariable, relevant for $\forall x\,(v\to w)$
$\vdash \forall x\,(v\to w) \to (\exists x\, v \to \exists x\, w)$	by (QD)

The comparition with the earlier deductions shows that, as in the propositional case, the employment of (QD) avoids the use of various propositional tautologies which, sometimes, may have appeared artificial. On the other hand, it requires the careful bookkeeping of eigenvariables.

6. Tautologies of Positive Quantifier Logic

The formulas listed below are deducible from the empty set in the calculi $cgqt_i$, i.e. making use of only the positive part of propositional logic. As it did before, $\simeq$ shall denote the relation of interdeducibility.

(A1)	$\forall x\,(u\to w) \simeq u \to \forall x\, w$	if x is not in fr(u)
(A2)	$\forall x\,(w\to u) \simeq \exists x\, w \to u$	if x is not in fr(u)
(B1)	$\exists x\,(u\to w) \to (u \to \exists x\, w)$	if x is not in fr(u)
(B2)	$\exists x\,(w\to u) \to (\forall x\, w \to u)$	if x is not in fr(u)
(A3)	$\exists x\,(u \wedge w) \simeq u \wedge \exists x\, w$	if x is not in fr(u)
(B3)	$(u \vee \forall x\, w) \to \forall x\,(u \vee w)$	if x is not in fr(u)
(A4)	$\forall x\,(u \wedge w) \simeq \forall x\, u \wedge \forall x\, w$	
(A5)	$\exists x\,(u \vee w) \simeq \exists x\, u \vee \exists x\, w$	
(B4)	$\exists x\,(u \wedge v) \to (\exists x\, u \wedge \exists x\, w)$	
(B5)	$\forall x\, u \vee \forall x\, w \to \forall x\,(u \vee w)$.	

In (A1) and (A2), the implications to the right are (a12), (a13). I then continue as follows :

(A1b) $(u \to \forall x\, w) \to (u \to w)$ by (a00)
$(u \to \forall x\, w) \to \forall x\, (u \to w)$ by (R00)

(A2b) $(\exists x\, w \to u) \to (w \to u)$ from (a01) propositionally
$(\exists x\, w \to u) \to \forall x\, (w \to u)$ by (R00)

(B1) $(u \to w) \to (u \to \exists x\, w)$ from (a01)
$\exists x\, (u \to w) \to (u \to \exists x\, w)$ by (R01)

(B2) $(w \to u) \to (\forall x\, w \to u)$ from (a00)
$\exists x\, (w \to u) \to (\forall x\, w \to u)$ by (R01)

(A3a) $(u \wedge w) \to (u \wedge \exists x\, w)$ from (a01)
$\exists x\, (u \wedge w) \to (u \wedge \exists x\, w)$ by (R01)

(A3b) $(u \wedge w) \to \exists x\, (u \wedge w)$ by (a01)
$w \to (u \to \exists x\, (u \wedge w))$ propositionally
$\exists x\, w \to (u \to \exists x\, (u \wedge w)$ by (R01)
$u \wedge \exists x\, w \to \exists x\, (u \wedge w)$ propositionally

(B2) $(u \vee \forall x\, w) \to (u \vee w)$ from (a00)
$(u \vee \forall x\, w) \to \forall x\, (u \vee w)$ by (R00)

(A4a) $\forall x\, (u \wedge w) \to u$ from (a00)
$\forall x\, (u \wedge w) \to \forall x\, u$ by (R00)
$\forall x\, (u \wedge w) \to \forall x\, w$ analogously
$\forall x\, (u \wedge w) \to \forall x\, u \wedge \forall x\, w$ propositionally

(A4b) $(\forall x\, u \wedge \forall x\, w) \to u \wedge w$ by (a00)
$(\forall x\, u \wedge \forall x\, w) \to \forall x\, (u \wedge w)$ by (R00)

(A5a) $u \to \exists x\, u$ from (a01)
$u \to \exists x\, u \vee \exists x\, w$ propositionally
$w \to \exists x\, u \vee \exists x\, w$ analogously
$(u \vee w) \to \exists x\, u \vee \exists x\, w$ propositionally
$\exists x\, (u \vee w) \to \exists x\, u \vee \exists x\, w$ by (R01)

(A5b) $u \to (u \vee w)$ propositionally
$\forall x\, (u \to (u \vee w))$ by (R1)
$\exists x\, u \to \exists x\, (u \vee w)$ by (a15) and MP
$\exists x\, w \to \exists x\, (u \vee w)$ analogously
$\exists x\, u \vee \exists x\, w \to \exists x\, (u \vee w)$ propositionally

(B4) $u \wedge w \to \exists x\, u$ from (a01)
$u \wedge w \to \exists x\, w$ from (a01)
$u \wedge w \to (\exists x\, u \wedge \exists x\, w)$ propositionally
$\exists x\, (u \wedge v) \to (\exists x\, u \wedge \exists x\, w)$ by (R01)

(B5)	$\forall x\, u \to (u \vee w)$	from (a00)
	$\forall x\, w \to (u \vee w)$	analogously
	$\forall x\, u \vee \forall x\, w \to (u \vee w)$	propositionally
	$\forall x\, u \vee \forall x\, w \to \forall x\,(u \vee w)$	by (R01) .

7. Tautologies of Minimal Quantifier Logic

I recall from Chapter **7** that the three formulas

(dm1) $v \to \neg\neg v$
(dm2) $(v \to w) \to (\neg w \to \neg v)$
(dm5) $(v \to \neg w) \to ((v \to w) \to \neg v)$

can be derived in the minimal propositional calculus. Making use of (QD) and (dm5), I obtain the metarule of *reductio ad absurdum*:

> If S: $C, v \vdash w$ and S' $C, v \vdash \neg w$ both are deductive for v, then a deduction S": $C \vdash \neg v$ can be found such that E(S") = E(S)∪E(S').

The statements listed below are deducible from the empty set in the calculi $cmqt_i$ in which the propositional part is restricted to minimal propositional logic. I shall use an abbreviating notation: $\exists \to \neg\forall\neg$, say, expresses that for every variable x and for every formula v there is a deduction of the formula $\exists x\, v \to \neg\forall x\, \neg v$, and $\forall\neg \simeq \neg\exists$ expresses that for every x and every v the formulas $\forall x\, \neg v$, $\neg\exists x\, v$ are interdeducible.

(M1) $\exists \to \exists\neg\neg \to \neg\neg\exists \simeq \neg\neg\exists\neg\neg \simeq \neg\forall\neg$

(M2) $\forall \to \neg\neg\forall \to \forall\neg\neg \simeq \neg\neg\forall\neg\neg \simeq \neg\exists\neg$

(M3) $\forall\neg \simeq \neg\exists \simeq \neg\neg\forall\neg \simeq \neg\exists\neg\neg$

(M4) $\exists\neg \to \neg\neg\exists\neg \simeq \neg\forall\neg\neg \to \neg\forall$

I begin the proof by deducing from (a00) and (a01) with help of (dm2) the properties

0. $\neg \to \neg\forall$ 1. $\neg\exists \to \neg$.

Applying (R01) and (R00), there follow

2. $\exists\neg \to \neg\forall$ 3. $\neg\exists \to \forall\neg$.

From (a00), applied to $\neg v$, there follows $v \to \neg\neg v \to \neg\forall x\, \neg v$, hence with (R01) also

4. $\exists \to \neg\forall\neg$.

Applying (R1) to (dm1), I obtain $\forall x\,(v \to \neg\neg v)$, from where by (a14) and (a15) also

5. $\forall \to \forall\neg\neg$ 6. $\exists \to \exists\neg\neg$.

The rest now is pure combinatorics, applying (dm2), (dm1) and their consequence $\neg\neg\neg \simeq \neg$:

7.	$\exists\neg \rightarrow \neg\forall\neg\neg$	by 4
8.	$\neg\neg\forall\neg\neg \rightarrow \neg\exists\neg$	by 1, (dm2)
9.	$\forall\neg\neg \rightarrow \neg\neg\forall\neg\neg$	by (dm1)
10.	$\forall\neg\neg \rightarrow \neg\exists\neg$	by 9, 8
11.	$\forall \rightarrow \neg\exists\neg$	by 5, 10
12.	$\forall\neg \rightarrow \neg\exists\neg\neg$	by 11
13.	$\neg\exists\neg\neg \rightarrow \neg\exists$	by 6, (dm2)
14.	$\forall\neg \rightarrow \neg\exists$	by 12, 13
15.	$\forall\neg \simeq \neg\exists$	by 3, 14
16.	$\forall\neg\neg \simeq \neg\exists\neg$	by 15
17.	$\forall\neg\neg \simeq \neg\neg\forall\neg\neg \simeq \neg\exists\neg$	by 9, 8, 16
18.	$\neg\neg\forall \rightarrow \neg\neg\forall\neg\neg$	by 5, (dm2) twice
19.	$\neg\forall\neg \simeq \neg\neg\exists$	by 15
20.	$\neg\forall\neg \simeq \neg\neg\exists\neg\neg$	by 19 and $\neg \simeq \neg\neg\neg$
21.	$\exists\neg\neg \rightarrow \neg\neg\exists\neg\neg$	by 6, (dm2) twice
22.	$\neg\exists \simeq \neg\neg\forall\neg$	by 14 and $\neg \simeq \neg\neg\neg$
23.	$\neg\exists \simeq \neg\exists\neg\neg \simeq \neg\forall$	by 12, 13, 15
24.	$\exists\neg \rightarrow \neg\neg\exists\neg$	by (dm1)
25.	$\neg\forall\neg\neg \simeq \neg\neg\exists\neg$	by 17
26.	$\neg\forall\neg\neg \rightarrow \neg\forall$	by 5, (dm2) .

Now (M2) follows from 18, 17; (M1) follows from 6, 21, 20; (M3) follows from 16, 22, 23; (M4) follows from 24, 25, 26.

8. Tautologies of Classical Quantifier Logic

Classically, $\neg\neg v$ is equivalent to v. Thus (M1)–(M4) simplify to

$$\exists \simeq \neg\forall\neg \qquad \forall \simeq \neg\exists\neg \qquad \exists\neg \simeq \neg\forall \qquad \forall\neg \simeq \neg\exists \ .$$

But the consequences of classicity reach farther: the implications (B1), (B2), (B3) become equivalences:

(C1) $\exists x\,(u \rightarrow w) \simeq u \rightarrow \exists x\, w$ if x is not in fr(u)

(C2) $\exists x\,(w \rightarrow u) \simeq \forall x\, w \rightarrow u$ if x is not in fr(u)

(C3) $(u \vee \forall x\, w) \simeq \forall x\,(u \vee w)$ if x is not in fr(u) .

Observe first that, for x not in fr(u), $u \simeq \exists x\, u$ follows from $u \rightarrow u$ by (R01). Hence (A5) becomes $\exists x\,(u \vee w) \simeq u \vee \exists x\, w$. Replacing $u \rightarrow v$ by $\neg u \vee v$, I thus obtain (C1). As for (C2) and (C3), I find

$$\begin{aligned}
\forall x\, w \to u \;&\simeq\; \neg u \to \neg\forall x\, w\\
&\simeq\; \neg u \to \exists x\, \neg w\\
&\simeq\; \exists x\,(\neg u \to \neg w) \quad \text{by (C1)}\\
&\simeq\; \exists x\,(w \to u)
\end{aligned}$$

$$\begin{aligned}
\neg\forall x\,(u \vee w) \;&\simeq\; \exists x\, \neg(u \vee w)\\
&\simeq\; \exists x\,(\neg u \wedge \neg w)\\
&\simeq\; \neg u \wedge \exists x\, \neg w \quad \text{by (A3)}\\
&\simeq\; \neg(u \vee \neg\exists x\, \neg w)\\
&\simeq\; \neg(u \vee \forall x\, w)
\end{aligned}$$

whence $\forall x\,(u \vee w) \;\simeq\; \neg\neg\forall x\,(u \vee w) \;\simeq\; \neg\neg(u \vee \forall x\, w) \;\simeq\; u \vee \forall x\, w$.

I shall show in Chapter **10** that the various implications proved here for classical logic, but not proved for minimal or intuitionistic logic, *cannot* be deduced in intuitionistic logic.

It now can also be stated that, for the classical calculi, the relation of interdeducibility is *prenex* in the sense of Chapter **1.14**. That $\simeq$ is a congruence relation also for quantifiers follows from Lemma 1 (i), and of the defining properties of prenex relations the first, (pn0), follows from Lemma 1 (ii). Property (pn1) follows from the classically strengthened (M3), (M4); (pn2), (pn3), (pn4) follow for $Q = \forall$ from (A1), (C2) and (A4) for $\wedge$, (C3) for $\vee$, and they follow for $Q = \exists$ from (C1), (A2) and (A3) for $\wedge$, (A5) for $\vee$. Consequently, every formula is classically interdeducible with its prenex normal form.

The classical relations between the quantifiers $\forall$ and $\exists$ offer the possibility to set up calculi which use only one of them in their structural axioms and rules, and to introduce then the other by a suitable relationship taken as its definition. In order to choose such definitions, observe first that already propositionally the the four properties

(k1) $\neg\forall\neg \to \exists$ (k2) $\neg\exists \to \forall\neg$ (k3) $\neg\exists\neg \to \forall$ (k4) $\neg\forall \to \exists\neg$

are equivalent, since contraposition leads from (k1) to (k2) and from (k3) to (k4), and inserting negations leads from (k2) to (k3) and from (k4) to (k1). In the same way, the four properties

(k5) $\exists \to \neg\forall\neg$ (k6) $\forall\neg \to \neg\exists$ (k7) $\forall \to \neg\exists\neg$ (k8) $\exists\neg \to \neg\forall$

are propositionally equivalent.

Let me now choose $\forall$ as the quantifier to be described structurally, and $\exists$ as the one to be defined. I start from a pure MP-calculus cc for classical propositional logic and define a *basic* $\forall$-calculus with the axiom (a00) and the rule (R00).

Taking a property from the first group, I can derive (a01) in the basic calculus, because the (a00)-instance $\forall x\, \neg w \to rep(x,t \mid \neg w)$, preceded by (k2), leads to $\neg\exists x\, w \to \neg rep(x,t \mid w)$ from where (a01) by contraposition. Taking a property from the second group, I see that (R01) is admissible in the basic calculus, because if x is not in fr(u) then

$$
\begin{array}{rcll}
w \rightarrow u & \vdash & \neg u \rightarrow \neg w & \text{propositionally} \\
\neg u \rightarrow \neg w & \vdash & \neg u \rightarrow \forall x \neg w & \text{by (R00)} \\
\neg u \rightarrow \forall x \neg w & \vdash & \neg \forall x \neg w \rightarrow u & \text{propositionally} \\
\neg u \rightarrow \forall x \neg w & \vdash & \exists x\, w \rightarrow u & \text{by (k5)}\ .
\end{array}
$$

There are, consequently, 16 possible definitions for the quantifier $\exists$ over the basic $\forall$-calculus, obtained by adjoining one of the (ki) with $i<5$ together with one of the (kj) with $j>4$.

Chapter 10. Selected Topics in Sequential Quantitifier Logic

1. Translating Between Sequential and Modus Ponens Calculi

For propositional logic, I have translated, in the several variants of Theorem 7.2, between derivations in sequential calculi and deductions in modus ponens calculi. Let now XQ be the sequential quantifier calculus based upon one of these propositional calculi, which shall at least be positive, and let cxqt, cxqs be quantifier calculi from the preceding Chapter which are based upon the corresponding propositional modus ponens calculus cx.

THEOREM 1 Let C consist of sentences, and let M consist of formulas. There is a linear operator **P** transforming cxqt–deductions S of $C \vdash u$ into XQC–derivations **P**(S) of $C \Longrightarrow u$, as well as cxqs–deductions S of $M \vdash_s u$ into XQC–derivations **P**(S) of $M \Longrightarrow u$. There is a linear operator **Q** transforming XQC–derivations D of $M \Longrightarrow u$ into cxqs–deductions **Q**(D) of $M \vdash_s u$.

I first consider a $cxqt_0$–deduction S from C. As in the proof of Theorem 7.2 I construct, by recursion in the tree of S, for every node e and for the formula v_e at e, an XQ–derivation D_e of the the sequent $D(e): C \Longrightarrow v_e$. If v_e is an axiom (a00) or (a01) then I derive D(e) as

$$\begin{array}{rl} \mathrm{rep}(x,t\,|\,w) \Longrightarrow & \mathrm{rep}(x,t\,|\,w) \\ \forall x\ w \Longrightarrow & \mathrm{rep}(x,t\,|\,w) \\ \Longrightarrow & \forall x\ w \rightarrow \mathrm{rep}(x,t\,|\,w) \end{array} \qquad \begin{array}{rl} \mathrm{rep}(x,t\,|\,w) \Longrightarrow & \mathrm{rep}(x,t\,|\,w) \\ \mathrm{rep}(x,t\,|\,w) \Longrightarrow & \exists x\ w \\ \Longrightarrow & \mathrm{rep}(x,t\,|\,w) \rightarrow \exists x\ w\,. \end{array}$$

If D(e) is conclusion from the premiss D(e') under (R00) then I find a derivation D' of D(e'): $A \Longrightarrow u \rightarrow w$ by inductive hypothesis. After a first transformation ensuring that D' it becomes eigen, I then apply the inversion rule (JI→) and obtain a derivation of $C, u \Longrightarrow w$. Since $w = \mathrm{rep}(x,x\,|\,w)$ and since x is neither free in u nor in the sentences C, the rule (I∀) produces a derivation of $C, u \Longrightarrow \forall x\, w$, and continuing that with (I→), I then obtain D(e). The case of (R01) analogously makes use of (E∃).

Observe here that, in view of the freeness of the involved maps $\zeta(x,t)$ and $\zeta(x,x)$, the Q–rules may employ rep as well as sub. Observe also that the transformation of instances of MP leads to cuts. Employing Q–rules with sub, these cuts can be eliminated; employing Q–rules with rep, they may resist elimination if the formula u is not pure.

Consider now a $cxqs_0$–deduction S of $M \vdash_s u$ which I view as a propositional cx–deduction S_0 from M and from formulas u_n in $cxqt_0(0)$, together with $cxqt_0$–deductions S_n leading to the maximal nodes n of S_0 which carry the

u_n. Again I proceed as in the propositional case: I replace the formula v_e at e by the sequent D(e): $M \Longrightarrow v_e$ and construct by recursion the derivation D_e of D(e). There only remain the maximal nodes of S_0 which carry the u_n. Now the S_n can be transformed into XQ-derivations D_n'' of $\blacktriangle \Longrightarrow u_n$. If I consider them as employing Q-rules with sub, then I can can transform them by Lemma **8**.8 into derivations D_n' of the same endsequents which uses only eigenvariables not free in M. If I consider them as employing Q-rules with rep, then repeated applications of the Lemma **8**.2 will also produce such D_n'. Weakening their antecedents with M, I then obtain the desired derivations D_n of $M \Longrightarrow u_n$. - This completes the construction of **P**, and its linearity follows as in the propositional case.

Let D be an XQC-derivation. Again it suffices to expand the proof of Theorem **7**.2 by handling the instances of the new Q-rules, and here I shall consider the case that they employ sub, that case of those with rep being even simpler. After an application of Lemma **8**.8 I may assume that D is eigen. So let (R) be a Q-rule ending D. If (R) is (E$\forall$) with the principal formula $\forall x\, w$ and if the endsequent is $\forall x\, w, M_0 \Longrightarrow u$, then there is a (shorter) subderivation H of the premiss $\mathrm{sub}(x,t \mid w), \forall x\, w, M_0 \Longrightarrow u$ for which I find a deduction **Q**(H). But $\forall x\, w \to \mathrm{sub}(x,t \mid w)$ is an axiom of cxqs by (ao0), and so I obtain **Q**(H) by implanting

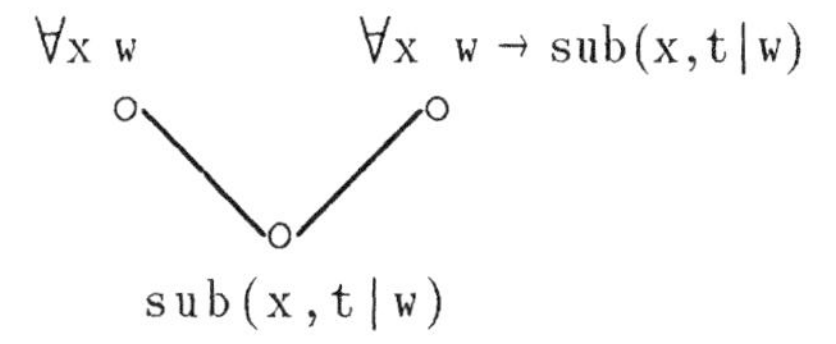

at every maximal node of **Q**(H) carrying $\mathrm{sub}(x,t \mid w)$. If (R) is (I$\exists$) then I use the axiom (a01) and obtain **Q**(D) by prolonging **Q**(H):

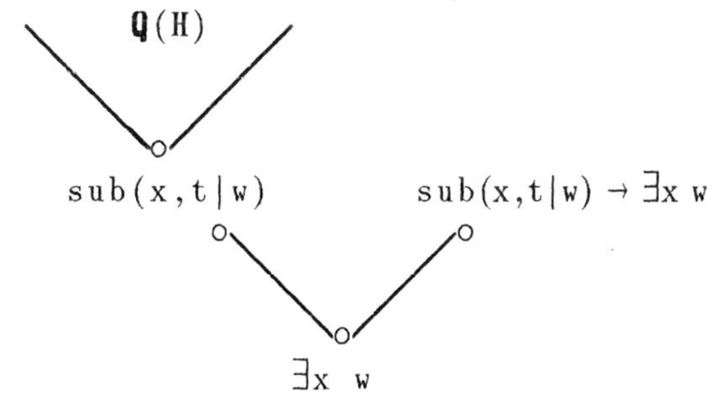

If (R) is (E$\exists$) then the premiss is $\mathrm{sub}(x,z \mid w), \exists x\, w, M_0 \Longrightarrow u$, and its subderivation H leads to a cxqs-deduction **Q**(H): $\mathrm{sub}(x,z \mid w), \exists x\, w, M_0 \vdash_s u$, i.e. $\mathrm{rep}(x,z \mid \mathrm{tot}(x,z \mid w)), \exists x\, w, M_0 \vdash_s u$. The derivation D being eigen, the eigenvariable z occurs neither in M_0 nor in $\exists x\, w$; hence z is neither free in M nor in w, and $\zeta(x,z)$ is free for $\mathrm{tot}(x,y \mid w)$. Therefore I can apply the second operator from Lemma **9**.3 and transform **Q**(H) into a deduction H*: $\exists x\, \mathrm{tot}(x,z \mid w), M_0 \vdash_s u$. By **8**.(3) the formula $w \to \mathrm{tot}(x,z \mid w)$ belongs to cxqt(0), whence by **9**.(a15) and the rule (R1) also $\exists x\, w \to \exists x\, tot(x,z \mid w)$ is in cxqt(0). Thus I obtain **Q**(D) by implanting

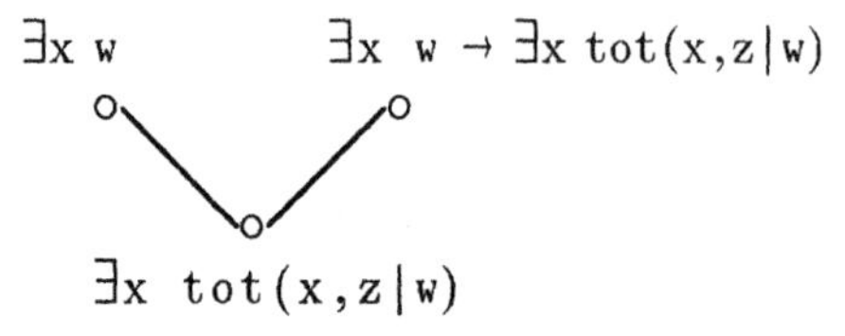

at every maximal node of H* carrying $\exists x\, tot(x,z \mid w)$.

If (R) is (I∀) then the premiss is $M \Longrightarrow sub(x,z \mid w)$, and its subderivation H leads to a cxqs-deduction **Q**(H) of $sub(x,z \mid w) = rep(x,z \mid tot(x,z \mid w))$ from M. D being eigen, the eigenvariable z occurs neither in M nor in $\forall x\, w$; hence z is neither free in M nor in w, and $\zeta(x,z)$ is free for $tot(x,y \mid w)$. I apply the operator from Lemma **9**.3 and transform **Q**(H) into a deduction $H^*\colon M \vdash_s \forall x\, tot(x,z \mid w)$. By **8**.(3) then $tot(x,z \mid w) \to w$ is in cxqt(0), whence by **9**.(a14) and the rule (R1) also $\forall x\, tot(x,z \mid w) \to \forall x\, w$ belongs to cxqt(0), and so I obtain **Q**(D) by prolonging H* with MP.

Forgetting the operators **P** and **Q**, the Theorem says in particular that

$M \Longrightarrow u$ is derivable if, and only if, $M \vdash_s u$ is deducible

and that for sets C of sentences

$C \Longrightarrow u$ is derivable if, and only if, $C \vdash u$ is deducible.

As a consequence, I obtain a new proof of Lemma **9**.2: if C consists of sentences then a cxqt-deduction $C \vdash u$ transforms into $C \Longrightarrow u$ which transforms into a cxqs-deduction $C \vdash_s u$.

It follows from the Theorem together with Theorem $\mathbf{9}.2_c$ that for classical logic the derivability of a sequent $M \Longrightarrow u$ is equivalent to $u \in cs(M)$; in so far, the sequent calculus describes the consequence operation cs. More generally, for a logic based at least on positive logic and a class **X** of algebras, it follows that for sets C of sentences the derivability of a sequent $C \Longrightarrow u$ is equivalent to $u \in cs^{\mathbf{X}}(C)$ and to $ct^{\mathbf{X}}(C)$.

When setting up sequent calculi, the derivability of $M \Longrightarrow v$ was understood to express that v be deducible from the assumptions M. It is a familiar practice to fix a certain set C of sentences as an *axiom system* and to consider derivabilities *from* this axiom system. In that case, it will be convenient to read a sequent

$$C, M \Longrightarrow v ,$$

as saying that v is deducible from the assumptions M *under the axiom system* C. The derivability of this sequent is equivalent to $v \in cs(C \cup M)$, and as C consists of sentences, this means that for every model A of C: every valuation in A satisfying M also satisfies v .

2. Relations between Classical and Intuitionistic Derivability

During the discussion of modus ponens calculi in the last Chapter, I established various deducibilities $\vdash u \rightarrow v$ for positive and intuitionistic logic, and in view of the translations between Hilbert type and sequential calculi, these deducibilities give rise to deducibilities $u \Longrightarrow v$ in the corresponding sequential calculi (which actually are much simpler to derive). In the list of tautologies 9.(M1)–(M4) for minimal logic, there was number of implications for which the converse implications were not deduced. This was no accident:

LEMMA 1 The missing implications

(i1) $\neg\forall \rightarrow \exists\neg$
(i2) $\neg\neg\forall \rightarrow \forall$
(i3) $\exists\neg\neg \rightarrow \exists$
(i4) $\forall\neg\neg \rightarrow \neg\neg\forall$
(i5) $\neg\neg\exists \rightarrow \exists\neg\neg$
(i6) $\neg\forall \rightarrow \neg\forall\neg\neg$
(i7) $\neg\neg\exists\neg \rightarrow \exists\neg$

cannot be derived in intuitionistic logic (but can easily be derived in the classical calculi with multiple sequents).

An intuitionistic derivation of (i1) would necessarily have to employ the premisses in the left attempt

$$\begin{array}{rl} v(t) \Longrightarrow & \forall x \; v(x) \\ v(t), \neg\forall x\, v(x) \Longrightarrow & \blacktriangle \\ \neg\forall x\, v(x) \Longrightarrow & \neg v(t) \\ \neg\forall x\, v(x) \Longrightarrow & \exists x \neg v(x) \end{array} \qquad \begin{array}{rl} v(x) \Longrightarrow & v(x) \\ \blacktriangle \Longrightarrow & \neg v(x),\ v(x) \\ \blacktriangle \Longrightarrow & \exists x \neg v(x),\ v(x) \\ \blacktriangle \Longrightarrow & \exists x \neg v(x),\ \forall x \; v(x) \\ \neg\forall x\, v(x) \Longrightarrow & \exists x \neg v(x) \end{array}$$

and if v is atomic then the topmost sequent could only come from $v(t) \Longrightarrow v(y)$ with an eigenvariable not in t. Classically, however, I have the right derivation with the eigenvariable x.

An intuitionistic derivation of (i2) would necessarily have to employ the premisses in the left attempt

$$\begin{array}{rl} \neg\neg\forall x\, v(x) \Longrightarrow & v(y) \\ \neg\neg\forall x\, v(x) \Longrightarrow & \forall x \; v(x) \end{array} \qquad \begin{array}{rl} \forall x\, v(x) \Longrightarrow & \forall x\, v(x) \\ \blacktriangle \Longrightarrow & \forall x\, v(x),\ \neg\forall x\, v(x) \\ \neg\neg\forall x\, v(x) \Longrightarrow & \forall x\, v(x) \end{array}$$

and if v is atomic no further premiss can be found. Classically I have the right derivation.

(i3) can be reduced intuitionistically only to $\neg\neg v(x) \Longrightarrow v(x)$ which cannot be proved intuitionistically if v(x) is atomic. In (i4), the right quantifier will remain critical after removing the double negation and thus will have to be resolved before the the uncritical left quantifier. This again leads to a premiss $\neg\neg v(t) \Longrightarrow v(x)$ which cannot be derived intuitionistically if $v(x)$ is atomic. In (i5), the only intuitionistically possible premiss is $\neg\neg\exists x v(x) \Longrightarrow \neg\neg v(t)$ which again would force the resolution of the uncritical quantifier before the critical one.

(i6) requires the premisses $\neg\forall x v(x), \forall\neg\neg v(x) \Longrightarrow \blacktriangle\ , \forall\neg\neg v(x) \Longrightarrow \forall x v(x)$. Intuitionistically this reduces to $\neg\neg v(x) \Longrightarrow v(x)$ which cannot be proved intuitionistically if $v(x)$ is atomic. (i7) requires intuitionistically the premiss $\neg\neg\exists\neg v(x) \Longrightarrow v(t)$ which prevents a later resolution of the critical quantifier on the left; classically I can use the premisses $\blacktriangle \Longrightarrow \neg\exists x\neg v(x)$, $\exists x\neg v(x)$ and then $\exists x\neg v(x) \Longrightarrow \exists x\neg v(x)$.

LEMMA 2 The implications

$$
\begin{array}{ll}
\text{(i8)} & u \rightarrow \exists x w \Longrightarrow \exists x (u \rightarrow w) \\
\text{(i9)} & \forall x w \rightarrow u \Longrightarrow \exists x (w \rightarrow u) \\
\text{(i10)} & \forall x (u \vee w) \Longrightarrow u \vee \forall x w \ ,
\end{array}
$$

with x not in fr(u), *cannot* be derived in intuitionistic logic.

If $\blacktriangle \Longrightarrow u$ is not derivable then an intuitionistic derivation of (i8) with w atomic would require the premisses

$$
\begin{array}{ll}
u \Rightarrow u & u,\ \exists x\ w \Rightarrow w(t) \\
\hline
\multicolumn{2}{l}{u,\ u \rightarrow \exists x w \Longrightarrow w(t)} \\
\multicolumn{2}{l}{u \rightarrow \exists x w \Longrightarrow u \rightarrow w(t)} \\
\multicolumn{2}{l}{u \rightarrow \exists x w \Longrightarrow \exists x\ (u \rightarrow w)}
\end{array}
$$

where the right upper premiss cannot be resolved to an axiom in view of the criticality of the quantifier. Classically in MK I have

$$
\begin{array}{ll}
 & u,\ u, w(x) \Longrightarrow w(x),\ w(t) \\
 & u, w(x) \Longrightarrow u \rightarrow w(x), w(t) \\
u \Longrightarrow u, \exists x (u \rightarrow w) & u, w(x) \Longrightarrow \exists x (u \rightarrow w),\ w(t) \\
u \Longrightarrow u, \exists x (u \rightarrow w) & u, \exists x\ w \Longrightarrow \exists x (u \rightarrow w),\ w(t) \\
\hline
\multicolumn{2}{l}{u, u \rightarrow \exists x\ w \Rightarrow \exists x (u \rightarrow w),\ w(t)} \\
\multicolumn{2}{l}{u \rightarrow \exists x\ w \Rightarrow \exists x (u \rightarrow w),\ u \rightarrow w(t)} \\
\multicolumn{2}{l}{u \rightarrow \exists x\ w \Rightarrow \exists x (u \rightarrow w)}
\end{array}
$$

If $\blacktriangle \Longrightarrow \forall x w$ is not derivable then an intuitionistic derivation of (i9) with w atomic would require the premisses

$$
\begin{array}{ll}
w(t) \Longrightarrow \forall x\ w & w(t),\ u \Longrightarrow u \\
\hline
\multicolumn{2}{l}{w(t),\ \forall x\ w \rightarrow u \Longrightarrow u} \\
\multicolumn{2}{l}{\forall x\ w \rightarrow u \Longrightarrow w(t) \rightarrow u} \\
\multicolumn{2}{l}{\forall x\ w \rightarrow u \Longrightarrow \exists x\ (w \rightarrow u)}
\end{array}
$$

where the left upper premiss cannot be resolved to an axiom in view of the criticality of the quantifier. Classically, however, I have

$$\begin{array}{ll} w(x) \Longrightarrow w(x),\, u & \\ \Longrightarrow w(x),\, w(x) \to u & w(t),\, u \Longrightarrow u \\ \Longrightarrow w(x),\, \exists x\,(w \to u) & u \Longrightarrow w(t) \to u \\ \Longrightarrow \forall x\, w,\, \exists x\,(w \to u) & u \Longrightarrow \exists x\,(w \to u) \\ \multicolumn{2}{c}{\forall x\, w \to u \Longrightarrow \exists x\,(w \to u)} \end{array} .$$

In a derivation of (i10) the critical quantifier on the right has to be resolved below the one on the left. Classically, this is done in

$$\begin{array}{ll} u \Longrightarrow u,\, w(x) & w(x) \Longrightarrow u,\, w(x) \end{array}$$

$$\begin{array}{l} u \vee w(x) \Longrightarrow u,\ w(x) \\ \forall x\,(u \vee w) \Longrightarrow u,\ w(x) \\ \forall x\,(u \vee w) \Longrightarrow u,\ \forall x\, w \\ \forall x\,(u \vee w) \Longrightarrow u \vee \forall x\, w \end{array} ,$$

while an intuitionistic proof, permitting only either u or $\forall xw$ on the right, would not be possible.

In the case of propositional logic, I studied in Chapter 6 the three translations maps ϑ, κ, γ sending classical into intuitionistic derivations. The maps κ and γ can be extended to the case of quantifier logic by setting

$$(\forall x v)^{\kappa} = \neg\neg\forall x\, v^{\kappa} \quad , \quad (\exists x v)^{\kappa} = \neg\neg\exists x\, v^{\kappa} \quad ,$$

$$(\forall x v)^{\gamma} = \forall x\, v^{\gamma} \quad , \quad (\exists x v)^{\gamma} = \neg\forall x \neg v^{\gamma} \quad .$$

The map ϑ now must be defined with help of an auxiliary map φ by

$a^{\varphi} = a$ if a is atomic , $(\neg v)^{\varphi} = \neg v^{\varphi}$, $(v \% w)^{\varphi} = v^{\varphi} \% w^{\varphi}$ if % is a propositional connective,

$(\forall x v)^{\varphi} = \forall x \neg\neg v^{\varphi}$, $(\exists x v)^{\varphi} = \exists x v^{\varphi}$

and then $v^{\vartheta} = \neg\neg v^{\varphi}$. Observe that for each of these maps τ the formulas v and v^{τ} are classically interdeducible. Also, v^{τ} and $\neg\neg v^{\tau}$ are intuitionistically interdeducible: this is clear in call cases different from $(\forall x v)^{\gamma}$ since then v^{τ} starts with a negation, and if $v^{\gamma} \simeq \neg\neg v^{\gamma}$ then $\forall x\, v^{\gamma} \simeq \forall x \neg\neg v^{\gamma}$, $\neg\neg\forall x\, v^{\gamma} \simeq \neg\neg\forall x \neg\neg v^{\gamma}$, but $\neg\neg\forall x \neg\neg v^{\gamma} \simeq \forall x \neg\neg v^{\gamma}$ by 9.(M2) whence $\neg\neg\forall x\, v^{\gamma} \simeq \forall x \neg\neg v^{\gamma}$. Now the statement of Lemma 6.4 remains in effect without changes:

LEMMA 3 For each of the classical quantifier calculi LK_xQ, arising from the calculi LK_x, and for each of the maps ϑ, κ, γ there is an operator G_x^{τ} transforming an LKQ-derivation D of $M \Longrightarrow u$ into an LJQ-derivation of $M^{\tau} \Longrightarrow u^{\tau}$.

The function *length*$(G_x^{\kappa}(D))$ depends linearly on *length*(D) for $x = n,d$, and *length*$(G_0^{\kappa}(D))$ depends linearly on *length*(D) when computed in the calculus obtained from LJQ by adding the admissible rule (M_0). In all other cases, the lengths may grow exponentially due to the occurrence of cuts.

Observe that, by definition of my maps, $(\mathrm{rep}(x,t\,|\,v))^{\tau} = \mathrm{rep}(x,t\,|\,v^{\tau})$ holds for each of them, and the same for sub. Consider first the case of κ; then I translate rules (IQ) by

$$
\begin{array}{l}
M^{\kappa} \Longrightarrow \mathrm{rep}(x,t\,|\,v^{\kappa}) \\
M^{\kappa} \Longrightarrow Qx\, v^{\kappa} \\
M^{\kappa} \Longrightarrow \neg\neg Qx\, v^{\kappa}
\end{array}
$$

and rules (EQ) by

$$
\begin{array}{rll}
\mathrm{rep}(x,t\,|\,v^{\kappa}),\ M^{\kappa} \Longrightarrow & u^{\kappa} & \\
\mathrm{rep}(x,t\,|\,v^{\kappa}),\ \neg u^{\kappa},\ M^{\kappa} \Longrightarrow & \blacktriangle & \\
\mathrm{rep}(x,t\,|\,v^{\kappa}),\ \neg\neg t,\ M^{\kappa} \Longrightarrow & \blacktriangle & u^{\kappa} = \neg t \\
\mathrm{rep}(x,t\,|\,v^{\kappa}),\ t,\ M^{\kappa} \Longrightarrow & \blacktriangle & \mathbf{JS}\neg\neg \\
Qx\, v^{\kappa},\ t,\ M^{\kappa} \Longrightarrow & \blacktriangle & \\
\neg\neg Qx\, v^{\kappa},\ t,\ M^{\kappa} \Longrightarrow & \blacktriangle & \\
\neg\neg Qx\, v^{\kappa},\ M^{\kappa} \Longrightarrow & u^{\kappa}\ . &
\end{array}
$$

So here the growth of the lengths is linear again. For the map γ the case of the quantifier $\forall$ is trivial; for the quantifier $\exists$ there holds

$$
\begin{array}{rl}
M^{\gamma} \Longrightarrow & \mathrm{rep}(x,t\,|\,v^{\gamma}) \\
\neg\,\mathrm{rep}(x,t\,|\,v^{\gamma}),\ M^{\gamma} \Longrightarrow & \blacktriangle \\
\mathrm{rep}(x,t\,|\,\neg v^{\gamma}),\ M^{\gamma} \Longrightarrow & \blacktriangle \\
\forall x\,\neg v^{\gamma},\ M^{\gamma} \Longrightarrow & \blacktriangle \\
M^{\gamma} \Longrightarrow & \neg\forall x\,\neg v^{\gamma}\ ,
\end{array}
$$

$$
\begin{array}{rl}
\mathrm{rep}(x,y\,|\,v^{\gamma}),\ M^{\gamma} \Longrightarrow & u^{\gamma} \\
\mathrm{rep}(x,y\,|\,v^{\gamma}),\ \neg u^{\gamma},\ M^{\gamma} \Longrightarrow & \blacktriangle \\
\neg u^{\gamma},\ M^{\gamma} \Longrightarrow & \mathrm{rep}(x,y\,|\,\neg v^{\gamma}) \\
\neg u^{\gamma},\ M^{\gamma} \Longrightarrow & \forall x\neg v^{\gamma} \\
\neg\forall x\neg v^{\gamma},\ M^{\gamma} \Longrightarrow & \neg\neg u^{\gamma} \qquad\qquad \neg\neg u^{\gamma} \Longrightarrow u^{\gamma} \quad (u^{\gamma} \text{ is regular})
\end{array}
$$

$$
\neg\forall x\neg v^{\gamma},\ M^{\gamma} \Longrightarrow \neg u^{\gamma}\ .
$$

As for the map ϑ, I find

$$
\begin{array}{l}
M^{\vartheta} \Longrightarrow \mathrm{rep}(x,y\,|\,v^{\varphi}) \\
M^{\vartheta} \Longrightarrow \mathrm{rep}(x,y\,|\,\neg\neg v^{\varphi}) \\
M^{\vartheta} \Longrightarrow \forall x\,\neg\neg v^{\varphi} \\
M^{\vartheta} \Longrightarrow \neg\neg\forall x\,\neg\neg v^{\varphi}\ ,
\end{array}
$$

$$
\begin{array}{l}
M^{\vartheta} \Longrightarrow \mathrm{rep}(x,t\,|\,\neg\neg v^{\varphi}) \\
M^{\vartheta} \Longrightarrow \exists x\,\neg\neg v^{\varphi} \qquad \exists x\,\neg\neg v^{\varphi} \Longrightarrow \neg\neg\exists x\, v^{\varphi}
\end{array}
$$

$$
M^{\vartheta} \Longrightarrow \neg\neg\exists x\, v^{\varphi}
$$

where the right premiss of the cut comes from **8.**(M1), further

$$\frac{\neg\neg\forall x\neg\neg v^{\varphi} \Longrightarrow \forall x\neg\neg v^{\varphi} \qquad \dfrac{\mathrm{rep}(x,t\,|\,\neg\neg v^{\varphi}),\ M^{\vartheta} \Longrightarrow \neg\neg u^{\varphi}}{\forall x\neg\neg v^{\varphi},\ M^{\vartheta} \Longrightarrow \neg\neg u^{\varphi}}}{\neg\neg\forall x\neg\neg v^{\varphi},\ M^{\vartheta} \Longrightarrow \neg\neg u^{\varphi}}$$

where the left premiss of the cut comes from **8**.(M2), and

$$\begin{array}{rll}
\neg\neg\mathrm{rep}(x,y\,|\,v^{\varphi}),\ M^{\vartheta} & \Longrightarrow \neg\neg u^{\varphi} & \\
\mathrm{rep}(x,y\,|\,v^{\varphi}),\ M^{\vartheta} & \Longrightarrow \neg\neg u^{\varphi} & \mathbf{JS}\neg\neg \\
\neg\neg\neg u^{\varphi},\ \mathrm{rep}(x,y\,|\,v^{\varphi}),\ M^{\vartheta} & \Longrightarrow \blacktriangle & \\
\neg u^{\varphi},\ \mathrm{rep}(x,y\,|\,v^{\varphi}),\ M^{\vartheta} & \Longrightarrow \blacktriangle & \mathbf{JS}\neg\neg \\
\neg u^{\varphi},\ \exists x\, v^{\varphi},\ M^{\vartheta} & \Longrightarrow \blacktriangle & \\
\neg u^{\varphi},\ \neg\neg\exists x\, v^{\varphi},\ M^{\vartheta} & \Longrightarrow \blacktriangle & \\
\neg\neg\exists x\, v^{\varphi},\ M^{\vartheta} & \Longrightarrow \neg\neg u^{\varphi} & .
\end{array}$$

3. Equality Logic

At the end of Section **1**, I explained deducibility from axiom systems. A particular axiom system, consisting of sentences, is the set EA_u of equality axioms, studied already in Chapter **1.11** and **1.12** :

$\forall x_0\,(x_0 \equiv x_0)$,
$\forall x_0\,\forall x_1\,(x_0 \equiv x_1 \to x_1 \equiv x_0)$,
$\forall x_0\,\forall x_1\,\forall x_2\,(x_0 \equiv x_1 \wedge x_1 \equiv x_2 \to x_0 \equiv x_2)$
$\forall \xi\,\forall \xi'\,(\xi \equiv \xi' \to f(\xi) \equiv f(\xi')$ for every function symbol f [from ≡
$\forall \xi\,\forall \xi'\,(\xi \equiv \xi' \to (p(\xi) \to p(\xi'))$ for every predicate symbol p different

where $\xi = \langle x_0, \ldots, x_{n-1}\rangle$, $\xi' = \langle x_0', \ldots, x_{n-1}'\rangle$ are of appropriate lengths and

$\forall \xi$, $\forall \xi'$ abbreviate $\forall x_0 \ldots \forall x_{n-1}$ and $\forall x_0' \ldots \forall x_{n-1}'$ respectively,
$\xi \equiv \xi'$ abbreviates $(x_0 \equiv x_0' \wedge \ldots \wedge x_{n-1} \equiv x_{n-1}')$.

At the end of Chapter **9**, Section **1**, I mentioned the calculi $ccqt_=$ and $ccqs_=$, for which I shall also write $M \vdash_= v$ for $EA_u, M \vdash v$ and $M \vdash_{s=} v$ for $EA_u, M \vdash_s v$.

It follows from Theorem 1 that a sequent $EA_u, M \Longrightarrow v$ is derivable if, and only if, $M \vdash_{s=} v$; if M is a set of sentences then this is equivalent to $M \vdash_= v$. I thus shall say that a sequent $M \Longrightarrow v$ is derivable *from the axioms of equality logic* if the sequent $EA_u, M \Longrightarrow v$ is derivable in the usual sense. The substitution instances of the unquantified formulas from EA_u I shall occasionally call *equality atoms*.

A second way to study derivabilities from particular axioms consists in defining a suitable extension of the sequential calculus – the form of which, of course, will depend on these axioms. In the case of equality axioms, such extension is easily formed: all it requires is to enlarge the set of axioms (axiom sequents), say

$$p(\lambda), M \Longrightarrow p(\lambda) ,$$

of a given sequential calculus X by additional *equality axiom sequents*: first for all terms r, s, t the sequents

$(E_0 0)$ $\qquad M \Longrightarrow t \equiv t$

$(E_0 1)$ $\qquad s \equiv t\ ,\ M \Longrightarrow t \equiv s$

$(E_0 2)$ $\qquad r \equiv s\ ,\ s \equiv t\ ,\ M \Longrightarrow r \equiv t\ ,$

next, for every function symbol f of arity n, and for all sequents μ, μ' of terms of length n, the sequents

$(E_0 3)$ $\qquad E(\mu \equiv \mu'), M \Longrightarrow f(\mu) \equiv f(\mu')\ ,$

and then, for every predicate symbol p, different from $\equiv$, of arity m, and for all sequents μ, μ' of terms of length n, the sequents

$(E_0 4)$ $\qquad E(\mu \equiv \mu'), p(\mu), M \Longrightarrow p(\mu')$

where $E(\mu \equiv \mu')$ abbreviates the sequence $\mu(0) \equiv \mu'(0), \ldots, \mu(n-1) \equiv \mu'(n-1)$.

I shall write XE_0 for the calculus arising this way from the given calculus X. Distinguishing between derivabilities in XE_0 and in X, I shall speak of derivabilities in *equational logic* and of derivabilities in *pure logic*. Let XC, XE_0C be now the extensions of X and XE_0 by (CUT). There holds the

LEMMA 4 $\quad$ Every XC-derivation from the axioms EA_u can be transformed into an XE_0C-derivation of the same endsequent; every XE_0-derivation can be transformed into an X-derivation of the same endsequent from the axioms EA_u.

Let D be a derivation of EA_u, $M \Longrightarrow v$ in X or XC. In XE_0, I can derive

$$\mu(0) \equiv \mu'(0), \ldots, \mu(n-1) \equiv \mu'(n-1), p(\mu) \Longrightarrow p(\mu')$$
$$\mu(0) \equiv \mu'(0), \ldots, \mu(n-1) \equiv \mu'(n-1) \Longrightarrow p(\mu) \to p(\mu')$$
$$\mu(0) \equiv \mu'(0) \wedge \ldots \wedge \mu(n-1) \equiv \mu'(n-1) \Longrightarrow p(\mu) \to p(\mu')$$
$$\Longrightarrow \mu \equiv \mu' \to (p(\mu) \to p(\mu'))$$
$$\Longrightarrow \forall\xi\forall\xi' (\xi \equiv \xi' \to (p(\xi) \to p(\xi')) ,$$

and in the same way I find derivations for sequents with the other axioms of EA_u. Applying (CUT) to the endsequents of D and of these derivations, I obtain an XE_0C-derivation of $M \Longrightarrow v$. Conversely, it suffices to show that I can derive the new axiom sequents of XE_0 from the axioms EA_u in X, e.g.

$$\mu \equiv \mu', E(\mu \equiv \mu'), p(\mu), M \Longrightarrow p(\mu) \qquad \mu \equiv \mu', p(\mu'), E(\mu \equiv \mu'), p(\mu), M \Longrightarrow p(\mu')$$
$$\mu \equiv \mu'\ ,\ p(\mu) \to p(\mu')\ ,\ E(\mu \equiv \mu'), p(\mu), M \Longrightarrow p(\mu')$$
$$\mu \equiv \mu' \wedge (p(\mu) \to p(\mu'))\ ,\ E(\mu \equiv \mu'), p(\mu), M \Longrightarrow p(\mu')$$
$$\forall\xi\forall\xi' (\xi \equiv \xi' \wedge (p(\xi) \to p(\xi'))), E(\mu \equiv \mu'), p(\mu), M \Longrightarrow p(\mu')$$
$$EA_u, E(\mu \equiv \mu'), p(\mu), M \Longrightarrow p(\mu') \ .$$

There is an unpleasantness about XE_0 in that without (CUT) sequents obviously valid cannot be derived, for instance $r \equiv s, s \equiv t, t \equiv u \Longrightarrow r \equiv u$. On the other hand, inspection of the elimination algorithm shows immediately that the one, and only, obstacle preventing it to work is that, in contrast to the reflexivity axiom sequents, the equality axiom sequents are not closed under cuts.

The obvious way to overcome this difficulty is, therefore, to enforce this closedness: let XE_1 be the calculus whose axiom sequents are the sequents belonging to the smallest set ES of sequents which contains the axiom sequents of XE_0 and is closed under instances of structural rules, including cuts. Then the cut elimination algorithm remains in effect for XE_1, and thus the statement of Lemma 4 will hold with XE_1 replacing both XE_0 and XE_0C.

Cut free calculi obey a subformula property. This I wish to use for the calculus XE_1 in order to conclude that the extension of X by equality axioms is *conservative:* if a sequent, not containing the predicate $\equiv$, can be derived in equality logic then it can be derived already in pure logic. In order to perform this argument, however, I need to know that ES contains a sequent without the predicate symbol $\equiv$ only if it is already a reflexive axiom.

The candidate to produce sequents without $\equiv$ is, obviously, (E_04), used as the right premiss of a cut with left premiss (E_00) where the parameter set M does not contain $\equiv$. But while a cut with *that* kind of left premiss only produces a reflexive axiom again, it might be just conceivable that more complicated iterations of cuts give rise to sequents $M \Longrightarrow s \equiv t$ in ES for which $s \neq t$, and *such* sequents as left premisses of a cut with (E_04) would indeed be fatal. I thus have to show that such sequents *do not* arise in ES, and to this end a different description of ES, due to TAKEUTI 75 , is required.

Let ET be the smallest set of sequents containing (E_00)-(E_03) and closed under instances of structural rules, including cut. All sequents in ET have a right side of the form $s \equiv t$, and I write them in the form $E, M \Longrightarrow s \equiv t$ where E shall contain *only* the predicate symbol $\equiv$ and where M shall *not* contain it. I call a sequence E *identical* if all its members $p \equiv q$ are such that $p = q$. I first observe

(1e) If $E, M \Longrightarrow s \equiv t$ is in ET and if E is identical then $s = t$.

This is obvious for (E_00)-(E_03) and remains trivially preserved under structural rules distinct from cut. It remains also preserved under cut: if in the conclusion of

$$\frac{E_0, M_0 \Longrightarrow s \equiv t \qquad s \equiv t, E_1, M_1 \Longrightarrow u \equiv v}{E_0, E_1, M_0, M_1 \Longrightarrow u \equiv v}$$

the sequence E_0, E_1 is identical then E_0 is identical, hence $s = t$ by inductive hypothesis. But also E_1 is identical, hence the inductive hypothesis also gives $u = v$. - As a particular case of (1e) I remark

(2e) If $M \Longrightarrow s \equiv t$ is in ET and M does not contain $\equiv$ then $s = t$.

LEMMA 5 The set ES consists precisely of

(i) the reflexive axioms,

(ii) the sequents in ET,

(iii) the sequents $E, M, p(\mu) \Longrightarrow p(\mu')$ where p is different from $\equiv$ and such that, for every $i<n$, the sequent $E, M \Longrightarrow \mu(i) \equiv \mu'(i)$ is in ET.

Clearly, ET and the reflexive axioms are in ES, and the sequents (iii) arise by cuts from $(E_0 4)$:

$$\frac{E, M \Longrightarrow \mu(0) \equiv \mu'(0) \qquad \mu(0) \equiv \mu'(0), \ldots, \mu(n-1) \equiv \mu'(n-1), p(\mu), M \Longrightarrow p(\mu')}{E, M, \mu(1) \equiv \mu'(1), \ldots, \mu(n-1) \equiv \mu'(n-1), p(\mu), M \Longrightarrow p(\mu')}$$

$$\cdots\cdots\cdots$$

$$\cdots\cdots\cdots$$

$$\frac{E, M \Longrightarrow \mu(n-1) \equiv \mu'(n-1) \qquad E, M, \mu(n-1) \equiv \mu'(n-1), p(\mu), M \Longrightarrow p(\mu')}{E, M, p(\mu) \Longrightarrow p(\mu')}\ .$$

In order to see that the sequents (i)-(iii) make up all of ES, it suffices to show that they are closed under cuts. Obviously, both the sequents (i) and (ii) are closed with respect to cuts among themselves. A sequent (ii) or (iii), cut as left premiss with a reflexive axiom, will give a reflexive axiom if the cut formula is not characteristic for the axiom, and otherwise it will give a sequent (ii) respectively (iii) again. A sequent (ii) or (iii), cut as right premiss with a reflexive axiom, will give a sequent (ii) respectively (iii).

If a sequent (ii) is cut with a sequent (iii),

$$\frac{E_0, M_0 \Longrightarrow s \equiv t \qquad s \equiv t, E_1, M_1, p(\mu) \Longrightarrow p(\mu')}{E_0, E_1, M_0, M_1, p(\mu) \Longrightarrow p(\mu')}\ ,$$

then the conclusion is a sequent (iii) again since another cut

$$\frac{E_0, M_0 \Longrightarrow s \equiv t \qquad s \equiv t, E_1, M_1 \Longrightarrow \mu(i) \equiv \mu'(i)}{E_0, E_1, M_0, M_1 \Longrightarrow \mu(i) \equiv \mu'(i)}$$

shows that this sequent is in ET for every $i<n$. If a sequent (iii) is cut with a sequent (ii) then the conclusion is a sequent (ii) again. If a sequent (iii) is cut with a sequent (iii) and the cut formula is the distinguished formula of the right premiss

$$\frac{E_0, M_0, p(\mu) \Longrightarrow p(\mu') \qquad E_1, M_1, p(\mu') \Longrightarrow p(\mu'')}{E_0, E_1, M_0, M_1, p(\mu) \Longrightarrow p(\mu'')}$$

then the conclusion is a sequent (ii) again, since the further cuts

$$\dfrac{E_0, M_0 \Longrightarrow \mu(i) \equiv \mu'(i) \qquad \mu(i) \equiv \mu'(i), \mu'(i) \equiv \mu''(i) \Longrightarrow \mu(i) \equiv \mu''(i)}{E_1, M_1 \Longrightarrow \mu'(i) \equiv \mu''(i) \qquad \mu'(i) \equiv \mu''(i), E_0, M_0 \Longrightarrow \mu(i) \equiv \mu''(i)}$$

$$E_0, E_1, M_0, M_1 \Longrightarrow \mu(i) \equiv \mu''(i)$$

show that this sequent is in ET for every $i<n$. If the cut formula is not the distinguished formula of the right premiss but a formula in M_1, then the conclusion $E_0, E_1, M_0, M_1', p(\mu), p(\mu') \Longrightarrow p(\mu'')$ is trivially a sequent (iii). This concludes the proof of Lemma 5.

It follows from the observation (2e) that a for sequent (iii), in which E is empty, there holds $\mu = \mu'$. Consequently, the only sequents in ES which not contain $\equiv$ are reflexivity axioms. Summing up, I thus obtain the

THEOREM 2 Every XC-derivation from the axioms EA_u of equality logic can be transformed into an XE_1-derivation of the same endsequent, and vice versa.

An XE_1-derivation, of an endsequent without the predicate symbol $\equiv$, is already an X-derivation.

An XC-derivation from the axioms EA_u of equality logic, of an endsequent without the predicate symbol $\equiv$, can be transformed into an X-derivation of that endsequent.

Because if D is an XC-derivation from the axioms EA_u then by Lemma 4 it can be transformed into an XE_0C-derivation D' of its endsequent. But D' is also an XE_1C-derivation; hence it can be transformed into a cut free XE_1-derivation D'' of its endsequent. If this endsequent does not contain $\equiv$ then D'' is an X-derivation.

I wish to add that an explicit description of the sequents in ET (in contrast to the generic one of ET's definition) appears to be difficult. For instance, if f and g are of arities 2 and 3 respectively then the sequent

$$x \equiv x', y \equiv y', z \equiv z', u \equiv u' \Longrightarrow f(u, g(x,y,z)) \equiv f(u', g(x',y',z'))$$

will be in ET, but the only way to relate its left side to its right one seems to write down the history of its origin, viz. the cut between two particular instances of $(E_0 3)$.

It follows from Theorem 2 that also the calculi $cxqt_=$ and $cxst_=$ are conservative over the calculi cxqt and csqs respectively: if M and v do not contain the predicate symbol $\equiv$ then a deduction $M \vdash_{s=} v$ by equality logic can be transformed into a deduction $M \vdash_s v$ by pure logic, and if M consists of sentences then also $M \vdash_= v$ can be transformed into $M \vdash v$. In particular, equality logic (or the axiom system EA_u) is consistent if pure logic is consistent.

I conclude with a few simple, but useful observations.

$L_0(v)$ If not $y \in var(v)$ then there are derivations of both
$x \equiv y, v \Longrightarrow rep(x, y \mid v)$ and $x \equiv y, rep(x, y \mid v) \Longrightarrow v$.

For an atomic formula v, the first sequent is a special cases of $(E_0 4)$, and as $rep(x,y \mid v)$ is an atomic formula as well, also the second sequent is such special case because $rep(y,x \mid rep(x,y \mid v) = v$ since y is not in v. If $L_0(v)$ holds for v then a derivation of the first [second] sequent of $L_0(v)$ can be extended to a derivation of the second [first] sequent of $M(\neg v)$. Also, if $L_0(v)$ and $L_0(w)$ then $L_0(v \to w)$ since e.g.

$$x \equiv y, rep(x,y \mid v) \Longrightarrow v \qquad x \equiv y, rep(x,y \mid v), w \Longrightarrow rep(x,y \mid v)$$

$$x \equiv y, rep(x,y \mid v), v \to w \Longrightarrow rep(x,y \mid v)$$
$$x \equiv y, v \to w \Longrightarrow rep(x,y \mid v) \to rep(x,y \mid w) .$$

The cases of $v \wedge w$ and $v \vee w$ are even simpler. Further, derivations of the sequents in $L_0(v)$ can be extended by

$$\begin{array}{ll}
x \equiv y, v \Longrightarrow rep(x,y \mid v) & x \equiv y, rep(x,y \mid v) \Longrightarrow v \\
x \equiv y \Longrightarrow v \to rep(x,y \mid v) & x \equiv y \Longrightarrow rep(x,y \mid v) \to v \\
x \equiv y \Longrightarrow \exists z\, (v \to rep(x,y \mid v)) & x \equiv y \Longrightarrow \exists z\, (rep(x,y \mid v) \to v) \\
x \equiv y \Longrightarrow \exists z v \to \exists z\, rep(x,y \mid v) & x \equiv y \Longrightarrow \exists z\, rep(x,y \mid v) \to \exists z v \\
x \equiv y, \exists z v \Longrightarrow \exists z\, rep(x,y \mid v) & x \equiv y, \exists z\, rep(x,y \mid v) \Longrightarrow \exists z v .
\end{array}$$

If $z \neq x$ then $\exists z\, rep(x,y \mid v)) = rep(x,y \mid \exists z v)$, and so I have $L_0(\exists z v)$. If $z = x$ then $rep(x,y \mid \exists x v) = \exists x v$ and so a derivation of $x \equiv y, \exists z v \Longrightarrow \exists z v$ gives both the derivations of $L_0(\exists z v)$. The case of $L_0(\forall z v)$ is analogous.

LEMMA 6 (i) $\vdash_= x \equiv t \to (v \leftrightarrow sub(x, t \mid v))$.

(ii) $\vdash_= t_0 \equiv t_1 \to (sub(x, t_0 \mid v) \leftrightarrow sub(x, t_1 \mid v))$.

(iii) $\vdash_= sub(x, t \mid v) \leftrightarrow \exists x\, (x \equiv t \wedge v)$ if not $x \in occ(t)$.

Assume first that $\zeta(x, t)$ is free for v. Let y be not in var(v). Then $\zeta(y, t)$ is free for $rep(x, y \mid v)$, hence also for $x \equiv y \to (v \to rep(x, y \mid v))$. By $L_0(v)$ this formula is provable in ccqs$_=$, hence in ccqt$_=$. Thus $\vdash_= rep(y, t \mid x \equiv y \to (v \to rep(x,y \mid v)))$, and this is $x \equiv t \to (v \to rep(y, t \mid rep(x,y \mid v)))$. But $rep(y,t \mid rep(x,y \mid v)) = rep(x,t \mid v)$ by Lemma **1.13**.9(iii). The case of $x \equiv t \to (rep(x, t \mid v) \to v)$ is analogous.

In the general case, $\zeta(x, t)$ is free for $tot(x, t \mid v)$, thus now $\vdash_= x \equiv t \to (tot(x, t \mid v) \leftrightarrow sub(x, t \mid v))$. But $\vdash v \leftrightarrow tot(x, t \mid v)$ by Lemma **9**.1(iii), and this implies (i).

The axiom $\forall\xi\ \forall\xi'\ (\xi \equiv \xi' \to (p(\xi) \to p(\xi'))$ in EA_u implies (ii) in the case of an atomic formula v; if both $\zeta(x, t_0)$, $\zeta(x, t_1)$ are free for v then (ii) follows by a straightforward induction on v. In the general case, let Y_0 contain all variables and set $Y_1 = occ(t_0) \cup occ(t_1)$. Thus $tot(Y_0, Y_1, \zeta(x, t_0) \mid v)$ has its bound variables also outside $occ(t_1)$, and so *both* $\zeta(x, t_0)$, $\zeta(x, t_1)$ are free for $tot(Y_0, Y_1, \zeta(x, t_0) \mid v)$. Hence

$$(x_0) \quad \vdash_= \ t_0 \equiv t_1 \to (rep(x, t_0 \mid tot(Y_0, Y_1, \zeta(x, t_0) \mid v)) \longleftrightarrow rep(x, t_1 \mid tot(Y_0, Y_1, \zeta(x, t_0) \mid v))).$$

But $\vdash v \longleftrightarrow tot(Y_0, Y_1, \zeta(x, t_0) \mid v)$, $\vdash v \longleftrightarrow tot(x, t_0 \mid v)$ by Lemma 9.1(iii), hence $\vdash tot(Y_0, Y_1, \zeta(x, t_0) \mid v) \longleftrightarrow tot(x, t_0 \mid v)$ and so

$$\vdash rep(x, t_0 \mid tot(Y_0, Y_1, \zeta(x, t_0) \mid v)) \longleftrightarrow rep(x, t_0 \mid tot(x, t_0 \mid v)),$$

i.e.

$$(x_1) \quad \vdash rep(x, t_0 \mid tot(Y_0, Y_1, \zeta(x, t_0) \mid v)) \longleftrightarrow sub(x, t_0 \mid v)$$

By the same argument also

$$(x_2) \quad \vdash rep(x, t_1 \mid tot(Y_0, Y_1, \zeta(x, t_1) \mid v)) \longleftrightarrow sub(x, t_1 \mid v).$$

So from (x_1) and (x_0)

$$(x_3) \quad \vdash_= \ t_0 \equiv t_1 \to (sub(x, t_0 \mid v) \longleftrightarrow rep(x, t_1 \mid tot(Y_0, Y_1, \zeta(x, t_0) \mid v))).$$

Now $\vdash v \longleftrightarrow tot(Y_0, Y_1, \zeta(x, t_0) \mid v)$, $\vdash v \longleftrightarrow tot(Y_0, Y_1, \zeta(x, t_1) \mid v)$ by Lemma 9.1(iii), hence $\vdash tot(Y_0, Y_1, \zeta(x, t_0) \mid v) \longleftrightarrow tot(Y_0, Y_1, \zeta(x, t_1) \mid v)$ and

$$\vdash rep(x, t_1 \mid tot(Y_0, Y_1, \zeta(x, t_0) \mid v)) \longleftrightarrow rep(x, t_1 \mid tot(Y_0, Y_1, \zeta(x, t_1) \mid v).$$

Hence from (x_2)

$$\vdash rep(x, t_1 \mid tot(Y_0, Y_1, \zeta(x, t_0) \mid v)) \longleftrightarrow sub(x, t_1 \mid v)$$

which together with (x_3) proves (ii).

As for (iii), $\vdash_= sub(x, t \mid v) \to (t \equiv t \wedge sub(x, t \mid v))$ holds trivially, i.e. $\vdash_= sub(x, t \mid v) \to sub(x, t \mid x \equiv t \wedge v)$. This implies $\vdash_= sub(x, t \mid v) \to \exists x\ (x \equiv t \wedge v)$.

On the other hand, $\vdash_= x \equiv t \to (v \to sub(x, t \mid v))$ by (i), i.e. $\vdash_= x \equiv t \wedge v \to sub(x, t \mid v)$. Thus $\vdash_= \exists x\ (x \equiv t \wedge v) \to sub(x, t \mid v)$ because x is not free in $sub(x, t \mid v)$

Statement (iii) offers the possibility to eliminate replacement and substitution from equality logic.

Reference

G.Takeuti: Proof Theory. Amsterdam 1975. 2nd. edition 1987.

4. Language Extensions with Predicate Symbols

Consider a language L and an extension L_p of L by a predicate symbol p of arity n, n>0. Let $v(\xi)$ be an L-formula which has its free variables in $\xi = \langle x_0, \ldots, x_{n-1}\rangle$. The L-sentence

$$u^p = \forall\xi\, w(\xi) \quad \text{with} \quad w(\xi) = (v(\xi) \to p(\xi)) \wedge (p(\xi) \to v(\xi))$$

I call *the definition* of $p(\xi)$ by v. If C is a set of L-sentences then u^p, C^p is called the *definitorial extension* of C by u^p. In the case of equality logic the equality axioms $EA_u{}^p$ of L_p consist of the equality axioms EA_u of L and the axiom $\forall\xi\,\forall\xi'\ (\xi \equiv \xi' \to (p(\xi) \to p(\xi')))$ which I denote as e^p.

It is trivial that every definitorial extension u^p, C^p is semantically *conservative* over C: if v is an L-sentence then $v\epsilon ct(\{u^p\}\cup C)$ implies $v\epsilon ct(C)$. Because every L-structure A which is a model of C can be made into an L_p-structure B which is a model of $\{u^p\}\cup C$: define p^B to be the set of all α in $u(A)^n$ which satisfy $u(\xi)$.

I shall now prove that such extensions are also *conservative* with respect to derivability. To this end, I consider a sequential calculus which may be intuitionistic or classical (an in the former case the sequences on the right sides will consist of at most one formula); for reasons of convenience I assume that X is either L_tJ or M_tK. In the case of equational logic I use the axioms EA_u and e^p. Let C_0, C_1 are sequences of L-sentences, and let E the subset of EA_u containing the equality axioms for predicate symbols used in C_0, C_1 or $v(\xi)$.

THEOREM 3 (A) There is an operator transforming every L_p-derivation D of $u^p, C_0 \Longrightarrow C_1$ into an L-derivation of $C_0 \Longrightarrow C_1$.

(B) There is an operator transforming every L_p-derivation D of $u^p, C_0, E, e^p \Longrightarrow C_1$ into an L-derivation of $C_0, E \Longrightarrow C_1$.

Since C_0, C_1 and u^p (and e^p) are sentences, the endsequent of D is pure. Hence I can restrict myself to calculi working with rep and can assume that D itself is pure and cut free. I further may assume that a separation between free and bound variables has been performed with respect to D. In D, the precursors of u^p arise first by dissolving the (uncritical) quantifiers $\forall$ of u^p. When the last of these quantifiers has been dissolved, the precursors are still in the antecendent and have the form

$$\mathrm{rep}(x_{n-1}, t_{n-1} \mid \ldots\ \mathrm{rep}(x_0, t_0 \mid w(\xi))\ \ldots)\ ,$$

and setting $\tau = \langle t_0, \ldots, t_{n-1}\rangle$, I can write them as $w(\tau)$, because successive replacements may also be performed simultaneously in view of the separation between free and bound variables. These $w(\tau)$ I call the *critical* precursors of u^p. All further precursors of $w(\tau) = (v(\tau) \rightarrow p(\tau) \wedge (p(\tau) \rightarrow v(\tau))$ then are $v(\tau) \rightarrow p(\tau)$, $p(\tau) \rightarrow v(\tau)$, the precursors of $v(\tau)$, and $p(\tau)$ itself. None of the precursors containing p will occur as precursor of a formula C_0 or C_1, i.e. as a side formula of a rule not leading to the next precursor of u^p. In the same manner, the precursors of e^p lead to *critical* precursors

$$g(\tau, \tau') = \bigwedge < \tau_0 \equiv \tau_0' \mid i<n> \rightarrow (p(\tau) \rightarrow p(\tau')) .$$

In a first step, I transform D into an object D_1 by removing all quantified precursors of u^p and e^p, including u^p and e^p themselves, in all sequents of D; I also remove all instances of Q–rules which produce them as principal formulas and all instances of (RC) in which one of them was contracted. This preserves all instances of propositional and of Q–rules in D, with the exception of those instances of (E$\forall$) in which a critical $w(\tau)$ or $g(\tau, \tau')$ was quantified:

$$\begin{array}{c} w(t_0, t_1, \dots, t_{n-1}), M_0 \Longrightarrow M_1 \\ \forall x w(x_0, t_1, \dots, t_{n-1}), M_0 \Longrightarrow M_1 \end{array} \quad \text{becomes} \quad \begin{array}{c} w(\tau), M_{01} \Longrightarrow M_1 \\ M_{01} \Longrightarrow M_1 \end{array}$$

$$\begin{array}{c} M_0, g(t_0, t_1, \dots, t_{n-1}, \tau') \Longrightarrow M_1 \\ M_0, \forall x\, g(x_0, t_1, \dots, t_{n-1}, \tau') \Longrightarrow M_1 \end{array} \quad \text{becomes} \quad \begin{array}{c} M_{01}, g(\tau, \tau') \Longrightarrow M_1 \\ M_{01} \Longrightarrow M_1 \end{array}$$

where M_{01} arises from M_0 by omitting *all* quantified precursors of u^p or e^p. Thus the new object D_1 is almost a derivation, with the exception of these pairs of sequents; these I call *critical*.

In a second step, I replace, in every sequent of D_1, every precursor $p(\tau)$ of $w(\tau)$ by $v(\tau)$. Logical axioms depending on $p(\tau)$ then become reflexive axioms depending on $v(\tau)$ (for which I can implant derivations). Instances of logical rules are preserved since $p(\tau)$ cannot be a principal formula and since eigenvariable conditions cannot be violated as no new variables are introduced. Thus I arrive at an object D_{1+} which still is almost a derivation, with the exception of the situation at critical pairs. Those for $w(\tau)$ are replaced by formulas $w'(\tau) = (v(\tau) \rightarrow v(\tau) \wedge (v(\tau) \rightarrow v(\tau))$, such that

$$\begin{array}{c} w(\tau), M_{01} \Longrightarrow M_1 \\ M_{01} \Longrightarrow M_1 \end{array} \quad \text{becomes} \quad \begin{array}{c} w'(\tau), M_{01+} \Longrightarrow M_{1+} \\ M_{01+} \Longrightarrow M_{1+} \end{array}$$

where M_{01+}, M_{1+} arise from M_{01}, M_1 under the replacement of $p(\tau)$. But for each $w'(\tau)$ I have a trivial derivation of $\blacktriangle \Longrightarrow w'(\tau)$, and connecting it with a cut to the upper sequent of my pair, I obtain a derivation of $M_{01+} \Longrightarrow M_{1+}$. Implanting all these derivations into D_{1+}, and then eliminating the cuts, I arrive at derivations for the lower sequents of these critical pairs.

There remain the critical precursors $g(\tau, \tau')$ and their precursors $p(\tau)$, $p(\tau')$. Again, the separation between free and bound variables ensures that the terms in τ, τ' do not contain variables bound in $v(\xi)$; hence the maps η

and η' with $\eta \cdot \xi = \tau$, $\eta' \cdot \xi = \tau'$ are free for $v(\xi)$ and so give rise to $v(\tau) = rep(\eta \mid v)$, $v(\tau') = rep(\eta' \mid v)$. In a third step, I replace, in every sequent of D_{1+}, every precursor $p(\tau)$, $p(\tau')$ of $g(\tau,\tau')$ by $v(\tau)$, $v(\tau')$. Again, logical axioms and instances of rules are preserved. So I arrive at an object D_{1++} which still is almost a derivation, with the exception of the situation at critical pairs for the $g(\tau,\tau')$. These are replaced by formulas $g'(\tau,\tau') = \bigwedge < \tau_0 \equiv \tau_0' \mid i<n> \rightarrow (v(\tau) \rightarrow v(\tau'))$, such that

$$\frac{M_{01+}, g(\tau , \tau') \Longrightarrow M_{1+}}{M_{01+} \Longrightarrow M_{1+}} \quad \text{becomes} \quad \frac{M_{01++}, g'(\tau , \tau') \Longrightarrow M_{1++}}{M_{01++} \Longrightarrow M_{1++}} .$$

It follows by repeated applications of Lemma 6(ii) that $\vdash_= g'(\tau,\tau')$, hence $EA_u \vdash g'(\tau,\tau')$ and $EA_u \vdash_s g'(\tau,\tau')$ which translates into a derivation of $E \Longrightarrow g'(\tau,\tau')$ by definition of E. Connecting this with a cut to the upper sequent of my pair, I obtain a derivation of $E, M_{01++} \Longrightarrow M_{1++}$. Implanting all these derivations into D_{1++}, and then eliminating the cuts, I arrive at derivations for the lower sequents, weakened with E, of these critical pairs. Carrying the sentences E along in the antecedents, I arrive at a derivation of $E, C_0, E \Longrightarrow C_1$ and by contractions at the desired derivation of $C_0, E \Longrightarrow C_1$. This completes the proof of Theorem 3.

Another method to translate situtations from L_p to L is the use of *translation maps*. I define the *direct* translation a^π of an L_p-formula a as $a^\pi = a$ if a is an L-formula, as

$$p(\lambda)^\pi = sub(\eta \mid v) \text{ where } \eta(x_i) = \lambda(i) \text{ for the } x_i \text{ in } \xi ,$$

and as $(\neg a)^\pi = \neg a^\pi$, $(a \% b)^\pi = a^\pi \% b^\pi$ for propositional connectives %, $(Qxa)^\pi = Qx\, a^\pi$ for quantifiers. Then

(x) $\vdash u^p \rightarrow (a^\pi \leftrightarrow a)$ and $fr(a) = fr(a^\pi)$

holds for every L_p-formula a. Because

$$\vdash u^p \rightarrow (sub(\eta \mid v) \rightarrow p(\lambda)) \wedge (p(\lambda) \rightarrow sub(\eta \mid v)),$$

shows the first part of (x) for the formulas $p(\lambda)$, and then induction proves all other cases, e.g. for $(\forall xa)^\pi$ since

$\vdash u^p \rightarrow (a^\pi \rightarrow a)$
$\vdash u^p \rightarrow \forall x\, (a^\pi \rightarrow a)$ as u^p is a sentence
$\vdash u^p \rightarrow \forall x a^\pi \rightarrow \forall x a$ by 9.(a14).

The second part of (x) follows from the fact that all free variables of $v(\xi)$ occur among the members of ξ.

COROLLARY 1 Let S be a $cxqt_0$-deduction of $C \vdash a$, let S^π arise from S replacing each of its formulas a by a^π. Then S^π can be extended to a deduction of $C^\pi \vdash a^\pi$.

To see this, I first define the Y-variant $^{\pi}{}_{Y}$ of $^{\pi}$ for any finite subset Y of the set X of all variables. The only difference occurs for $p(\lambda)$ where I set

$$p(\lambda)^{\pi}{}_{Y} = sub(X,Y,\eta \mid v) = rep(\eta \mid tot(X,Y,\eta \mid v)) \;\; ;$$

I then continue as in the definition of $^{\pi}$. Since v and $tot(X,Y,\eta \mid v)$ are interdeducible in L by Lemma **9**.1.(ii), also $p(\lambda)^{\pi}$ and $p(\lambda)^{\pi}{}_{Y}$ are interdeducible, and induction shows that for every L_p-formula a also (a and) the formulas a^{π} and $a^{\pi}{}_{Y}$ are interdeducible in L. – Next I observe that

(y) $\quad \vdash rep(x,t \mid a^{\pi}{}_{Y}) \longleftrightarrow rep(x,t \mid a)^{\pi}{}_{Y}$.

If $a = p(\lambda)$ and $\eta(x_i) = \lambda(i)$ then $tot(X,Y,\eta \mid v)))$ and $tot(X,Y,h_{\zeta(x,t)} \cdot \eta \mid v)$ are interdeducible with v and, therefore, between themselves. Thus

$$rep(x,t \mid p(\lambda)^{\pi}{}_{Y}) = rep(x,t \mid rep(\eta \mid tot(X,Y,\eta \mid v))) \quad \text{and}$$
$$rep(x,t \mid rep(\eta \mid tot(X,Y,h_{\zeta(x,t)} \cdot \eta \mid v)))$$

are interdeducible. As η is free for $tot(X,Y,h_{\zeta(x,t)} \cdot \eta \mid v)$, it follows from Lemma **1.13**.6 that the last formula is

$$rep(h_{\zeta(x,t)} \cdot \eta \mid tot(X,Y,h_{\zeta(x,t)} \cdot \eta \mid v) = p(h_{\zeta(x,t)} \cdot \lambda)^{\pi}{}_{Y} = rep(x,t \mid p(\lambda))^{\pi}{}_{Y} .$$

Now (y) follows by straightforward induction, e.g. for $(\forall z\, a)^{\pi}{}_{Y}$ since

$$rep(x,t \mid (\forall z\, a)^{\pi}{}_{Y}) = rep(x,t \mid \forall z\; a^{\pi}{}_{Y}) = \forall z\; rep(\zeta(x,t)_z \mid a^{\pi}{}_{Y})$$
$$rep(x,t \mid \forall z\, a)^{\pi}{}_{Y} = (\forall z\; rep(\zeta(x,t)_z \mid a))^{\pi}{}_{Y} = \forall z\; (rep(\zeta(x,t)_z \mid a))^{\pi}{}_{Y} .$$

Further I observe

(z) If t is a term with $occ(t) \subseteq Y$ and if $\zeta(x,t)$ is free for a then $\zeta(x,t)$ is free for $a^{\pi}{}_{Y}$.

Because if a is atomic and $a \neq a^{\pi}{}_{Y}$ then $a^{\pi}{}_{Y}$ is $sub(X,Y,\eta \mid v)$ and, therefore, has no bound variables occurring in t. If a is not atomic then (z) follows by straightforward induction on a.

Given my deduction S of $C \vdash a$, let Y be the set of all variables occurring in S as eigenvariables or in terms t used in axioms (a00), (a01). Let $S^{\pi}{}_{Y}$ arise from S by replacing each of its formulas a by $a^{\pi}{}_{Y}$. Then $S^{\pi}{}_{Y}$ can be extended to a deduction of $C^{\pi}{}_{Y} \vdash a^{\pi}{}_{Y}$. Because the only logical axioms not directly preserved are (a00) and (a01). As for the (a00),

$$(\forall x\, a \rightarrow rep(x,t \mid a))^{\pi}{}_{Y} \;=\; \forall x\, a^{\pi}{}_{Y} \rightarrow rep(x,t \mid a)^{\pi}{}_{Y} .$$

By (y) this is interdeducible with $\forall x\, a^{\pi}{}_{Y} \rightarrow rep(x,t \mid a^{\pi}{}_{Y})$, and by the choice of Y it follows from (z) that $\zeta(x,t)$ is free for $a^{\pi}{}_{Y}$. Thus (a00) is replaced by a formula equivalent to an axiom, and the case of axioms (a01) is analogous. In the same manner I see that quantifier rules e.g.

$$\frac{a \rightarrow rep(x,y \mid b)}{a \rightarrow \forall x\;\; b} \quad \text{can be replaced by} \quad \frac{a^{\pi}{}_{Y} \rightarrow rep(x,y \mid b^{\pi}{}_{Y})}{a^{\pi}{}_{Y} \rightarrow \forall x\;\; b^{\pi}{}_{Y}}$$

where y is not free in a or in $a^{\pi}{}_{Y}$. – Finally, the a^{π} and $a^{\pi}{}_{Y}$ being interdeducible in L, the extension of $S^{\pi}{}_{Y}$ also extends S^{π}.

These developments remain in effect if L_p is obtained from L by extending it with not only one but with a whole set of new predicate symbols; let U^p be the set of the definitions u^p of these predicate symbols. Consider now an axiom system G consisting of

a set G_0 of L-sentences
the definitions U^p
a set G_1 of L_p-sentences.

Assume that for every g in G_1 an L-sentence g^σ is determined such that $U^p, G_0 \Longrightarrow g \leftrightarrow g^\sigma$ is derivable, for instance $g^\sigma = g^\pi$; let $G_1{}^\sigma$ be the set of these g^σ. Let w be an L_p-sentence.

COROLLARY 2 Every L_p-derivation of $G \Longrightarrow w$ can be transformed into an L-derivation of $G_0, G_1{}^\sigma \Longrightarrow w^\pi$.

The equivalences $U^p, G_0 \Longrightarrow g \leftrightarrow g^\sigma$ permit to transform the L_p-derivation of $G_0, U^p, G_1 \Longrightarrow w$ into an L_p-derivation of $G_0, U^p, G_1{}^\sigma \Longrightarrow w$, and the equivalences $U^p \Longrightarrow a \leftrightarrow a^\pi$ permit to transform that into an L_p-derivation of $G_0, U^p, G_1{}^\sigma \Longrightarrow w^\pi$. By Theorem 1 this can be transformed into an L-derivation of $G_0, G_1{}^\sigma \Longrightarrow w^\pi$.

5. Language Extensions with Function Symbols 1

I shall now turn to the analogous questions for extensions of languages by function symbols. In a semantical setting, I treated these matters already in Chapter **1.14** when discussing Skolem extensions of L and Skolem expansions of L-structures.

Consider a language L and an extension L_f of L by a function symbol f of arity n, $n \geq 0$. Let $v = v(\xi, y)$ be an L-formula with y as a free variable and its remaining free variables in $\xi = \langle x_0, \dots, x_{n-1} \rangle$; I may assume that none of the free variables in v occurs bound. I then form the L-sentence

$$u = \forall\xi\, \exists y\, v(\xi, y)$$

and the L_f-sentence

$$u^f = \forall\xi\, v(\xi, f(\xi))$$

where $v(\xi, f(\xi))$ abbreviates $rep(y, f(\xi) \mid v)$. I call u^f the *definition* of $f(\xi)$ by u. If C is a set of L-sentences then u^f, $C^f = \{u^f\} \cup C$ is called *the definitorial extension* of C by u^f.

It is easy to see that everything which can be derived from u can also be derived from u^f. Because I can derive $u^f \Longrightarrow u$ by

$$v(\xi, f(\xi)) \Longrightarrow v(\xi, f(\xi))$$
$$v(\xi, f(\xi)) \Longrightarrow \exists y\ v(\xi, y)$$
$$\forall\xi\ v(\xi, f(\xi)) \Longrightarrow \exists y\ v(\xi, y)$$
$$\forall\xi\ v(\xi, f(\xi)) \Longrightarrow \forall\xi\ \exists y\ v(\xi, y) \quad .$$

Thus an L-derivation of $u, C_0 \Longrightarrow C_1$ can be transformed into an L_f-derivation of u^f, $C_0 \Longrightarrow C_1$ by performing a cut with $u^f \Longrightarrow u$ which afterwards can be eliminated.

It will require some more effort to establish the reverse result that u^f, C^f_0 is *conservative* over u, C_0 if an algorithmic transformation of derivations is desired.

I again consider a sequential calculus which may be classical as well as intuitionistic, and I choose again L_tJ or M_tK. I shall consider derivations D with endsequents consisting of sentences only. Hence they will be pure, and I so can restrict myself to calculi working with rep, and then can assume that D itself is pure and cut free. I shall further assume that a separation between free and bound variables has been performed with respect to D.

In the case of intuitionistic equality logic I shall have to make the following hypothesis on the *decidability of equalities:*

(DC) If a is a equality atom and if q is a conjunction of equality atoms then $q \vee (q \rightarrow a)$ is derivable.

From now on, let C_0, C_1 be sequences of L-sentences. Let E be a sufficiently large subset of L-equality axioms; I denote by

$$e^f = \forall\xi\forall\xi\,(\xi \equiv \xi' \rightarrow f(\xi) \equiv f(\xi'))$$

the equality axiom for f, where $\xi = <x_0, \ldots, x_{n-1}>$, $\xi' = <x_0', \ldots, x_{n-1}'>$ are two sequences of pairwise distinct variables.

THEOREM 4 (A) There is an operator transforming an L_f-derivation of u^f, $C_0 \Longrightarrow C_1$ into an L-derivation of u, $C_0 \Longrightarrow C_1$.

(B) There is an operator transforming an L_f-derivation of u^f, C_0, E, $e^f \Longrightarrow C_1$ into an L-derivation of u, C_0, E $\Longrightarrow C_1$, where in the case of intuitionistic logic (DC) is assumed for L .

The proof will be arranged in such a manner that the algorithm in (A) is a special case of that in (B). Let D be a cut free L_f-derivation of the sequent given in (B). The algorithm transforming D consists in replacing each of its L_f-sequents

(c0) $M_0 \Longrightarrow M_1$

by an L-sequent

(c1) $\quad$ $u, A(\mu), M_0^{\dagger}, E \Longrightarrow M_1^{\dagger}$

and in showing that the tree of L-sequents arising this way can be expanded into an L-sequent of the new sequent listed in (B). Before introducing the concepts necessary to define (c1), I shall describe the idea underlying this construction.

A term, starting with the new operation symbol f, I shall call an f-*term* . If D is to be transformed into an L-derivation then, in the sequents (c0), the f-terms must be removed or be replaced by terms already belonging to L. A convenient way to achieve this will be to replace the f-terms by new variables, where *new* means those which not occur in the finitely many formulas in sequents of D. In order not to create new f-terms (containing the new variables), this replacement shall start with f-terms of maximal degree (such that none of the other f-terms can be affected) and then continue with the f-terms of next lower degree. As a matter of fact, I shall perform this replacement procedure under certain hypotheses, because it is easily seen that it will preserve instances of propositional rules, and it will preserve instances of Q-rules, provided I have not met a *wicked* situation in which the quantified variable in the principal formulas itself occurs within an f-term. In that wicked case, however, instances of Q-rules will not be preserved. And the wicked case *will* be met, because the endsequence of D does contain the formulas u^f and e^f which contain f-terms with bound variables. Moreover, the example of u^f makes it evident that the results of the replacement process may not always be of desirable form: if $f(\xi)$ is replaced by a new variable z then u^f becomes $\forall\xi\, v(\xi,z)$, and while in antecedents this formula might be replaced by the stronger formula $\exists y \forall\xi\, v(\xi,y)$, it hardly could be replaced by the weaker formula u itself.

In such wicked situations, therefore, additional changes will become necessary, and in order to describe them I shall call *wicked* any occurrence of a formula in a sequent of D which contains an f-term with at least one bound variable. Since D is cut free, occurrences of wicked formulas in the premiss of a rule will have wicked successors in the conclusion. But the only wicked formulas in the endsequent of D are u^f and e^f; hence all wicked formulas are precursors of u^f and e^f. Such precursors arise by dissolving the (uncritical) quantifiers $\forall$ of u^f and e^f; they thus remain in the antecedents. When the last of these quantifiers $\forall$ has been dissolved, I arrive at formulas

$$\mathrm{rep}(x_{n-1}, t_{n-1} \mid \ldots\, \mathrm{rep}(x_0, t_0 \mid v(\xi, f(\xi))\, \ldots) \ ,$$

$$\mathrm{rep}(x_{n-1}', t_{n-1}' \mid \ldots\, \mathrm{rep}(x_0', t_0' \mid \mathrm{rep}(x_{n-1}, t_{n-1} \mid \ldots\, \mathrm{rep}(x_0, t_0 \mid e^f)\, \ldots))\, \ldots) \ ,$$

and setting $\tau = \langle t_0, \ldots, t_{n-1} \rangle$, $\tau' = \langle t_0', \ldots, t_{n-1}' \rangle$, I can write them as

$$v(\tau, f(\tau)) \quad \text{and} \quad \tau \equiv \tau' \rightarrow f(\tau) \equiv f(\tau')$$

because the successive replacements may also be performed simultaneously in view of the separation between free and bound variables. It follows from the shape of u^f and of e^f that *these* formulas are *not* wicked anymore; hence they are the *lowest* non-wicked precursors of u^f and e^f. So it is precisely the places at which the exterior quantifiers $\forall$ of u^f and e^f are dissolved which give rise to the wicked situations. Observe, in particular, that wicked formulas only occur on the left.

My aim is the transformation of D into an L-derivation in which u^f and e^f have disappeared from the endsequents (and u may have taken the place of u^f). I shall circumvent the problem of wicked situations by removing, in a first step, besides u^f and e^f, *also all* their wicked precursors; all not wicked formulas I shall subject to the replacement procedure for f-terms. It then remains to discharge also all those formulas which arose by the replacement procedure from the lowest non-wicked predecessors of u^f and e^f. Of them, for instance

$$v(\tau, f(\tau)) \quad \text{becomes} \quad v(\tau^\dagger, z)$$

if $f(\tau)$ is replaced by z and if $\tau^\dagger$ consists of the replaced terms from τ. In order to remove this formula, it suggests itself to use the critical rule (E$\exists$) to obtain $\exists y\, v(\tau^\dagger, y)$ and then to proceed by uncritical instances of (E$\forall$) to $\forall y\, \exists y\, v(\xi, y)$, i.e. to the formula u. However, to do so it must be secured that z can serve as the eigenvariable for (E$\exists$), meaning that it does not occur in any of the terms in $\tau^\dagger$.

But it may well happen that in the derivation D an exterior quantifier $\forall$ of u^f will be dissolved more than once. In that case, on the same branch of D, there may appear *several* of such lowest non-wicked precursors of u^f, and if besides $v(\tau, f(\tau))$ another one, $v(\rho, f(\rho))$, occurs in the same sequent and then is transformed into $v(\rho^\dagger, z')$, then an eigenvariable z must not occur in the terms of $\rho^\dagger$ either. It is this the reason that not only the replacement of f-terms, but also the removal of the formulas $v(\tau^\dagger, z)$ must be performed in a *particular order,* permitting the execution of the critical (E$\exists$) instances. *That* order, however, may be quite different from the one in which the terms from τ appeared during the dissolution of the exterior quantifiers of u^f in D, and for this reason care must be taken also with respect to possible *other* instances of critical Q-rules, in order not disturb *their* conditions on eigenvariables by putting terms, say from $\tau^\dagger$ into their transformed conclusions when originally no terms from τ had been present there.

Having described the *idea* of the construction, I now begin with its details. I consider the f-terms occurring in D which contain exclusively free variables, and I order them in a sequence

$$(1) \qquad f(\tau_0)\ ,\ f(\tau_1)\ ,\ f(\tau_2)\ ,\ \ldots$$

with descending degrees: $|f(\tau_i)| \geq |f(\tau_{i+1})|$ for $i>0$. I then choose a sequence

$$(2) \qquad z_0\ ,\ z_1\ ,\ z_2\ ,\ \ldots$$

of new free variables, and perform the term replacements

$f(\tau_0)$ replaced by z_0 ,
$f(\tau_1)$ replaced by z_1 ,

....

on all terms t occurring in sequents of D. The result after the performance of *all* these replacements upon some term t, or some formula w, I denote as $t^\dagger$ and $w^\dagger$ respectively, and for the argument sequences $\tau_k = <t_0^k, \dots, t_{n-1}^k>$ from (1) I set $\tau_i^\dagger = <t_0^{k\dagger}, \dots, t_{n-1}^{k\dagger}>$. Making use of the notion of wicked (occurrences) of formulas as defined above, I then obtain:

(3) A formula $w^\dagger$ contains an f-term if, and only if, w is wicked ,

(4) The variables different from z_i in $fr(w^\dagger)$ occur already in fr(w) ,

(5) If z_j occurs in a term from $\tau_i^\dagger$ then $f(\tau_j)$ is subterm of the corresponding term in τ_i, hence $|f(\tau_j)| < |f(\tau_i)|$ and thus $i<j$.

(6) If $j \leq i$ then z_j does not occur in any term from $\tau_i^\dagger$.

This describes the replacement process. I now introduce the abbreviations

$$a(i) = v(\tau_i^\dagger, z_i) \quad , \quad a(i,j) = \tau_i^\dagger \equiv \tau_j^\dagger \rightarrow z_i \equiv z_j$$

and define, for every sequence μ of indices, $def(\mu) = k$, the sequence $A(\mu)$ of formulas to consist of

$$\text{all } a(\mu(j)) , j<k, \quad \text{and} \quad \text{all } a(\mu(j), \mu(h)) , \quad h<j<k$$

(where for the construction (A) the $a(\mu(j), \mu(h))$ are omitted). For every sequent (c0) I then define the sequent (c1) by setting for sequences M of formulas

M^- to be the set of all not wicked w from M ,

$M^\dagger$ to be the set of all $w^\dagger$ for w in M^- ,

and defining for $M_0 \Longrightarrow M_1$

μ to be the sequence of numbers of f-terms listed in (1) which occur in $M^- \cup M_1$.

It follows from this definition that, in particular, the endsequent of D is transformed into the desired sequent $u, C_0, E \Longrightarrow C_1$: there are no f-terms in C_0, C_1 or E, hence $(u^f, C_0, E, e_f)^\dagger$ is C_0, E, and the sequence μ is empty. It also follows that logical axioms are transformed into logical axioms again, because their atomic formulas w cannot be wicked and thus will be replaced everywhere by $w^\dagger$. By the same reason, instances of propositional rules are preserved. It remains to study the transformation of instances of Q-rules. To this end, and writing $\mu+\pi$ for the concatenation of two sequences μ and π , I state the

LEMMA 7 Let H, K, $A(\mu)$ be sequences of formulas. Let π be a sequence of indices, $\mathrm{def}(\pi) = q$, such that the new variables $z_{\pi(p)}$, $p<q$, from (2) do not occur in those formulas. Then a derivation of

(a1) $A(\mu+\pi)$, H, E $\Longrightarrow$ K

can be extended to one of

(a2) u, $A(\mu)$, H, E $\Longrightarrow$ K .

I defer its proof and use it to proceed in proving Theorem 4. I shall show that a derivation of the transformed premiss of a Q-rule can be extended to a derivation of the transformed conclusion. I distinguish the following possibilities:

1. The principal formula is wicked.

1.1. The side formula is wicked.

In that case premiss and conclusion have the same transform, because both principal and side formula disappear, and none of them contributes to the sequence μ .

1.2. The side formula is not wicked.
There then are precisely the two following subcases:

1.2.1. The side formula is $v(\tau_i, f(\tau_i))$ for a suitable i.

Then the premiss has the form

$$v(\tau_i, f(\tau_i)),\ M_0 \Longrightarrow M_1\ .$$

Let μ be the sequence of numbers of f-terms in (1) which occur in $M_0^{-} \cup M_1$; let π be the sequence of numbers of f-terms in (1) which occur in $v(\tau_i, f(\tau_i))$ and do *not* occur in $M_0^{-} \cup M_1$. Then the premiss is transformed into

$$u,\ A(\mu+\pi),\ M_0^{\dagger},\ E \Longrightarrow M_1^{\dagger}$$

because the index i is a member of μ or of π and so $(v(\tau_i, f(\tau_i)))^{\dagger} = a(i)$ is already contained in $A(\mu+\pi)$. By *definition* of π, the variables $z_{\pi(p)}$ cannot occur in $M_0^{\dagger}$ or in $M_1^{\dagger}$. Nor can they occur in $A(\mu)$. Because then they would have to occur in of the formulas $a(\mu(j))$ or $a(\mu(j), \mu(h))$, hence in one of the terms from $\tau_{\mu(j)}^{\dagger}$, $\tau_{\mu(h)}^{\dagger}$. It would follow from (5) that $f(\tau_{\pi(p)})$ is subterm of a term from $\tau_{\mu(j)}$ or $\tau_{\mu(h)}$, hence also subterm of $f(\tau_{\mu(j)})$ or $f(\tau_{\mu(h)})$, and thus, together with these terms, also $z_{\pi(p)}$ would occur already in $M_0 \cup M_1$. Hence the hypotheses of Lemma 7 are satisfied, and a derivation of the transformed premiss can be extended to one of

$$u,\ A(\mu),\ M_0^{-\dagger},\ E \Longrightarrow M_1^{\dagger}$$

which this is the transform of the conclusion.

1.2.2. The side formula is $\tau_i \equiv \tau_h \rightarrow f(\tau_i) \equiv f(\tau_h)$ for suitable i, h .

This case is handled in analogy to the preceding one. The only additional consideration required is caused by the fact that the transform $a(i,h)$ of the side formula will occur in $A(\mu+\pi)$ only if, in the particular sequence $\mu+\pi$, the number h appears before the number i. If that is not the case, then $a(h,i)$ will occur also, and $a(i,h)$ disappears by a cut with $A(\mu+\pi), E \Longrightarrow a(i,h)$; if i equals h then $a(i,i)$ disappears by a cut with $E \Longrightarrow a(i,i)$.

2. The principal formula $Qx\,w$ is not wicked.

Then the the bound variable x of $Qx\,w$ cannot occur in an f-term; hence it will be in $fr(w^\dagger)$ if, and only if, it is in $fr(w)$. So a replacement $rep(x,y \mid w)$ cannot put y into an f-term whence $(rep(x,y \mid w))^\dagger = rep(x,y \mid w^\dagger)$. And in the same way, a replacement $rep(x,t \mid w)$ cannot put the term t into a larger f-term, whence also $(rep(x,t \mid w))^\dagger = rep(x,t^\dagger \mid w^\dagger)$. For the following, it will not matter whether the principal formula is on the right or on the left; I restrict myself to the left case. However, I distinguish

2.1. The Q-rule is not critical.

Let $rep(x,t \mid w), \forall x\,w, M_0 \Longrightarrow M_1$ and $\forall x\,w, M_0 \Longrightarrow M_1$ be premiss and conclusion. Let μ be the sequence of numbers of f-terms in (1) which occur in $M_0^- \cup M_1$ or in w; let π be the sequence of numbers of f-terms in (1) which occur in t but *not* in $M_0^- \cup M_1$ or in w. Then the premiss is transformed into

$$u,\ A(\mu+\pi),\ rep(x,t^\dagger \mid w^\dagger),\ \forall x\,w^\dagger,\ M_0^\dagger,\ E \Longrightarrow M_1^\dagger$$

from where $(E\forall)$ leads to

$$u,\ A(\mu+\pi),\ \forall x\,w^\dagger,\ M_0^\dagger,\ E \Longrightarrow M_1^\dagger\ .$$

It follows, as it did in case 1.2.1, that the $z_{\pi(p)}$ occur neither in $A(\mu)$, $M_{01}^\dagger$, $M_1^\dagger$ nor in $\forall x\,w^\dagger$, and applying Lemma 7, I obtain a derivation of

$$u,\ A(\mu),\ \forall x\,w^\dagger,\ M_0^\dagger, E \Longrightarrow M_1^\dagger$$

which is the transform of the conclusion.

2.2. The Q-rule is critical.

Let $rep(x,y \mid w), \exists x\,w, M_0 \Longrightarrow M_1$ and $\exists x\,w, M_0 \Longrightarrow M_1$ be premiss and conclusion. These sequents determine the same sequence $\mu(0), \ldots, \mu(k-1)$, and so their transforms are

$$u,\ A(\mu),\ rep(x,y \mid w)^\dagger,\ \exists x\,w^\dagger,\ M_0^\dagger,\ E \Longrightarrow M_1^\dagger$$
$$u,\ A(\mu),\ \exists x\,w^\dagger,\ M_0^\dagger,\ E \Longrightarrow M_1^\dagger\ .$$

It remains to be shown that here y can serve as the eigenvariable of an instance of $(E\exists)$. Since y was not free in the original conclusion, it cannot be free in $M_0^\dagger$, $M_1^\dagger$ or in $\exists x\,w^\dagger$. If y were free in $A(\mu)$, hence in some $a(\mu(j))$ or $a(\mu(j), \mu(h))$, then it would occur in some term from $\tau_{\mu(j)}{}^\dagger$ or $\tau_{\mu(h)}{}^\dagger$, hence also in the corresponding term from $\tau_{\mu(j)}$ or $\tau_{\mu(h)}$ and, therefore, in the f-term $f(\tau_{\mu(j)})$ or $f(\tau_{\mu(h)})$. By definition of μ, these terms occur in

$M_0 \cup M_1$ or in rep(x,y | w) – and in this latter case then also in w – and thus y would have been free already in the original conclusion.

This concludes the proof of the Theorem; it remains to prove Lemma 7. For sequences ρ and σ of terms, $\mathrm{def}(\rho) = \mathrm{def}(\sigma) = n$, and for variables z, y, I introduce the abbreviations

$$b(\rho,\sigma,z,y) = (\rho \equiv \sigma \to z \equiv y) .$$

In the case of equality logic, I shall need a further Lemma in which I use the sequence $\tau_0, \tau_1, \tau_2, \ldots$ of sequences of terms, each of length n, which is determined by the sequence (1) of terms.

LEMMA 8 Let H, K, $A(\mu)$ be sequences of formulas. Let z be a free variable, not free in their formulas, and let ρ be a sequence of n terms in which z does not occur either. Then a derivation of

$$\text{(b1)} \quad b(\rho, \tau_{\mu(0)}{}^{\dagger}, z, z_{\mu(0)}), \ldots, b(\rho, \tau_{\mu(k-1)}{}^{\dagger}, z, z_{\mu(k-1)}), v(\rho,z), A(\mu), H, E \Longrightarrow K$$

can be extended to one of

$$\text{(b2)} \quad v(\rho,z), A(\mu), H, E \Longrightarrow K .$$

It is the proof of this Lemma which, in the intuitionistic case, requires the hypothesis (DC). I start from the fact that the equality axioms permit, for every further sequence π of n terms, derivations of

$$E, \rho \equiv \pi, b(\pi,\sigma,z,y) \Longrightarrow b(\rho,\sigma,z,y) ,$$

$$E, \rho \equiv \pi, v(\pi,z) \Longrightarrow v(\rho,z) .$$

Let k be $\mathrm{def}(\mu)$. I fix a number r such that $r<k$ and set $\pi = \tau_{\mu(r)}{}^{\dagger}$. Choosing as ρ and z successively the $\tau_{\mu(j)}{}^{\dagger}$ and the $z_{\mu(j)}$ for $j<k$, the derivation of (b1) gives rise, after 2k cuts, to a derivation of

$$\rho \equiv \tau_{\mu(r)}{}^{\dagger}, b(\tau_{\mu(r)}{}^{\dagger}, \tau_{\mu(0)}{}^{\dagger}, z, z_{\mu(0)}), \ldots, b(\tau_{\mu(r)}{}^{\dagger}, \tau_{\mu(k-1)}{}^{\dagger}, z, z_{\mu(k-1)}), v(\tau_{\mu(r)}{}^{\dagger}, z), A(\mu), H, E \Longrightarrow K .$$

Since z shall not be free in $A(\mu)$, it occurs in none of the sequences $\tau_{\mu(j)}{}^{\dagger}$, $\tau_{\mu(r)}{}^{\dagger}$. Making use of the Replacement Lemma, I replace z by $z_{\mu(r)}$ and thus obtain a derivation of

$$\rho \equiv \tau_{\mu(r)}{}^{\dagger}, b(\tau_{\mu(r)}{}^{\dagger}, \tau_{\mu(0)}{}^{\dagger}, z_{\mu(r)}, z_{\mu(0)}), \ldots$$
$$\ldots, b(\tau_{\mu(r)}{}^{\dagger}, \tau_{\mu(k-1)}{}^{\dagger}, z_{\mu(r)}, z_{\mu(k-1)}), v(\tau_{\mu(r)}{}^{\dagger}, z_{\mu(r)}), A(\mu), H, E \Longrightarrow K .$$

But $b(\tau_{\mu(r)}{}^{\dagger}, \tau_{\mu(j)}{}^{\dagger}, z_{\mu(r)}, z_{\mu(j)}) = a(\mu(r), \mu(j))$ and $v(\tau_{\mu(r)}{}^{\dagger}, z_{\mu(r)}) = a(\mu(r))$, and so these formulas belong to $A(\mu)$. Hence I have a derivation of

$$\rho \equiv \tau_{\mu(r)}{}^{\dagger}, A(\mu), H, E \Longrightarrow K \;;$$

Let me abbreviate this sequent as $\Sigma(r)$. Weakening the derivation of $\Sigma(0)$ on the left, I obtain a derivation of

$$\rho \equiv \tau_{\mu(0)}{}^{\dagger},\ b(\rho,\ \tau_{\mu(1)}{}^{\dagger},\ z,\ z_{\mu(1)}),\ \dots\ ,\ b(\rho,\ \tau_{\mu(k-1)}{}^{\dagger},\ z,\ z_{\mu(k-1)}),$$
$$v(\rho, z),\ A(\mu),\ H,\ E \Longrightarrow K\ .$$

Taking this derivation, together with that of (b1), I can apply the rule (Ev), and so obtain a derivation of

$$(\rho \equiv \tau_{\mu(0)}{}^{\dagger}) \vee b(\rho,\ \tau_{\mu(0)}{}^{\dagger},\ z,\ z_{\mu(0)})\ ,\ b(\rho,\ \tau_{\mu(1)}{}^{\dagger},\ z,\ z_{\mu(1)}),\ \dots$$
$$\dots\ ,\ b(\rho,\ \tau_{\mu(k-1)}{}^{\dagger},\ z,\ z_{\mu(k-1)}),\ v(\rho, z),\ A(\mu),\ H,\ E \Longrightarrow K\ .$$

Writing q for $\rho \equiv \tau_{\mu(0)}{}^{\dagger}$ and p for $z \equiv z_{\mu(0)}$, I see that the first formula in this sequent has the form $q \vee (q \rightarrow p)$. It follows from (DC) that is has a derivation, and so a cut leads to a derivation of

$$b(\rho,\ \tau_{\mu(1)}{}^{\dagger},\ z,\ z_{\mu(1)}),\ \dots\ ,\ b(\rho,\ \tau_{\mu(k-1)}{}^{\dagger},\ z,\ z_{\mu(k-1)}),$$
$$v(\rho, z),\ A(\mu),\ H,\ E \Longrightarrow K\ .$$

I thus have removed the first of the formulas of (b1), and now induction shows that I can analogously remove the next k−1 formulas, arriving at a derivation of (b2). This concludes the proof of Lemma 8.

I now turn to the proof of Lemma 7. I set $k = \mathrm{def}(\mu)$ and $q = \mathrm{def}(\pi)$ and further $\pi' = \pi \upharpoonright q-1$. It will be no restriction of generality if I assume that for $p<q-1$ there holds $\pi(p) > \pi(p+1)$. I shall construct a derivation of the sequent which arises from the given (a1) if I replace

$$A(\mu+\pi) \quad \text{by} \quad u\ ,\ A(\mu+\pi')\ ;$$

then the Lemma will follow by induction on q. Here the first sequence differs from the second by containing also the formulas

$$a(\pi(q-1)) = v(\tau_{\pi(q-1)}{}^{\dagger},\ z_{\pi(q-1)})\ ,$$
$$a(\pi(q-1), m) = b(\tau_{\pi(q-1)}{}^{\dagger},\ \tau_{m}{}^{\dagger},\ z_{\pi(q-1)},\ z_{m}) \qquad \text{for m in } \mu+\pi'\ .$$

Writing ρ for $\tau_{\pi(q-1)}{}^{\dagger}$ and z for $z_{\pi(q-1)}$, the sequent (a1) takes the form

$$b(\rho,\ \tau_{\mu(0)}{}^{\dagger},\ z,\ z_{\mu(0)}),\ \dots\ ,\ b(\rho,\ \tau_{\mu(k-1)}{}^{\dagger},\ z,\ z_{\mu(k-1)}),$$
$$b(\rho,\ \tau_{\pi(0)}{}^{\dagger},\ z,\ z_{\pi(0)}),\ \dots\ ,\ b(\rho,\ \tau_{\pi(q-2)}{}^{\dagger},\ z,\ z_{\pi(q-2)}),$$
$$v(\rho, z),\ A(\mu+\pi'),\ H,\ E \Longrightarrow K\ .$$

About the occurrences of the variable z, I observe:

1. z does not occur in H, K, $A(\mu)$

   as follows from the hypothesis of the Lemma ,

2. z does not occur in the sequences $\tau_{\mu(j)}{}^{\dagger}$ for $j<k$

   as follows from 1. and the form of the formulas in $A(\mu)$,

3. z does not occur in the sequences $\tau_{\pi(p)}{}^{\dagger}$ for $p<q-1$, nor in ρ

   as follows from (6) because $\pi(q-1) \leq \pi(p)$ for $p \leq q-1$,

4. z does not occur in any formula $v(z_{\pi(p)}, \tau_{\pi(p)}^{\dagger})$,
$b(\tau_{\pi(p)}^{\dagger}, \tau_m^{\dagger}, z_{\pi(p)}, z_m)$ for $p<q-1$ and m in $\mu+\pi'$

as follows from 2. and 3. ,

5. z does not occur in $A(\mu+\pi')$

since this sequence is $A(\mu)$ together with the formulas from 4. .

Thus the sequent (a1) satisfies the hypotheses of Lemma 8. Applying it, I obtain a derivation of

$$v(\rho, z),\ A(\mu+\pi'),\ H,\ E \Longrightarrow K\ .$$

It follows from 1. and 5. that z here occurs only in $v(\rho, z)$ and at the indicated position. Thus it can serve as the eigenvariable for an (E$\exists$)-instance with the conclusion

$$\exists y\, v(\rho,z),\ A(\mu+\pi'),\ H,\ E \Longrightarrow K\ .$$

Proceeding with n instances of (E$\forall$), I thus obtain a derivation of

$$\forall\xi\, \exists y\, v(\xi,y),\ A(\mu+\pi'),\ H,\ E \Longrightarrow K\ .$$

This concludes the proof of Lemma 7 and, thereby, the proof of Theorem 4. It is essentially due to MAEHARA 55 .

In classical logic there will be occasion to study the situation dual to that of Theorem 4. Given an L-formula $v = v(y,\xi)$ with y and $\xi = \langle x_0, \dots, x_{n-1}\rangle$ as its free variables, I form the L-sentence

$$w = \exists\xi\, \forall y\, v(\xi,y)$$

and the L_f-sentence

$$w^f = \exists\xi\, v(\xi, f(\xi))$$

where $v(\xi, f(\xi))$ abbreviates $\mathrm{sub}(y, f(\xi) \mid v)$. If v does not contain quantifiers then w^f is the Herbrand normal form NH(h) of w, for which a semantical treatment was given in Chapter **1.14**.

It is easy to see that every L-derivation of $C_0 \Longrightarrow w, C_1$ can be transformed into an L_f-derivation of $C_0 \Longrightarrow w^f, C_1$, because I can derive $w \Longrightarrow w^f$ by

$$\begin{array}{r}
v(\xi,f(\xi)) \Longrightarrow v(\xi,\ f(\xi)) \\
\forall y\ v(\xi,y) \Longrightarrow v(\xi,\ f(\xi)) \\
\forall y\ v(\xi,y) \Longrightarrow \exists\xi\ v(\xi,\ f(\xi)) \\
\exists\xi\ \forall y\ v(\xi,y) \Longrightarrow \exists\xi\ v(\xi,\ f(\xi))
\end{array}\ .$$

The reverse result concerns the classical calculus with multiple sequents :

COROLLARY 3 (A) There is an operator transforming an L_f-derivation of $C_0 \Longrightarrow w^f, C_1$ into an L-derivation of $C_0 \Longrightarrow w, C_1$.

(B) There is an operator transforming an L_f-derivation of $C_0, E, e^f \Longrightarrow w^f, C_1$ into an L-derivation of $C_0, E \Longrightarrow w, C_1$.

The construction of the algorithms from Theorem 4 can easily be modified to cover this new situation; observe, however, that while the lowest wicked precursors $\tau \equiv \tau' \to f(\tau) \equiv f(\tau')$ of e^f will continue to appear on the left, the lowest wicked precursors $v(\tau, f(\tau))$ of w^f now will be on the right sides. Instead of one sequence $A(\mu)$, therefore, I will have to use a sequence $A_0(\mu)$ containing the $a(\mu(j), \mu(h))$, which appears on the left sides, and a sequence $A_1(\mu)$ containing the $a(\mu(j))$, which appears on the right sides, such that the transformed sequents (c1) now become

$$A_0(\mu_0), M_0^\dagger, E \Longrightarrow w, A_1(\mu_1), M_1^\dagger$$

where μ_i is the sequence of f-terms listed in (1) which occur in M^-_1. Lemma 7 then has to be stated as saying that, under the hypotheses made, both a derivation

$$A_0(\mu_0+\pi_0), H, E \Longrightarrow A_1(\mu_1), K \quad \text{and} \quad A_0(\mu_0), H, E \Longrightarrow A_1(\mu_1+\pi_1), K$$

can be extended to one of $w, A_0(\mu_0), H, E \Longrightarrow A_1(\mu_1), K$. In the same manner, Lemma 8 has to say that a derivation of

$$b(\rho, \tau_{\mu(0)}{}^\dagger, z, z_{\mu(0)}), \ldots, b(\rho, \tau_{\mu(k-1)}{}^\dagger, z, z_{\mu(k-1)}),$$
$$A_0(\mu_0), H, E \Longrightarrow v(\rho, z), A_1(\mu_1), K$$

can be extended to one of $A_0(\mu_0), H, E \Longrightarrow v(\rho, z), A_1(\mu_1), K$.

The proofs given above carry over immediately to these situations.

A more convenient way, however, is that of a direct reduction of the Corollary, making use of classical derivabilities:

$$C_0, E, e^f \Longrightarrow \exists\xi\, v(\xi, f(\xi)), C_1$$
$$\neg\exists\xi\, v(\xi, f(\xi)), C_0, E, e^f \Longrightarrow C_1$$
$$\forall\xi\, \neg v(\xi, f(\xi)), C_0, E, e^f \Longrightarrow C_1 \qquad \text{since } \forall\neg \Longrightarrow \neg\exists$$
$$\forall\xi\exists y\, \neg v(\xi, y), C_0, E \Longrightarrow C_1 \qquad \text{algorithm in Thm. 2}$$
$$\exists\xi\forall y\, \neg\neg v(\xi, y), C_0, E \Longrightarrow C_1 \qquad \text{since } \neg\exists \Longrightarrow \forall\neg$$
$$C_0, E \Longrightarrow \neg\neg\exists\xi\forall y\, \neg\neg v(\xi, y), C_1$$
$$C_0, E \Longrightarrow \exists\xi\forall y\, v(\xi, y), C_1 \qquad \text{since } \exists \Longrightarrow \neg\neg\exists\neg\neg$$

Theorem 4 is often applied in situations in which a set G of *sentences* is given as an *axiom system* and the classical consequences u of G are studied.

For sentences u the semantical statements $u \epsilon ct(G)$ and $u \epsilon cs(G)$ are equivalent, and by Theorems $\mathbf{9}.1_c$ and $\mathbf{9}.2_c$ they are equivalent to $u \epsilon ccqt(G)$ and $u \epsilon ccqs(G)$; hence by Theorem 1 they are equivalent to the classical derivability of $G \Longrightarrow u$.

Let now u and u^f be as before. Then an L-sentence c deducible, or derivable, from the definitorial extension $\{u^f\} \cup G$ is already deducible and derivable from G alone; this follows as $G \Longrightarrow c$ by a cut from the derivations of $G, u^f \Longrightarrow c$, $G, u \Longrightarrow c$ and $G \Longrightarrow u$.

This purely syntactical result contains an algorithm producing a derivation of $G \Longrightarrow c$ from a derivation of $G, u^f \Longrightarrow c$; it thus is much stronger than the trivial semantical observation that every L-model of G can be expanded into an L_f-model of u^f. Moreover, if G admits infinite models then the semantical argument would require the strong assumption that every such model can be well ordered - unless additional knowledge such as

$$u^n : v(\xi, x) \wedge v(\xi, y) \rightarrow x \equiv y$$

can be deduced from G.

Reference

S.Maehara: The Predicate Calculus with Epsilon-Symbol. J.Math.Soc.Japan **7** (1955) 323-344

6. Language Extensions with Function Symbols 2

If g is an m-ary function symbol, $g(\mu)$ a term, and if $\vartheta = \langle z_0, \ldots, z_{m-1} \rangle$ is a sequence of new variables of the same length m as μ, then by equality logic

$$\vdash z \equiv g(\mu) \longrightarrow z \equiv g(\mu) \wedge \mu(m-1) \equiv \mu(m-1)$$
$$\vdash z \equiv g(\mu) \longrightarrow \exists z_{m-1} (z \equiv g(\mu(0), \ldots, \mu(m-2), z_{m-1}) \wedge z_{m-1} \equiv \mu(m-1))$$
$$\cdots\cdots$$
$$\vdash z \equiv g(\mu) \longrightarrow \exists \vartheta (z \equiv g(\vartheta) \wedge \bigwedge\!\!\!\bigwedge \langle z_i \equiv \mu(i) \mid i < m \rangle)$$

as well as

$$z \equiv g(\vartheta) \wedge \bigwedge\!\!\!\bigwedge \langle z_i \equiv \mu(i) \mid i < m \rangle \longrightarrow g(\mu)$$
$$\exists \vartheta (z \equiv g(\vartheta) \wedge \bigwedge\!\!\!\bigwedge \langle z_i \equiv \mu(i) \mid i < m \rangle) \longrightarrow g(\mu)$$

whence

(7a) $\vdash z \equiv g(\mu) \longleftrightarrow \exists \vartheta (z \equiv g(\vartheta) \wedge \bigwedge\!\!\!\bigwedge \langle z_i \equiv \mu(i) \mid i < m \rangle)$.

In the same way, if $p(\mu)$ is an atomic formula, and if ϑ is a sequence of new variables, then

(7b) $\vdash p(\mu) \longleftrightarrow \exists \vartheta (p(\vartheta) \wedge \bigwedge\!\!\!\bigwedge \langle \vartheta(i) \equiv \mu(i) \mid i < m \rangle)$.

Consider as before the L–sentence u: $\forall\xi\,\exists y\, v(\xi,y)$. I may assume that no variable from ξ occurs bound in $v(\xi,y)$ because otherwise an application of a suitable map $tot(Y_1|-)$ will transform u into a sentence interdeducible with u which does have that property; thus $\zeta(y, f(\xi))$ is free for $v(\xi,y)$. Consider the extended language L_f and the definition u^f: $\forall\xi\, v(\xi, f(\xi))$. Assume from now on that G is set of L–sentences which G proves both u and

$$u^n : v(\xi,x) \wedge v(\xi,y) \rightarrow x\equiv y \quad .$$

Then $G \vdash u^f \leftrightarrow u^{ff}$ for

$$u^{ff} : \forall\xi\,\forall y\,(\, v(\xi,y) \leftrightarrow y\equiv f(\xi)\,)\ ,$$

because $G, u^f \vdash v(\xi, f(\xi))$, hence $G, u^f, u^n \vdash v(\xi,y) \rightarrow y\equiv f(\xi)$, and further $\vdash y\equiv f(\xi) \rightarrow (v(\xi, f(\xi)) \rightarrow v(\xi,y))$ by equality logic; thus $G,\ u^f,\ u^n \vdash y\equiv f(\xi) \rightarrow v(\xi,y)$. Conversely, $G,\ u^{ff} \vdash y\equiv f(\xi) \rightarrow v(\xi,y)$ whence $G,\ u^{ff} \vdash v(\xi, f(\xi))$ by equality logic. – It follows from $G \vdash u^{ff}$ that now the equality axioms for f become provable from G: if ξ and ξ' are sequences of variables of the same length n such that $G \vdash \bigwedge\!\!\!\bigwedge\langle \xi(i)\equiv \xi'(i)\,|\,i<n\rangle$ then

$$\begin{array}{lll} G \vdash \bigwedge\!\!\!\bigwedge\langle \xi(i)\equiv \xi'(i)\,|\,i<n\rangle & \rightarrow\ v(\xi,x) \rightarrow v(\xi',x) & \text{by equality logic} \\ G \vdash \bigwedge\!\!\!\bigwedge\langle \xi(i)\equiv \xi'(i)\,|\,i<n\rangle & \rightarrow\ y\equiv f(\xi) \rightarrow y\equiv f(\xi') & \text{by } u^{ff}\ , \\ G \vdash \bigwedge\!\!\!\bigwedge\langle \xi(i)\equiv \xi'(i)\,|\,i<n\rangle & \rightarrow\ f(\xi)\equiv f(\xi')\ . & \end{array}$$

Next I observe that for distinct variables ξ, y, z there holds

$$(8)\qquad G \vdash z\equiv f(\xi) \leftrightarrow \exists y\,(v(\xi,y) \wedge z\equiv y)\ .$$

Because

$$\begin{array}{l} u^f,\ z\equiv f(\xi) \vdash v(\xi, f(\xi)) \wedge z\equiv f(\xi) \\ u^f,\ z\equiv f(\xi) \vdash \exists y\,(v(\xi,y) \wedge z\equiv y)\ , \end{array}$$

and by equality logic

$$\begin{array}{l} u^{ff},\ v(\xi,y) \wedge z\equiv y) \vdash y\equiv f(\xi) \wedge z\equiv y \\ u^{ff},\ v(\xi,y) \wedge z\equiv y) \vdash z\equiv f(\xi) \\ u^{ff},\ \exists y\,(v(\xi,y) \wedge z\equiv y) \vdash z\equiv f(\zeta)\ . \end{array}$$

For every L_f–term t and every formula a = (z≡ t) I define the formula a^* as follows. If t is a variable then I set $a^* = a$. If $t = g(\mu)$ and $g\neq f$ and $def(\mu) = m$ then I choose a sequence $\vartheta = \langle z_0,\ldots,z_{m-1}\rangle$ of new variables and set

$$a^* = \exists\vartheta\,(z\equiv g(\vartheta) \wedge \bigwedge\!\!\!\bigwedge\langle(\vartheta(i)\equiv \mu(i))^*\,|\,i<m\rangle).$$

If $t = f(\lambda)$ then I choose a sequence $\vartheta = \langle z_0,\ldots,z_{n-1}\rangle$ of new variables and set

$$a^* = \exists\vartheta\,(\,\exists y\,(v(\vartheta, y) \wedge z\equiv y) \wedge \bigwedge\!\!\!\bigwedge\langle(\vartheta(i)\equiv \lambda(i))^*\,|\,i<m\rangle).$$

Induction on t shows that $G \vdash z\equiv t \leftrightarrow (z\equiv t)^*$; if $t = g(\mu)$ then I use (7a), and if $t = f(\lambda)$ then I also use (8). – There always holds $fr(a) = fr(a^*)$.

The *translation* a^σ of an L_f-formula a I define as follows. If a is an atomic formula $z \equiv t$ then I set $a^\sigma = a^*$; if a is an atomic formula $r \equiv t$ where r is not a variable then I set $a^\sigma = \exists z\,((z\equiv r)^* \wedge (z\equiv t)^*)$. If a is an atomic formula $p(\nu)$ of arity k then I choose a sequence $\vartheta = <z_0,\ldots,z_{k-1}>$ of new variables and set

$$p(\nu)^\sigma = \exists\vartheta\,(p(\vartheta) \wedge \bigwedge\!\!\!\bigwedge <(\vartheta(i)\equiv \nu(i))^* \mid i<m>) \; .$$

If a is composite then I set $(\neg a)^\sigma = \neg a^\sigma$, $(a \,\%\, b)^\sigma = a^\sigma \,\%\, b^\sigma$ for propositional connectives %, $(Qx\,a)^\sigma = Qx\,a^\sigma$ for quantifiers. Induction on formulas shows the

LEMMA 9 For every L_f-formula a, the translation a^σ is an L-formula with $fr(a) = fr(a^\sigma)$ and such that $G \vdash a \leftrightarrow a^\sigma$.

The translation σ also preserves deductions, and in order to see this it is more convenient to consider the modus ponens calculi $cxqt_2$ from Chapter **9** :

THEOREM 5 Assume that $G \vdash u$ and $G \vdash u^n$. Let d be an L_f-formula and let D be a set of L_f-formulas. Every $cxqt_2$-deduction in L_f of G, D, $u^f \vdash d$ can be expanded into an $cxqt_2$-deduction in L of G, $D^\sigma \vdash d^\sigma$ if each of its formulas a is replaced by a^σ.

A formula g in G is transformed into g^σ, and Lemma 9 then permits to implant subderivations of $G \vdash g^\sigma$. As for u^f, observe that

$$\begin{aligned}
&G \vdash \forall\xi\forall z\,(\,v(\xi,z)^\sigma \leftrightarrow v(\xi,z)) \qquad \text{by Lemma 9}\\
&G \vdash \forall\xi\forall z\,(\,v(\xi,z)^\sigma \leftrightarrow \exists\vartheta\,((v(\vartheta,z) \wedge \bigwedge\!\!\!\bigwedge<(\vartheta(i)\equiv \xi(i))^* \mid i<m>))\\
&G \vdash \forall\xi\forall z\,(\,v(\xi,z)^\sigma \leftrightarrow\\
&\qquad\qquad \exists\vartheta\,(\,\exists y\,(v(\vartheta,y) \wedge z\equiv y) \wedge \bigwedge\!\!\!\bigwedge<(\vartheta(i)\equiv \xi(i))^* \mid i<m>))\\
&G \vdash \forall\xi\forall z\,(\,v(\xi,z)^\sigma \leftrightarrow (z\equiv f(\xi))^*\,)\\
&G \vdash (\forall\xi\forall z\,(\,v(\xi,z) \leftrightarrow (z\equiv f(\xi)))^\sigma\\
&G \vdash (\forall\xi\, v(\xi,\, f(\xi)))^\sigma
\end{aligned}$$

and so $G \vdash (u^f)^\sigma$. The transformation with σ preserves the structure of formulas; hence all propositional axioms as well as the quantifier axioms (a14)-(a17) are preserved, and so are the instances of (MP) and of the generalization rule (R1). It remains to consider the translations of the axioms (a00), (a01)

$$\forall x\, w \rightarrow rep(x,t \mid w) \quad \text{and} \quad rep(x,t \mid w) \rightarrow \exists x\, w \;,$$

e.g. in the first case $a = a_0 \rightarrow a_1$ where $a_0 = \forall x\, w$, $a_1 = rep(x,t \mid w)$. It then will suffice to show that

(9a) $\quad G \vdash a_0{}^\sigma \rightarrow a_1{}^\sigma$.

Now $\vdash a_1 \leftrightarrow \exists z\,(z \equiv t \wedge w)$ for a new variable z, and so G proves $a_1{}^\sigma \leftrightarrow \exists z\,(z \equiv t \wedge w)^\sigma$, i.e.

$$G \vdash a_1{}^\sigma \leftrightarrow \exists z\,((z \equiv t)^* \wedge w^\sigma)\ .$$

I shall show that

(9b) $\quad G \vdash a_0{}^\sigma \rightarrow \exists z\,((z \equiv t)^* \wedge w^\sigma)\ ,$

whence also (9a). In order to see (9b), it suffices to prove

(9c) $\quad G \vdash \forall x\, w^\sigma \rightarrow \exists z\,((z \equiv t)^* \wedge w^\sigma)\ ,$

and this certainly holds if t is a variable. If $t = g(\mu)$ and $def(\mu) = m$ then by inductive hypothesis for every $i<m$

$$G \vdash \forall x\, w^\sigma \rightarrow \exists \vartheta(i)\,((\vartheta(i) \equiv \mu(i))^* \wedge w^\sigma)\ ,$$
$$G \vdash \forall x\, w^\sigma \rightarrow \bigwedge\!\!\!\bigwedge <\exists \vartheta(i)\,((\vartheta(i) \equiv \mu(i))^* \wedge w^\sigma)\,|\,i<m>\ .$$

If z_i does not occur in b_i then $\vdash \exists z_0 \exists z_1 (b_0 \wedge b_1) \leftrightarrow \exists z_0 b_0 \wedge \exists z_1 b_1$ by **9**.(A3); as the variables in ϑ are new and distinct, there follows

$$G \vdash \forall x\, w^\sigma \rightarrow \exists \vartheta \bigwedge\!\!\!\bigwedge <(\vartheta(i) \equiv \mu(i))^*\,|\,i<m> \wedge w^\sigma\ .$$

Thus also

$$G \vdash \forall x\, w^\sigma \rightarrow g(\vartheta) \equiv g(\vartheta) \wedge \exists \vartheta \bigwedge\!\!\!\bigwedge <(\vartheta(i) \equiv \mu(i))^*\,|\,i<m> \wedge w^\sigma\ .$$

$$G \vdash \forall x\, w^\sigma \rightarrow \exists z\,(\exists \vartheta\,(z \equiv g(\vartheta) \wedge \bigwedge\!\!\!\bigwedge <(\vartheta(i) \equiv \mu(i))^*\,|\,i<m>) \wedge w^\sigma)\ .$$
$$G \vdash \forall x\, w^\sigma \rightarrow \exists z\,(\exists \vartheta\,(z \equiv g(\vartheta) \wedge \bigwedge\!\!\!\bigwedge <(\vartheta(i) \equiv \mu(i))^*\,|\,i<m>) \wedge w^\sigma)\ .$$

If $g \neq f$ then this is (9c) for $t = g(\mu)$. If $g = f$ and $\mu = \lambda$ then

$$G \vdash \forall x\, w^\sigma \rightarrow \exists z\,(\exists \vartheta\,(z \equiv f(\vartheta) \wedge \bigwedge\!\!\!\bigwedge <(\vartheta(i) \equiv \lambda(i))^*\,|\,i<m>) \wedge w^\sigma)$$

whence (8) gives

$$G \vdash \forall x\, w^\sigma \rightarrow$$
$$\exists z\,(\exists \vartheta\,(\exists y\,(v(\vartheta,y) \wedge z \equiv y) \wedge \bigwedge\!\!\!\bigwedge <(\vartheta(i) \equiv \lambda(i))^*\,|\,i<m>) \wedge w^\sigma)$$

$$G \vdash \forall x\, w^\sigma \rightarrow \exists z\,(\exists \vartheta\,(v(\vartheta,z) \wedge \bigwedge\!\!\!\bigwedge <(\vartheta(i) \equiv \lambda(i))^*\,|\,i<m>) \wedge w^\sigma)\ ,$$

and this is (9c) for $t = f(\lambda)$.

7. The Midsequent Theorem

It follows from the subformula principle that a derivable sequent $A \Longrightarrow v$, none of whose formulas contains quantifiers, has a derivation employing propositional rules only. While this consequence is obvious, it would require much more effort to establish the corresponding observation for derivabilities in modus ponens calculi *without* making use of the translations into cut free sequent calculi; as a matter of fact, this observation then is, essentially, the content of the *First ε-Theorem* of HILBERT-BERNAYS 39 (cf. also the Consistency Theorem of SHOENFIELD 67 , p.49)

GENTZENs 34 *Midsequent Theorem*, to be discussed now, says, essentially, that every classical MK-derivation of an endsequent, containing prenex formulas only, can be transformed in such a manner that all instantiations of propositional rules precede all instantiations of quantifier rules. The transformed derivation, therefore, splits into two parts: an upper one which is purely propositional and ends at a certain sequent called the *midsequent*, and a lower one employing only quantifier rules (and thus laying on a linear tree). The proof is, apart from some minor details, quite simple. For the time being, I admit both classical calculi MK and intuitionistic calculi LJ.

As a preparation, consider a quantified formula Qxw in the endsequent of a derivation D. Reading D upwards, Qxw may be *resolved* into precursors which are substitution instances sub(x,t | w), and if sub(x,t | w) is quantified or propositionally composed itself, then *this* formula may be resolved further. It follows that, in the arising *tree of precursors* of Qxw, a propositional resolution can appear only *above* the corresponding resolutions of the exterior quantifier Qx. Observe, however, the word *corresponding,* because a quantified formula may be resolved *repeatedly* and may appear in an upper sequent still together with already rather distant ones of its precursors (and in extremal calculi this will always be the case). For example, *horror vacui* might have lead to write a classical derivation of the formulas (i9) as

$$
\begin{array}{ll}
w(x),\ w(t) \Rightarrow w(x)\ ,\ u,\ u & \\
\quad w(t) \Rightarrow w(x)\ ,\ u,\ w(x) \rightarrow u & \\
\quad w(t) \Rightarrow w(x)\ ,\ u,\ \exists x\ (w \rightarrow u) & \\
\quad w(t) \Rightarrow \forall x\ w\ ,\ u,\ \exists x\ (w \rightarrow u) & \qquad w(t),\ u \Rightarrow \exists x\ (w \rightarrow u) \\
\end{array}
$$

$$
\begin{array}{l}
w(t),\ \forall x\ w \rightarrow u \Rightarrow u\ ,\ \exists x(w \rightarrow u) \\
\forall x\ w \rightarrow u \Rightarrow w(t) \rightarrow u\ ,\ \exists x\ (w \rightarrow u) \\
\forall x\ w \rightarrow u \Rightarrow \exists x(w \rightarrow u) \qquad ,
\end{array}
$$

where the formula ∃x(w→u) is resolved a first time into its precursor w(t)→u , but also is carried along upwards and, on the left branch, resolved a second time into w(x)→u ; the upper resolution of its quantifier then happens *above* the propositional resolution of its precursor w(t)→u .

Given a derivation D and a rule (R), I shall say that an instance I(R) in D is *movable* if (R) is propositional or (W), or if (R) is (RC) *and* the contraction formula does *not* begin with a quantifier. I shall say that an instance I(P) of a rule (P) is *located immediately* above an instance I(R) of a movable rule (R) if the only rules, instantiated between the conclusion of I(P) and a premiss of I(R), are (RP) and (RC) with a contraction formula which *does* begin with a quantifier.

I shall say that, in a derivation D, an instance I(Q) of a Q-rule (Q), located immediately above an I(R) instance of a movable rule (R), *can be permuted with* I(R) if I can transform D into a derivation D*, differing from D in the location of instances of logical rules *only* in that now I(R) appears immediately above I(Q). – An obvious necessary condition for this to be the case is:

(a) If (R) is propositional then the (occurrence of the) principal formula Qxw of I(Q) is *not* a (occurrence of a) side formula of I(R).

In the case of critical Q–rule (Q), t is an eigenvariable which must not occur in the other side formula of I(R). This may be violated if (R) is one of (I∧), (E∧), (I∨), (E∨) or (W); a sufficient condition to prevent such violation is that D be *eigen*.

I shall now show that these two conditions together are, in most cases, sufficient to ensure the permutability of I(Q) with I(R). In order to express the simplification achieved by a permutation of I(Q) and I(R), I denote, for any I(Q), by $\nu(I(Q))$ the number of movable instances I(R) below I(Q) in D. I then set

$$\nu(D) = \Sigma < \nu(I(Q)) \mid \text{all instances } I(Q) \text{ of Q–rules in } D > .$$

LEMMA 10 Let D be a derivation, let I(Q) be an instance of a Q–rule and let I(R) be a movable instance, located immediately below I(Q). Assume that D is derivation either in a classical calculus with multiple sequents, or in a intuitionistic calculus with (R) distinct from (E∨). Also, assume that D is eigen and that (a) holds.

Then I can construct a derivation D* of the same endsequent in which I(Q) and I(R) have been permuted. D* is eigen and satisfies (a); D and D* have the same length, and there holds $\nu(D^*)+1 = \nu(D)$.

I construct D* from D by explicitly replacing the instances of I(Q) and I(R), distinguishing the following cases.

1. (R) is one of (W), (RC), (E∧), (I∨), (I→), (I¬), (E¬), e.g.

$$\begin{array}{c} b_0, M \Longrightarrow sub(x,t \mid w) \\ b_0, M \Longrightarrow Qxw \\ b_0 \wedge b_1, M \Longrightarrow Qx\, w \end{array} \quad \text{becomes} \quad \begin{array}{c} b_0, M \Longrightarrow sub(x,t \mid w) \\ b_0 \wedge b_1, M \Longrightarrow sub(x,t \mid w) \\ b_0 \wedge b_1, M \Longrightarrow Qx\, w \end{array} ,$$

$$\begin{array}{c} sub(x,t \mid w), M, b_0 \Longrightarrow u \\ Qxw, M, b_0 \Longrightarrow u \\ b_0, Qx\, w, M \Longrightarrow u \\ b_0 \wedge b_1, Qxw, M \Longrightarrow u \end{array} \quad \text{becomes} \quad \begin{array}{c} sub(x,t \mid w), M, b_0 \Longrightarrow u \\ b_0, sub(x,t \mid w), M \Longrightarrow u \\ b_0 \wedge b_1, sub(x,t \mid w), M \Longrightarrow u \\ sub(x,t \mid w), b_0 \wedge b_1, M \Longrightarrow u \\ Qxw, b_0 \wedge b_1, M \Longrightarrow u \\ b_0 \wedge b_1, Qxw, M \Longrightarrow u \end{array} .$$

In the intuitionistic case, if Qxw is on the right then (I∨), (I→), (I¬), (E¬) will not occur. – From now on, I shall omit the indication of instances of the permutation rule.

2. (R) is (E∨), e.g.

$$\dfrac{b_0,\ Qxw,\ M \Longrightarrow u \qquad \dfrac{\text{sub}(x,t\,|\,w),\ b_1,\ M \Longrightarrow u}{Qxw,\ b_1,\ M \Longrightarrow u}}{b_0 \vee b_1,\ Qxw,\ M \Longrightarrow u}$$

becomes

$$\dfrac{\dfrac{\dfrac{b_0,\ \text{sub}(x,t\,|\,w),\ Qxw,\ M \Longrightarrow u \qquad b_1,\ \text{sub}(x,t\,|\,w),\ Qxw,\ M \Longrightarrow u}{b_0 \vee b_1,\ \text{sub}(x,t\,|\,w),\ Qxw,\ M \Longrightarrow u}}{b_0 \vee b_1,\ Qxw,\ Qxw,\ M \Longrightarrow u}}{b_0 \vee b_1,\ Qxw,\ M \Longrightarrow u}\quad .$$

As I am working with 2-ary rules in pure parameter form, I have to weaken with sub(x,t | w) every sequent above the left premiss. Moreover, for M_tK and L_tJ, I have to weaken with Qxw every sequent above the right premiss, and after applying (Q), I have to remove one copy of Qxw with (RP). If these calculi would use rules in mixed parameter form, none of these weakenings would be needed, because in that case

$$\dfrac{b_0,\ M_0 \Longrightarrow u \qquad \dfrac{\text{sub}(x,t\,|\,w),\ b_1,\ M_1 \Longrightarrow u}{Qx\ w,\ b_1,\ M_1 \Longrightarrow u}}{b_0 \vee b_1,\ Qxw,\ M_0,\ M_1 \Longrightarrow u}$$

becomes

$$\dfrac{\dfrac{b_0,\ M_0 \Longrightarrow u \qquad b_1,\ \text{sub}(x,t\,|\,w),\ M_1 \Longrightarrow u}{b_0 \vee b_1,\ \text{sub}(x,t\,|\,w),\ M_0,\ M_1 \Longrightarrow u}}{b_0 \vee b_1,\ Qxw,\ M_0,\ M_1 \Longrightarrow u}\quad .$$

Similarly

$$\dfrac{b_0,\ M \Longrightarrow Qxw \qquad \dfrac{b_1,\ M \Longrightarrow \text{sub}(x,t\,|\,w)}{b_1,\ M \Longrightarrow Qx\ w}}{b_0 \vee b_1,\ M \Longrightarrow Qxw}$$

becomes

$$\dfrac{\dfrac{\dfrac{b\ ,\ M \Longrightarrow Qxw,\ \text{sub}(x,t\,|\,w) \qquad b\ ,\ M \Longrightarrow Qxw,\ \text{sub}(x,t\,|\,w)}{b_0 \vee b_1,\ M \Longrightarrow Qxw,\ \text{sub}(x,t\,|\,w)}}{b_0 \vee b_1,\ M \Longrightarrow Qxw,\ Qx\ w}}{b_0 \vee b_1,\ M \Longrightarrow Qxw}\quad .$$

But here then I would need multiple sequents in any case and thus would have to employ classical logic.

3. (R) is (I∧). This is treated analogously. In the intuitionistic case, Qxw cannot appear on the right.

4. (R) is (E→), e.g.

$$\dfrac{Qxw,\ M \Longrightarrow b_0 \qquad \dfrac{\mathrm{sub}(x,t\,|\,w),\ b_1,\ M \Longrightarrow u}{Qxw,\ b_1,\ M \Longrightarrow u}}{Qxw,\ b_0 \rightarrow b_1,\ M \Longrightarrow u}$$

becomes

$$\dfrac{\dfrac{\mathrm{sub}(x,t\,|\,w),\ Qxw,\ M \Longrightarrow b_0 \qquad \mathrm{sub}(x,t\,|\,w),\ Qxw,\ b_1,\ M \Longrightarrow u}{\mathrm{sub}(x,t\,|\,w),\ Qx\ w,\ b_0 \rightarrow b_1,\ M \Longrightarrow u}}{\dfrac{Qx\ w\ ,\ Qx\ w,\ b_0 \rightarrow b_1,\ M \Longrightarrow u}{Qx\ w,\ b_0 \rightarrow b_1,\ M \Longrightarrow u}}$$

and

$$\dfrac{M \Longrightarrow b_0 \qquad \dfrac{b_1,\ M \Longrightarrow \mathrm{sub}(x,t\,|\,w)}{b_1,\ M \Longrightarrow Qx\ w}}{b_0 \rightarrow b_1,\ M \Longrightarrow Qxw}$$

becomes

$$\dfrac{\dfrac{M \Longrightarrow b_0 \qquad b_1,\ M \Longrightarrow \mathrm{sub}(x,t\,|\,w)}{b_0 \rightarrow b_1,\ M \Longrightarrow \mathrm{sub}(x,t\,|\,w)}}{b_0 \rightarrow b_1,\ M \Longrightarrow Qxw}$$

and

$$\dfrac{\dfrac{M \Longrightarrow b_0,\ \mathrm{sub}(x,t\,|\,w)}{M \Longrightarrow b_0,\ Qxw} \qquad b_1,\ M \Longrightarrow Qxw}{b_0 \rightarrow b_1,\ M \Longrightarrow Qxw,}$$

becomes

$$\dfrac{\dfrac{M \Longrightarrow b_0,\ Qx\ w,\ \mathrm{sub}(x,t\,|\,w) \qquad b_1,\ M \Longrightarrow Qx\ w.\ \mathrm{sub}(x,t\,|\,w)}{b_0 \rightarrow b_1,\ M \Longrightarrow Qxw,\ \mathrm{sub}(x,t\,|\,w)}}{\dfrac{b_0 \rightarrow b_1,\ M \Longrightarrow Qxw,\ Qxw}{b_0 \rightarrow b_1,\ M \Longrightarrow Qxw}}$$

where the latter case cannot occur intuitionistically. This concludes the definition of D* by cases.

In the cases 2–4, the instance I(R) in D has a *first* branch on which Qxw is indicated, and a *second* branch on which sub(x,t | w) is indicated. These branches have images for the corresponding instance I(R)* in D*. With the exception of Qxw on the right in the second subcase of (E→), the occurrence of the side formula sub(x,t | w) must be added to all sequents of the image of the first branch. But for Qxw on the right and (E∨), the formula Qxw is already present on the right of the sequent starting the first branch. In the intuitionistic case, therefore, the image of this sequent cannot be enlarged, and the permutation cannot be performed.

For extremal calculi, the indicated occurrence of Qxw is present also in all sequents above, except for Qxw on the right in the intuitionistic case.

For non extremal calculi, the indicated occurrence of Qxw is a new one. Being parametric in I(R), Qxw then also occurs in the sequents, starting the images of the first branches, with the exception of Qxw on the right and (E→). In order to permit the instance $I(R)^*$ in pure parameter form, Qxw must be added to all sequents of the image of the second branch. The image $I(Q)^*$ in D^*, therefore, will produce an additional copy of Qxw, which has to be removed by a following application of (RP). In the intuitionistic case, the enlargement by Qxw of the images of the second branches happens only on the left, since if Qxw is on the right then (E→) is the only possible case. For non extremal calculi with mixed parameter forms, none of these weakenings are necessary, and so also the concluding instance of (RC) will not be needed.

It follows from the definition of D^* that neither the endsequent nor the length has been changed. For all instances I(Q') of Q-rules different from I(Q), there holds $\nu(I(Q')) = \nu(I(Q')^*)$. For I(Q) there holds $\nu(I(Q)) = \nu(I(Q)^*) + 1$. This concludes the proof of Lemma 10.

In a cut free derivation D, a side formula Qxw of an instance of propositional rule will also occur as propositional subformula of a formula in the endsequent. Consequently, if the endsequent of D contains only prenex formulas then this event cannot happen, and all quantified formulas occurring in sequents of D will be prenex. Hence if D is eigen and has an endsequent of prenex formulas then in D every instance of a Q-rule can be permuted according to Lemma 10.

In that situation, therefore, if an arbitrary instance I(Q) in D with $\nu(I(Q)) > 0$ is given, then there is a (unique) instance I(Q') of a rule (Q'), I(Q') either equal to I(Q) or below I(Q) in D, such that there is a (unique) instance I(R) immediately below I(Q') which can be permuted with I(Q') and such that $\nu(I(Q)) = \nu(I(Q'))$. Performing the permutation, I obtain D^* such that $\nu(D^*) < \nu(D)$.

If $\nu(D) = 0$ then I set $D^\# = D$. If $\nu(D) > 0$ then I obtain a first algorithm, transforming D into a derivation $D^\#$ such that $\nu(D^\#) = 0$, by executing the above construction successively on the highest, and farthest to the left, instance I(Q) such that $\nu(I(Q)) > 0$. - Observing that the existence of an I(Q) such that $\nu(I(Q)) > 0$ implies the existence of an I(Q") (below I(Q)) such that $\nu(I(Q")) = 1$, I obtain another algorithm by executing the construction successively on the leftmost I(Q") such that $\nu(I(Q")) = 1$.

In $D^\#$, all instances of Q-rules occur below all instances of propositional rules and of weakenings. Let S_0 be the sequent which is the premiss of the topmost premiss of a Q-rule in $D^\#$, if there are any, and otherwise let S_0 be the endsequent. S_0 is called the *midsequent* of $D^\#$.

If there are instances of (RC) above S_0 with a contraction formula v beginning with a quantifier, then above S_0 the copies of v were never used as side- or principal formulas of logical rules. Hence I will obtain a derivation again if I remove all repeated copies of v above S_0, together with those contractions, and together with possible weakenings which may have introduced them earlier. This removal does not change $D^{\#}$ below S_0. Calling the new derivation $\mathbf{M}_0(D)$, I have proven the *Midsequent Theorem*

THEOREM 6 There exist operators $\mathbf{M}_0$, acting on classical derivations D, or on intuitionistic derivations D not employing (E∨), which are eigen and have an endsequent consisting of prenex formulas.

$\mathbf{M}_0(D)$ is a derivation of the same endsequent and of the same length as D. $\mathbf{M}_0(D)$ contains a sequent S_0 such that :

The only rules (R) instantiated above S_0 are propositional, (RP), (W) or (RC) with a contraction formula not beginning with a quantifier .

The only rules instantiated below S_0 are Q–rules, (RP) or (RC) with a contraction formula beginning with a quantifier.

While the subderivation D_s of the midsequent S_0 in $\mathbf{M}_0(D)$ is purely propositional, in an extremal calculus its sequents will still contain all the quantified formulas inherited upwards from the later uses of Q–rules. Of course, I could remove all the quantified formulas in sequents of D_s and then obtain a propositional derivation of the arising rarefied part of S_0.

Matters become formally simpler if I restrict myself to non extremal calculi and to derivations D starting from axioms (not from generalized axioms).

I call *dumb* any quantified formula u occurring in $D^{\#} = \mathbf{M}_0(D)$ above S_0. Then u cannot occur in an axiom. As it will still be prenex, u cannot have been produced by a Q–rule and, therefore, must have been introduced by an instance I(W) of (W). Since u is parametric in all the rules instantiated between the I(W) and S_0, is also occurs in S_0. Removing an occurrence of u from S_0, and removing all its predecessors above S_0 in $D^{\#}$, together with the I(W) which introduced them (and the instances of (RP) required for that), I obtain a sequent S_0^{-u} and a derivation $D_0^{\#-u}$ ending with S_0^{-u}. Performing this for *all* occurrences of dumb formulas in S_0, I obtain a sequent $S_0^{\S}$ and a derivation $D_0^{\#\S}$ ending with $S_0^{\S}$.

From $S_0^{\S}$, I can proceed to S_0 by a series of instances of (W), re–introducing the omitted occurrences of dumb formulas. Calling the resulting derivation $\mathbf{M}_1(D)$, I have shown the

COROLLARY 4 There exist operators $\mathbf{M}_1$, acting on classical or intuitionistic derivations D in M_tK or L_tJ, but then not employing (E∨), which start from axioms, are eigen and have an endsequent consisting of prenex formulas.

$\mathbf{M}_1(D)$ is a derivation of the same endsequent and of the same length as D. $\mathbf{M}_1(D)$ contains a sequent S_0, and above S_0 a sequent $S_0{}^{\S}$, such that:

There are no quantified formulas above $S_0{}^{\S}$. The only rules (R) instantiated above $S_0{}^{\S}$ are propositional, (RP), (W) or (RC) with a weakening or contraction formula not beginning with a quantifier.

The only rules instantiated between $S_0{}^{\S}$ and S_0 are (RP) and (W) with weakening formulas beginning with a quantifier.

The only rules instantiated below S_0 are Q–rules, (RP) or (RC) with contraction formulas beginning with a quantifier.

If so desired, the weakenings between $S_0{}^{\S}$ and S_0 with quantified formulas $\mathfrak{QQ}$ v can be replaced by weakenings with propositional formulas, viz. substitution instances v' of v replacing the formerly quantified variables by completely new ones. These then have to be followed by Q–rules restoring the formulas $\mathfrak{QQ}$ v from v'.

Returning to $D^{\#}$, every occurrence of a dumb formula u in S_0 determines a unique occurrence v^u_n of a formula v^u in the endsequent T of which it is a predecessor; let v^u_n, v^u_{n-1}, ... , v^u_0 be the ascending chain of predecessors ending with the occurrence v^u_0 of u in S_0. I now define which of these predecessors shall be called *dumb:*

v^u_0 is dumb ,

if v^u_m is dumb then v^u_{m+1} is dumb if it arises from v^u_m under (RP) or under a Q–rule ,

if v^u_m is dumb then v^u_{m+1} is dumb if it arises under (RC) by contracting v^u_m and a dumb occurrence v^w belonging to a (possibly different) dumb formula w .

In particular, if no contractions are applied on the v^u_m then the occurrence v^u_n of v^u in the endsequent T itself will be dumb.

Removing in the sequents between S_0 and T all dumb occurrences of dumb formulas, and removing all instances of (RC) contracting at least one dumb occurrence (and the instances of (RP) required for that), I arrive at a locally correct subderivation $D_1{}^{\#\S}$ starting from $S_0{}^{\S}$ and ending with a subsequent $T^{\S}$ of T.

Placing $D_0{}^{\#\S}$ on top of $D_1{}^{\#\S}$, I thus obtain a derivation $\mathbf{M}_2(D)$ of $T^{\S}$.

COROLLARY 5 There exist operators $\mathbf{M}_2$, acting on classical or intuitionistic derivations D in M_tK or L_tJ, but then not employing ($E\vee$), which start from axioms, are eigen and have an endsequent consisting of prenex formulas.

$\mathbf{M}_2(D)$ is a derivation of an endsequent $T^§$ which is a subsequent of the endsequent T of D, and is of the same length as D. Thus $\mathbf{M}_2(D)$ can be prolonged to a derivation of T by introducing the missing formulas of T by applications of (W).

$\mathbf{M}_2(D)$ contains a sequent $S_0{}^§$ such that :

There are no quantified formulas above $S_0{}^§$. The only rules (R) instantiated above $S_0{}^§$ are propositional, (RP), (W) or (RC) with a weakening or contraction formula not beginning with a quantifier .

The only rules instantiated below $S_0{}^§$ are Q-rules, (RP) or (RC) with contraction formulas beginning with a quantifier.

It should be noticed that, in general, the use of (RC) below the midsequent S_0 cannot be avoided. The following *Example 1*, essentially due to D. Miller (cf. GALLIER 86, p.338), is offered by the derivation

$$\begin{array}{rl}
p(y),\ p(z) \Longrightarrow p(x),\ p(z) & \\
\neg p(x)\wedge p(z),\ \neg p(z)\wedge p(y) \Longrightarrow \blacktriangle & \\
\neg p(x)\wedge p(z),\ \exists y\ (\neg p(z)\wedge p(y)) \Longrightarrow \blacktriangle & \text{y EV} \\
\neg p(x)\wedge p(z),\ \forall x\ \exists y\ (\neg p(x)\wedge p(y)) \Longrightarrow \blacktriangle & \\
\exists y\ (\neg p(x)\wedge p(y)),\ \forall x\ \exists y\ (\neg p(x)\wedge p(y)) \Longrightarrow \blacktriangle & \text{z EV} \\
\forall x\ \exists y\ (\neg p(x)\wedge p(y)),\ \forall x\ \exists y\ (\neg p(x)\wedge p(y)) \Longrightarrow \blacktriangle & \\
\forall x\ \exists y\ (\neg p(x)\wedge p(y)) \Longrightarrow \blacktriangle &
\end{array}$$

where p is 1-ary and x, y, z are three distinct variables. If (RC) were not available, the only possible premisses would lead to the attempt

$$\begin{array}{rl}
p(z) \Longrightarrow p(t) & \\
\neg p(t)\wedge p(z) \Longrightarrow \blacktriangle & \\
\exists y\ (\neg p(t)\wedge p(y)) \Longrightarrow \blacktriangle & \text{z EV} \\
\forall x\ \exists y\ (\neg p(x)\wedge p(y)) \Longrightarrow \blacktriangle &
\end{array}$$

which cannot succeed because z must not occur in t.

In order to discuss some applications of the Midsequent Theorem, let me introduce a trivial temporary definition. Given a sequent $C_0 \Longrightarrow C_1$ of prenex formulas, a sequent $S: A_0 \Longrightarrow A_1$ of substitution instances A_0, A_1, of the matrices of the formulas in C_0, C_1 respectively, will be called *suitable* for $C_0 \Longrightarrow C_1$ if (s0) $A_0 \Longrightarrow A_1$ is derivable propositionally,

(s1) $C_0 \Longrightarrow C_1$ is derivable from $A_0 \Longrightarrow A_1$ by Q-rules alone.

Now the Midsequent Theorem assures me that

(A) Given a derivation of a sequent $C_0 \Longrightarrow C_1$ of prenex formulas then I can construct a suitable sequent $A_0 \Longrightarrow A_1$.

It was this propositional reduction which was the original motive to establish (theorems of the type of) the Midsequent Theorem. Because it affords a reduction of the proof that an axiom system, given as a set of prenex formulas C_0, be *consistent,* i.e. does not imply an inconsistency such as $0 \equiv 1$ or $p(0) \wedge \neg p(0)$: if that should happen, then the derivability of $C_0 \Longrightarrow 0 \equiv 1$ would lead already to the propositional inconsistency of suitable substitution instances of the matrices of C_0.

It further follows trivially from the definition of suitable sequents that

(B) If a sequent S, suitable for $C_0 \Longrightarrow C_1$, can be found then I can construct a derivation of $C_0 \Longrightarrow C_1$.

(and here the Midsequent Theorem becomes involved only if I want to add the remark that, in this situation, *any* derivation of $C_0 \Longrightarrow C_1$ can be transformed into one of this kind).

Statement (B) often is viewed as suggesting a reduction of a *proof search* for $C_0 \Longrightarrow C_1$ to the search for a propositional proof of such S, and given an S of which I only know that is suitable, the search for a propositional proof of that sequent can be decided by algorithms terminating after a finitely many steps. Unfortunately, however, *neither is it known* which terms to choose in order to obtain a sequent satisfying (s0), *nor is it known* which terms and *how many* substitution instances to choose in order to obtain a sequent satisfying (s1). As a *general methodic device,* therefore, the statement (B) is just as empty as it is trivial - though it may find applications in special, simple situations..

The condition (s1) for suitability involves, in particular, that the terms chosen for the substitution instances in $A_0 \Longrightarrow A_1$ respect the eigenvariable conditions caused by instances of (I$\exists$) and (E$\forall$). Such instances will not occur if all the formulas in C_0 are universal and if all the formulas in C_1 are existential, and in that case, therefore, no eigenvariable conditions need to be respected. That this simplification can be achieved even for *arbitrary* prenex C_0, C_1 is the content of

8. Herbrand's Theorem for Prenex Formulas

I recall from Chapter **1.14** the definition of the Skolem and the Herbrand normal forms NS(v) and NH(v) for a prenex formula v:

if v does not contain quantifiers $\exists$ then NS(v) = v ,
if v does not contain quantifiers $\forall$ then NH(v) = v ,

if $v = \forall\xi\exists y w$ then NS(v) = NS(u) for $u = \forall\xi \mathrm{sub}(y, f(\vartheta+\xi) \mid w)$

where ξ may be empty and (a) ϑ is the canonical sequence of variables free in w and not in ξ, (b) $\vartheta+\xi$ is the concatenation of ϑ with ξ, and (c) f is a new function symbol whose arity is the length of $\vartheta+\xi$;

if $v = \exists\xi\forall y w$ then NH(v) = NH(u) for $u = \exists\xi \mathrm{sub}(y, f(\vartheta+\xi) \mid w)$

with the same stipulations.

Thus if $v = v(\vartheta, \xi, y, \xi', y', \xi'', y')$ is without quantifiers (and all variables are distinct) then

$$\mathrm{NH}(\exists\xi\forall y\exists\xi'\forall y'\exists\xi''\forall y'' v)$$
$$= \mathrm{NH}(\exists\xi\exists\xi'\forall y'\exists\xi''\forall y'' v(\vartheta, \xi, f_0(\vartheta,\xi), \xi', y, \xi'', y''))$$
$$= \mathrm{NH}(\exists\xi\exists\xi'\exists\xi''\forall y'' v(\vartheta, f_0(\vartheta,\xi), \xi', f_1(\vartheta,\xi,\xi'), \xi'', y''))$$
$$= \exists\xi\exists\xi'\exists\xi'' v(\vartheta, f_0(\vartheta,\xi), \xi', f_1(\vartheta,\xi,\xi'), \xi'', f_2(\vartheta,\xi,\xi',\xi''))$$

with *Herbrand terms* $f_0(\vartheta,\xi)$, $f_1(\vartheta,\xi,\xi')$, $f_2(\vartheta,\xi,\xi',\xi'')$, and

$$\mathrm{NS}(\exists\xi\forall y\exists\xi'\forall y'\exists\xi''\forall y'' v)$$
$$= \mathrm{NS}(\exists x_1\ldots\exists x_{k-1}\forall y\exists\xi'\forall y'\exists\xi''\forall y'' v(\vartheta, f_0(\vartheta), x_1, \ldots, x_{k-1}, y, \xi', y', \xi'', y''))$$
$$= \mathrm{NS}(\forall y\exists\xi'\forall y'\exists\xi''\forall y'' v(\vartheta, f_0(\vartheta), \ldots, f_{k-1}(\vartheta), y, \xi', y', \xi'', y''))$$
$$= \mathrm{NS}(\forall y\forall y'\exists\xi''\forall y'' v(\vartheta, f_0(\vartheta), \ldots, f_{k-1}(\vartheta), y, g_0(\vartheta,y), \ldots, g_{h-1}(\vartheta,y), y', \xi'', y''))$$
$$= \forall y\ \forall y'\ \forall y''\ v(\vartheta, f_0(\vartheta), \ldots, f_{k-1}(\vartheta), y, g_0(\vartheta,y), \ldots, g_{i-1}(\vartheta,y), y', h_0(\vartheta,y,y'), \ldots, h_{j-1}(\vartheta,y,y'), y''))$$

with *Skolem terms* $f_p(\vartheta)$, $g_q(\vartheta,y)$, $h_r(\vartheta,y,y')$, where k, i, j are the lengths of ξ, ξ', ξ'' and $p<k$, $q<i$, $r<j$. In particular, if $\exists\xi\forall y\exists\xi'\forall y'\exists\xi''\forall y''\ v$ is a sentence then ϑ disappears and the Skolem terms $f_p(\vartheta)$ become constants.

Consider a prenex sentence v, and let g(v) be the number of quantifiers $\forall$ in v. If g(v) = 0 then trivially NH(v) = v. Assume now that g(v)>0 and that, for formulas u with g(u)<g(v), I have constructed algorithms transforming derivations of

$$C_o, E_1, E_H{}^u \Longrightarrow \mathrm{NH}(u), C_1$$

into derivations of

$$C_0, E_1 \Longrightarrow u, C_1$$

where $E_H{}^u$ is the set of equality axioms for the Herbrand functions in NH(u) and E_1 is a set of other equality axioms, including those for the functions symbols in u.

If $v = \exists\xi\forall y w(\xi,y)$ then $NH(v) = NH(u)$ for $u = \exists\xi sub(y, f(\xi) \mid w) = \exists\xi w(\xi, f(\xi))$ for the Herbrand term belonging to $\forall y w(\xi,y)$. But $g(w) < g(v)$, and $g(u) = g(w)$ by definition of u. Thus the inductive hypothesis holds for u. Consequently, I have algorithms transforming derivations of

$$C_0, E_1, E_H{}^u \Longrightarrow NH(v), C_1$$

into derivations of

$$C_0, E_1 \Longrightarrow \exists x w(\xi, f(\xi)), C_1 \ .$$

In E_1 there is, in particular, the equality axiom e^f for f, and $E_H{}^v$ is $E_H{}^u$ together with e^f; let E be the remaining set of equality axioms in E_1 not containing the symbol f. In Corollary 3 I presented algorithms which transformed L_f-derivations of

$$C_0, E, e^f \Longrightarrow \exists\xi w(\xi, f(\xi)), C_1$$

into L-derivations of

$$C_0, E \Longrightarrow \exists\xi\forall y w(\xi,y), C_1$$

and vice versa. Since the symbols for Herbrand functions are taken to be successively new ones, the hypotheses for these algorithms are satisfied, and composing the two types of algorithms, I arrive at algorithms transforming derivations of

$$C_0, E, E_H{}^v \Longrightarrow NH(v), C_1$$

into derivations of

$$C_0, E \Longrightarrow v, C_1$$

and vice versa. - Employing Theorem 4, there holds, obviously, an analogous result for prenex formulas v and their Skolem normal forms NS(v) occurring in the antecedent of a sequent. Hence I have proved the

THEOREM 7 Let C_0, C_1 be sequences of prenex sentences in a language L. Let $NS(C_0)$, $NH(C_1)$ be the sequences of their Skolem and Herbrand normal forms respectively, and let L_F the extension of L by the function symbols for the Skolem and Herbrand terms in $NS(C_0)$ and $NH(C_1)$. Let E be the equality axioms for L, and let E_S, E_H be the sets of equality axioms for the Skolem and Herbrand terms .

Then there are algorithms transforming L-derivations of C_o, $E \Longrightarrow C_1$ into L_F-derivations of $NS(C_o)$, E, E_S, $E_H \Longrightarrow NH(C_1)$ and vice versa.

I now consider calculi of type 1 and apply the construction of the Midsequent Theorem and its Corollaries to a derivation D of a sequent $NS(C_o)$, E, E_S, $E_H \Longrightarrow NH(C_1)$. I observe that

(i) All formulas in the antecedent are of the form $\forall\xi\, u_0(\xi)$, and all formulas in the succedent are of the form $\exists\xi\, u_1(\xi)$.

(ii) If $\forall\xi\, u_0(\xi)$ is in $NS(C_o)$ then there is a formula $v_o = \mathfrak{QQ}\, w_o$ in C_o and u_o arises from w_o by substituting Skolem terms for the existentially quantified variables of v_o.

(iii) If $\exists\xi\, u_1(\xi)$ is in $NH(C_1)$ then there is a formula $v_1 = \mathfrak{QQ}\, w_1$ in C_1 and u_1 arises from w_1 by substituting Herbrand terms for the universally quantified variables of v_1.

Let S_o be the midsequent of a derivation $\mathbf{M}_1(D)$. In view of (i), there follows

(iv) All formulas in the antecedent of S_o are substitution instances either of equality axioms from E, E_S, E_H or of formulas $u_0(\xi)$ from (iii) ; all formulas in the succedent of S are substitution instances of formulas $u_1(\xi)$ from (iv).

I thus obtain *Herbrand's Theorem* for prenex formulas as the

COROLLARY 6 Every L-derivation of a sequent C_o, $E \Longrightarrow C_1$ determines

a sequent S_o: $A_o \Longrightarrow A_1$ such that A_o consists of substitution instances either of equality axioms from E, E_S, E_H or of matrices $u_o(\xi)$ of formulas from $NS(C_o)$, and A_1 consists of substitution instances of matrices $u_1(\xi)$ of formulas in $NH(C_1)$,

a (propositional) derivation of S_o in the open predicate logic of the language L_F,

an L_F-derivation of $NS(C_o)$, E, E_S, $E_H \Longrightarrow NH(C_1)$ from $A_o \Longrightarrow A_1$ by Q-rules alone ,

an L-derivation of C_o, $E \Longrightarrow C_1$ from $NS(C_o)$, E, E_S, $E_H \Longrightarrow NH(C_1)$ by the algorithm of Theorem 7.

The advantage which the midsequent S_o in Herbrand's Theorem offers over that of the Midsequent Theorem, consists in that it avoids the eigenvariable conditions, required from the substitution instances in the general case. The price to be paid for that, consists in having to admit in A_0, A_1 substitution terms which use the new functions symbols of L_F - and it should be noticed that these function may occur nested: it will *not* suffice to just substitute old terms from L into the matrices $u_0(\xi)$ and $u_1(\xi)$, as the Example 2 below will show.

As for the use of Herbrand's Theorem in order to reduce a proof search to a propositional situation, observe that now the sequent S_0 must be suitable for the sequent $NS(C_0), E, E_S, E_H \Longrightarrow NH(C_1)$. Thus the same limitations prevail which were discussed above in connection with the usefulness of the observation (B) relying on the Midsequent Theorem. In simple situations, though, the circumstances created by the presence of Skolem and Herbrand functions may be simpler than those presented by the Midsequent Theorem. For instance, the sequent derived in Example 1 in a special manner, can with general substitution terms t_0, t_1 be derived from a midsequent

(e0)

$$
\begin{array}{rl}
p(z_0),\ p(z_1) \Longrightarrow p(t_0),\ p(t_1) & \\
\neg p(t_0)\wedge p(z_0),\ \neg p(t_1)\wedge p(z_1) \Longrightarrow \blacktriangle & z_1 \text{ EV} \\
\neg p(t_0)\wedge p(z_0),\ \exists y\ (\neg p(t_1)\wedge p(y)) \Longrightarrow \blacktriangle & \\
\neg p(t_0)\wedge p(z_0),\ \forall x\ \exists y\ (\neg p(x)\wedge p(y)) \Longrightarrow \blacktriangle & z_0 \text{ EV} \\
\exists y\ (\neg p(t_0)\wedge p(y)),\ \forall x\ \exists y\ (\neg p(x)\wedge p(y)) \Longrightarrow \blacktriangle & \\
\forall x\ \exists y\ (\neg p(x)\wedge p(y)),\ \forall x\ \exists y\ (\neg p(x)\wedge p(y)) \Longrightarrow \blacktriangle & \\
\forall x\ \exists y\ (\neg p(x)\wedge p(y)) \Longrightarrow \blacktriangle &
\end{array}
$$

with the eigenvariable conditions that (1) not $z_0 \epsilon occ(t_0)$, (2) not $z_1 \epsilon occ(t_1)$, (3) $z_1 \neq z_0$, (4) not $z_1 \epsilon occ(t_0)$. Thus the choice $t_1 = z_0$ will make the first sequent into an axiom. Observe that this example also shows that *one* prenex formula may require *several* substitution instances of its matrix. - If now I transform every step of this derivation by introducing an 1-ary Skolem function s, then I obtain with substitution terms r_0, r_1

(h1)

$$
\begin{array}{r}
p(s(r_0)),\ p(s(r_1)) \Longrightarrow p(r_0),\ p(r_1) \\
\neg p(r_0)\wedge p(s(r_0)),\ \neg p(r_1)\wedge p(s(r_1)) \Longrightarrow \blacktriangle \\
\neg p(r_0)\wedge p(s(r_0)),\ \forall x\ (\neg p(x)\wedge p(s(x))) \Longrightarrow \blacktriangle \\
\forall x\ (\neg p(x)\wedge p(s(x))),\ \forall x\ (\neg p(x)\wedge p(s(x))) \Longrightarrow \blacktriangle \\
\forall x\ (\neg p(x)\wedge p(s(x))) \Longrightarrow \blacktriangle \\
\forall x\ \exists y\ (\neg p(x)\wedge p(y)) \Longrightarrow \blacktriangle
\end{array}
$$

such that r_0 may be arbitrary (e.g. the variable x) and r_1 must be chosen as $s(r_0)$.

The transformation of a quantificational derivation with eigenvariables (such as (e0)) into one with Skolem/Herbrand functions is obvious. The reverse transformation is also possible, provided I use different function symbols for every copy of a formula contracted at a later stage. For instance, instead of (h1) with *one* Skolem function s I can form the derivation (h0) with *two* Skolem functions s_0, s_1

$$
\begin{array}{rl}
\text{(h0)} \qquad p(s_0(r_0)),\ p(s_1(r_1)) & \Rightarrow p(r_0),\ p(r_1) \\
\neg p(r_0)\wedge p(s_0(r_0)),\ \neg p(r_0)\wedge\neg p(s_1(r_1)) & \Rightarrow \blacktriangle \\
\neg p(r_0)\wedge p(s_0(r_0)),\ \forall x\ (\neg p(x)\wedge p(s_1(x))) & \Rightarrow \blacktriangle \\
\forall x\ (\neg p(x)\wedge p(s_0(x))),\ \forall x\ (\neg p(x)\wedge p(s_1(x))) & \Rightarrow \blacktriangle \\
\forall x\ \exists y\ (\neg p(x)\wedge p(y)),\ \forall x\ \exists y\ (\neg p(x)\wedge p(y)) & \Rightarrow \blacktriangle \\
\forall x\ \exists y\ (\neg p(x)\wedge p(y)) & \Rightarrow \blacktriangle
\end{array}
$$

where I then have to choose $r_1 = s_0(r_0)$. Writing the eigenvariable conditions in (e0) as $z_0(t_0)$ and $z_1(t_1,z_0,t_0)$, I see that the Skolem terms $s_0(r_0)$, $s_1(t_1)$ of (h0) *encode* these conditions:

$s_0(t_0)$ corresponds to the eigenvariable z_0, different from the variables in t_0,

$s_1(t_1)$ corresponds to the eigenvariable z_1, different from the variables in t_1 and from (all) the lower indexed eigenvariables and from the variables in terms excluded for them.

Obviously, in order to make this observation into a formal algorithm, also the nestings of Skolem/Herbrand terms (not present in (h0)) will have to be taken into account.

As *Example 2,* I assume that my language contains a 2-ary predicate symbol $<$ and consider the sequent $\forall x\,\exists y\ x<f(f(y)) \Rightarrow \forall x\,\exists y\ x<f(y)$. Working with the Midsequent Theorem, the two straightforward attempts

$$
\begin{array}{rl}
t_0<f(f(y)) & \Rightarrow x<f(t_1) \\
t_0<f(f(y)) & \Rightarrow \exists y\ x<f(y) \\
t_0<f(f(y)) & \Rightarrow \forall x\ \exists y\ x<f(y) \\
\exists y\ t_0<f(f(y)) & \Rightarrow \forall x\ \exists y\ x<f(y) \\
\forall x\ \exists y\ x<f(f(y)) & \Rightarrow \forall x\ \exists y\ x<f(y)
\end{array}
$$

$$
\begin{array}{rl}
t_0<f(f(y)) & \Rightarrow x<f(t_1) \\
\exists y\, t_0<f(f(y)) & \Rightarrow x<f(t_1) \\
\forall x\ \exists y\ x<f(f(y)) & \Rightarrow x<f(t_1) \\
\forall x\exists y\, x<f(f(y)) & \Rightarrow \exists y\ x<f(y) \\
\forall x\exists y\, x<f(f(y)) & \Rightarrow \forall x\ \exists y\ x<f(y)
\end{array}
$$

will fail, because in the first one the eigenvariable condition *not* $x\epsilon occ(t_0)$ makes it impossible to choose t_0 as x in the first sequent such as to obtain an axiom, and in the second attempt the condition *not* $y\epsilon occ(t_1)$ makes it impossible to choose t_1 as $f(y)$. A third attempt, not imposing any eigenvariable conditions,

$$
\begin{array}{rl}
t_0<f(f(y)) & \Rightarrow x<f(t_1) \\
t_0<f(f(y)) & \Rightarrow \exists y\, x<f(y) \\
\exists y\, t_0<f(f(y)) & \Rightarrow \exists y\, x<f(y) \\
\forall x\exists y\, t_0<f(f(y)) & \Rightarrow \exists y\, x<f(y) \\
\forall x\exists y\, x<f(f(y)) & \Rightarrow \forall x\exists y\ x<f(y)
\end{array}
$$

succeeds with $t_0 = x$ and $t_1 = f(y)$. Working with Herbrand's Theorem, I use the 1-ary Skolem function s and the 0-ary Herbrand constant c and again obtain a derivation

$$\begin{array}{r} r_0 < f(f(s(r_0))) \Longrightarrow c < f(r_1) \\ r_0 < f(f(s(r_0))) \Longrightarrow \exists y\ c < f(y) \\ \forall x\ x < f(f(s(x))) \Longrightarrow \exists y\ c < f(y) \\ \forall x \exists y\ x < f(f(y)) \Longrightarrow \forall x \exists y\ x < f(y) \end{array}$$

by choosing $r_0 = c$ and $r_1 = f(s(r_0))$. On the other hand, the sequent $\forall x\ \exists y\ x < f(y) \Longrightarrow \forall x\ \exists y\ x < f(f(y))$ can have *no* derivation. Because *any* attempt

$$\begin{array}{r} r_0 < f(s(r_0)) \Longrightarrow c < f(f(r_1)) \\ \forall x\ x < f(s(x)) \Longrightarrow \exists y\ c < f(f(y)) \\ \forall x\ \exists y x < f(y) \Longrightarrow \forall x \exists y\ x < f(f(y)) \end{array}$$

is bound to fail since the symbols s and f are distinct and therefore $f(s(r_0))$ can never be $f(f(r_1))$.

This last example raises the general question whether, given a class V of formulas v, there is a terminating algorithm which decides, for every v in V, whether v, i.e. the sequent $\blacktriangle \Longrightarrow v$, does have a derivation or not. While it follows from results to be discussed in Book 3 that there is *no* algorithm performing this task for the class of *all* formulas, positive solutions are available for certain special classes.

A particular such V is the class V_B introduced by HERBRAND 30 (cf. HILBERT-BERNAYS 39, p.163). As a technical abbreviation, I denote, for every formula w, as $1 \circ w$ this w itself and as $0 \circ w$ the formula $\neg w$. Let V_B be the class of all prenex formulas $\mathbb{QQ}$ v for which there exists a number n such that

there exists a sequence $< p^m(\lambda^m) \mid m < n >$ of atomic formulas $p^m(\lambda^m)$ and a sequence $< \varepsilon_m \mid m < n >$ of numbers ε_m either 0 or 1, such that v is the disjunction

$$\mathbb{W} < \varepsilon_m \circ p^m(\lambda^m) \mid m < n > .$$

Searching for a derivation of $\blacktriangle \Longrightarrow \mathbb{QQ}$ v, I shall extend my language L by Herbrand functions for $\mathbb{QQ}$ v into a language L_F, and in order to avoid more notation I will assume that this has been done already such that all quantifiers before v are existential. Thus $\blacktriangle \Longrightarrow \mathbb{QQ}$ v will have a derivation if, and only if, there is a sequence $< w_k \mid k < j >$ of substitution instances

$$v_k = rep(x_0, t_0^k \mid rep(x_1, t_1^k \mid \dots \mid v))$$

of $v = \mathbb{W} < \varepsilon_m \circ p^m(\lambda^m) \mid i < n >$ such that

$$v^* = \mathbb{W} < v_k \mid k < j >$$

is propositionally derivable. Being a disjunction of disjunctions, such a formula v^* is a disjunction itself.

Working in a calculus with rep, I can assume my substitution maps to be free for the formulas upon which they are performed. Hence for a formula $\exists x_0 \exists x_1 \exists x_2 \ldots v$, the term t_i^k replacing x_i in v_k does not contain any x_q with $q>p$, and so I may write $rep(x_0, t_0^k \mid rep(x_1, t_1^k \mid \ldots \mid v))$ as $rep(h^k \mid v)$ with $h^k(x_i) = t_i^k$. Hence the formulas v_k can be written as

$$v_k = \mathbb{W} < \varepsilon_m \circ p^m(h^k \cdot \lambda^m) \mid m<n>$$

whence

$$v^* = \mathbb{W} < \varepsilon_m \circ p^m(h^k \cdot \lambda^m) \mid m<n, k<j> .$$

But such a disjunction has a (classical) propositional derivation if, and only if, it is a tautology, and this is the case if, and only if there exist m, p below n and k, q below j such that

$$\varepsilon_m = 1 - \varepsilon_p \quad \text{and} \quad p^m = p^p \quad \text{and} \quad h^k \cdot \lambda^m = h^q \cdot \lambda^p .$$

Hence the decision algorithm for formulas $\mathbb{QQ}$ v from V_B is the following:

1. Introduce Herbrand functions and write the formula as $\exists\exists\exists \ldots \mathbb{W} < \varepsilon_m \circ p^m(\lambda^m) \mid m<n>$.

2. Check whether there is an uncancelled pair m, p below n such that $\varepsilon_m = 1 - \varepsilon_p$ and $p^m = p^p$. If not, then fail. If yes, then take the first such pair and continue.

3. Let r be the arity of the predicate symbol p^m. Check whether there are endomorphisms h^i and h^q of the term algebra of L_F which identify the corresponding members of λ^m, λ^p: $h^i(\lambda^m(s)) = h^q(\lambda^p(s))$ for every $s<r$. If not, then declare the pair m, p as cancelled and return to 2. If yes, then v^* becomes derivable where all the h^k with $k \neq i$, $k \neq q$ may be taken as the identity.

References

J.H.Gallier: Logic for Computer Science. New York 1986

G.Gentzen: Untersuchungen über das logische Schliessen I,II. Math.Z. **39** (1934/35), 176–210 and 405–431

J.Herbrand: Recherches sur la théorie de la démonstration. Trav.Soc.Sci.Varsovie, Cl. III, **33** (1930) 1–128 [Reprinted in: J.Herbrand: Ecrits logiques. Paris 1968]

D.Hilbert, P.Bernays: Grundlagen der Mathematik II . Berlin 1939

J.R.Shoenfield: Mathematical Logic. Reading 1967

9. Tableaux

The method of tableaux, developed for propositional logic in Chapters **1**.5 , **4**.6 and **5**.6 , extends to quantifier logic in a straightforward manner. Working again with signed formulas, the table of components becomes:

unramified				ramified			
$Q\ v\wedge w$	:	Qv	Qw	$P\ v\wedge w$	:	Pv	Pw
$P\ v\vee w$	:	Pv	Pw	$Q\ v\vee w$	:	Qv	Qw
$P\ v\rightarrow w$	:	Qv	Pw	$Q\ v\rightarrow w$	:	Pv	Qw
$Q\neg v$	:	Pv					
$P\neg v$	:	Qv	$P\blacktriangle$				
$Q\forall x\ v$	:	$Q\,sub(x,t\mid v)$		$P\forall x\ v$	:	$P\,sub(x,y\mid v)$	
$P\exists x\ v$	:	$P\,sub(x,t\mid v)$		$Q\exists x\ v$	:	$Q\,sub(x,y\mid v)$	

such that quantified formulas now are assigned *arbitrarily many* components, namely one for every term t or for every variable y respectively; in the case of classical logic the component $P\blacktriangle$ is again omitted.

A *tableau* again consists of a finite 2-ary tree T, a root piece R of T, and two functions t and ε such that

t assigns signed formulas to the nodes of T ,

ε is defined for the nodes e which are not maximal in E and which, should they belong to the root piece R , are the maximal node of R .

$\varepsilon(e)$ is a node such that $\varepsilon(e) \leq e$, $t(\varepsilon(e))$ is composite, and there holds:

(t1) If $t(\varepsilon(e))$ is unramified, but not by the connective $\neg$, or is ramified by a propositional connective and with identical components, then e has one upper neighbour e' only, and t(e') is one of the components of $t(\varepsilon(e))$. In addition, if $t(\varepsilon(e))$ is $P\ v\rightarrow w$ then also e' has one upper neighbour e'' only, and $t(e') = Qv$, $t(e'') = Pv$, $\varepsilon(e') = \varepsilon(e)$.

(t2) If $t(\varepsilon(e))$ is ramified by a propositional connective and with distinct components then e has two upper neighbours e', e'', and t(e'), t(e'') are the two components of $t(\varepsilon(e))$.

(t3) If $t(\varepsilon(e))$ is P-signed then $\varepsilon(e)$ is the P-predecessor of e.

(t4) If $t(\varepsilon(e))$ is $Q\neg v$ *and* if the P-predecessor of e carries $P\blacktriangle$ then t(e') is the component Pv .

(t5) If $t(\varepsilon(e))$ is $P\neg v$ then t(e') is the component Qv *and* e' has a further upper neighbour which carries $P\blacktriangle$.

(t7) If $t(\varepsilon(e))$ is ramified by a quantifier then e has one upper neighbour e' only, and t(e') is one of the components $R\,sub(x,y\mid v)$ of $t(\varepsilon(e))$, and y is not free in any t(f) with $f \leq e$.

In (t7) I call y the *eigenvariable* at e', and the last clause of (t7) again is called the *eigenvariable condition.* In the case of intuitionistic logic, the definition of ε is extended by permitting $t(\varepsilon(e))$ also to be P▲ (which is not composite), and the clause

> (t6) If $\varepsilon(e)$ is the P-predecessor of e (and $t(\varepsilon(e))$ is not already P▲) then e has one upper neighbour e' and t(e') is P▲ .

is added. In the case of classical logic, the component P▲ is omitted, (t3), (t4), (t5), (t6) are omitted, and (t1) is simplified to

> If $t(\varepsilon(e))$ is unramified, or is ramified by a propositional connective with identical components, then e has one upper neighbour e' only, and t(e') is one of the components of $t(\varepsilon(e))$.

A branch of a tableau is *closed* if it contains oppositely signed atomic formulas and, in case the logic is not classical, if the P-signed atomic formula is the lower, then it is the P-predecessor of the other; a tableau is *closed* if each of its branches is so.

Again, Lemma 1.4 and its proof remain in effect. However, in the case of non-classical logics and K- or L-sequents, the eigenvariable not occurring in the conclusion of a critical quantifier rule may well re-appear in lower sequents if there are instantiations of (I∧) or (I∨). In so far, then, the eigenvariable condition for tableaux is stronger than that for derivations, and a derivation H, from which a tableau is to be constructed, should first be transformed by Lemma 8.8 into a derivation H* which is eigen in that lemma's sense; for this H* then the construction in the proof of Lemma 1.4 works without changes. In the classical case with M-sequents , the rules of $M_u K$ prevent the re-appearance of eigenvariables, and so the transformation into H* is not necessary there.

The following are examples of closed tableaux, first for minimal and then for classical quantifier logic:

7	Q p(t)	t=y			
6	P p(y)	4, t(1)	6	Q p(t)	t=y
5	P ▲	3, t(4)	5	P ¬ p(t)	5, (t1)
4	Q ∀x v		4	P p(y)	2, (t1)
3	Q ¬ p(y)	2, t(5)	3	P ∀x p	3, (t7)
2	P ¬ ∀x p	1, t(7)	2	P ∃x ¬p	1, (t1)
1	Q ∃x ¬ p	0, t(1)	1	Q ¬ ∀x p	0, t(1)
0	P ∃x ¬ p → ¬ ∀x p	0, t(1)	0	P ¬ ∀x p → ∃x ¬p	0, t(1)

where on line n the comment m, t(p) abbreviates that line n+1 arises from dissolving line m with rule t(p) . In the left tableau, y on line 3 is an eigenvariable, and dissolving 4 into 7 the term t then is chosen as this y . Analogously, in the right tableau y in 4 is eigenvariable, and dissolving 2 into 5 the term t then is chosen as this y ; this tableau does not close minimally or intuitionistically, and the reference from 4 to 2 steps over the P-predecessor, as does the closing pair at 6 and 4. Moreover, no other closed minimal or intuitionistic tableau can be built from that root. Because 1 and 2 are necessarily so in view of the additional condition in (t1). If 2 is followed by dissolving 1 as above, then 2 becomes inaccessible to any dissolution itself, preventing an intuitionistic tableau to close through the lack of oppositely signed atomic formulas. If 2 is followed instead by dissolving 2 into $P \neg p(t)$ for some term t, leading later to $Q\,p(t)$, then an oppositely signed atomic formula would have to come via 1, hence via $P \forall x p$, and so it would be p(y) with an eigenvariable y. But if t had been chosen as y then the eigenvariable condition would be violated.

I shall now consider classical logic and shall show that the connection between closed tableaux and semantical consequence, expressed there in Corollary 5.4, remains in effect under the *hypothesis, made from now on,* that the language L has countably many variables and *at most countably* many function symbols. Consequently, it has countably many terms which I assume to be enumerated as t_i with i in ω . Moreover, it will become clear from the following that it is no restriction to also assume that L also has countably many predicate symbols. Consequently, it also has countably many formulas.

In order to abbreviate lenghty expressions, I shall denote the components of a quantified formula u as u_y and u_i : if u is $X\,Yx\,v$, where X is P or Q and Y is $\forall$ or $\exists$, then u_y shall be $X\,\mathrm{sub}(x,y \mid v)$ and u_i shall be $X\,\mathrm{sub}(x,t_i \mid v)$. Further, I shall abbreviate by UR the phrase 'unramified quantified' .

A tableau is said *to test* a node e if t(e) is not UR and is either a signed atomic formula or, for every branch B through e, both components of t(e) are on B in case t(e) is unramified, and at least one component of t(e) is on B in case e is ramified. A node for which t(e) is UR has arbitrarily many components and, therefore, cannot be tested in a finite tableau. Still, a *testing tower* of tableaux can be constructed in analogy to (the proof of) Theorem 5.4 .

To this end, I shall work right away with the choice function α mentioned on p.122 [which has the technical advantage that I need only *one* employment function and not a family of such]. So I shall consider tableaux T with an employment function δ which assigns to the nodes of E either a value ∞ or a value in ω such that (at least the) nodes carrying signed atomic formulas obtain the value ∞; the nodes with value ∞ then are *used*, the others are *open.* The function φ is defined on the set G of all those maximal nodes of T for which there are open nodes f with $f \leq e$, and $\varphi(e)$ shall be the smallest of them. A consequence of this particular choice is stated in the *confinality* condition:

$$\text{If } e,\ g \text{ are in G and } \varphi(e) \leq g \text{ then } \varphi(e) = \varphi(g)\ .$$

Given a finite tableau T together with δ ; I define its *direct extension* T' *with* δ' by adding at most two nodes on top of every node e of G. The node φ(e) being open, the formula $t\varphi(e)$ is composite, and if it is propositionally composed then I extend T, t, ε to T', t', ε' as in the propositional case. If $t\varphi(e)$ is a quantified formula which is ramified then I add one new upper neighbour n on top of e ; if it is unramified, i.e. UR, then I add a new upper neighbour n on top of e and a new upper neighbour e* on top of n. Extending t, if $t\varphi(e)$ is u, in the ramified case I set $t'(n) = u_y$ where y is the first variable not occurring in any t(f) for $f \lesssim e$; in the unramified case I set $t'(n) = u_i$ with $i = \delta\varphi(e)$, and $t'(e^*) = t\varphi(e)$. Extending ε, I set $\varepsilon'(e) = \varphi(e)$ and, in the unramified case, $\varepsilon'(n) = \varphi(e)$. This concludes the definition of T'. The function δ' shall declare φ(e) as used, and in view of the confinality condition this assignment does not depend on the node e with which φ(e) is described. It shall have the same values as δ for all other nodes of T' already in T; in all cases, δ' shall map the newly added nodes n, m to ∞ if they carry signed atomic formulas, and to 0 otherwise. Finally, in the UR case I set $\delta'(e^*) = \delta\varphi(e)+1$. – The following observations follow from this definition:

(d0) A node in T', but not in T, is open in T'.

(d1) If b is an upper bound for the heights of nodes in T then b+2 is an upper bound for the heights of nodes in T'.

(d2) If in T all nodes of height at most c are used, then in T' all nodes of height at most c+1 are used.

(d3) If a node in T has different values under δ and under δ' then it is used in T'.

(d4) If T tests all of its used nodes which do not carry an UR formula, then so does T'.

(d5) If g in T is open and is used in T' then g is a minimal open φ(e) for a maximal e of T.

(d6) If g is open in T' and is not in T, if $t'(g) = u$ is UR, and if $0 < \delta'(g)$, then there is an h in T with $h \lesssim g$, $t(h) = t'(g)$, $\delta(h) = \delta'(g)-1$.

(d7) If h is open in T and used in T', and if t(h) is UR, then on every branch of T' through h there is an e* in T' with $h \lesssim e^*$, $t'(e^*) = t(h)$, $\delta'(e^*) = \delta(h)+1$, and there is an h" between h and e* in T' with $t'(h") = u_i$ for $i = \delta(h)$.

Given now a sequent s, or equivalently a set of signed formulas, I recursively define a sequence $<T_n \mid n\epsilon\omega>$ of finite tableaux for s, together with a sequence $<\delta^n \mid n\epsilon\omega>$ of employment functions. T_0 shall be its own root piece: a linearly ordered set with a maximal node r, the nodes of which carry the signed formulas for s ; δ^0 declares as used only the nodes carrying signed atomic formulas, and assigns 0 to all other nodes. T_{n+1} with δ^{n+1} shall be the direct extension of T_n with δ^n.

So every T_n will have T_0 as its root piece, and T_{n+1} not only extends T_n, but also every T_m with $m \leq n$; the tree of T_m is an initial part of the tree of T_{n+1}, and t_m, ε_m are the restrictions of t_{n+1}, ε_{n+1} to that initial part.
In particular, the height and the value under t_m of a node e in T_m remains the same at every later stage T_n; thus I may write t(e) for $t_m(e)$. – In what follows, *node* shall always refer to a node of some T_m.

Let r+1 be the number of members of T_0; then the maximal node o of T_0 has height r. Consequently

(c1) The nodes in T_n have heights at most $r+2n$.

The node $t\varphi(o)$, with which T_1 is formed, has at least the height 0 ; thus the node of height 0 occurs in T_0 and is used in T_1. Consequently

(c2) All nodes of height at most n occur in T_n and are used in T_{n+1}.

Further, (d3) implies

(c3) If a node h is open in T_n and also open in T_a with $n < a$, then $\delta^n(h) = \delta^a(a)$.

In T_0 the only used nodes are those carrying signed atomic formulas; hence T_0 trivially tests all of its used nodes which do not carry a UR formula. Thus by (d4)

(c4) T_{n+1} tests all of its used nodes which do not carry an UR formula.

(c6) If g is open in T_{m+1} and is not in T_m, and if t(g) = u is UR, then for any j with $j \leq \delta^m(g)$ there is a stage $a \leq m$ and an h in T_a with $h \leq g$, $t(h) = t(g)$, $\delta^a(h) = \delta^m(g) - j$.

This is (d6) for j = 1 with a = m. Has it been proven for $j < \delta^m(g)$ then $\delta^a(h)$ is still positive, hence h is not in T_0 and $a > 0$. Thus I find $p < a$ such that h is in T_{p+1} and not in T_p. I obtain (c6) for j+1 by applying (d6) to h in T_{p+1} .

(c7) If h is open in T_n and used in T_{n+1}, and if t(h) = u is UR, then for any $j \geq 0$ there is a stage $b > n$ such that on every branch of T_b through h there is an h' with $h \leq h'$, $t(h') = t(h)$, $\delta^b(h') = \delta^n(h) + j$, and there is an h" between h and h' in T_b with $t(h") = u_i$ for $i = \delta^n(h) + j$.

This is (d7) for j = 0 with k = n+1. Has it been proven for j then h' is open in T_k, and so I find $p \geq k$ such that h remains open in T_p and becomes used in T_{p+1}. I obtain (c7) for j+1 by applying (d7) to h' in T_p .

The tower of the tableaux T_n I call the *testing tower* for the starting sequent s (or the starting set of signed formulas). The justification for this name is given in the

LEMMA 11 If a node g of the tower and of height n carries a composite formula which is not UR, then T_{n+1} (and hence every T_q with $n+1 \leq q$) tests g.

If a node g of the tower carries a UR formula u, then for every i there is stage b such that every branch through g of T_b (and hence of every T_q with $b \leq q$) carries the component u_i of u.

The first statement follows from (c2) and (c4). Assume now that g carries an UR formula. If g is in T_0, whence $\delta^0(g) = 0$, then I set $a = 0$ and $h = g$. If g is not in T_0 then there is an m such that g is not in T_m but is in T_{m+1}. Then g is open in T_{m+1} by (d0), and applying (c6) with $j = \delta^m(g)$ there is a stage a with $a \leq m$ and h in T_a with $h \leq g$, $t(h) = t(g)$, $\delta^a(h) = 0$. In any case, h is open in T_a, and as by (c2) it must become used at some later stage, there is an n with $n \geq a$ such that h is open in T_n and used in T_{n+1}. Now $\delta^n(h) = \delta^a(h) = 0$ by (c3), and applying (c7) there is, for every $i \geq 0$, a stage $b > n$ such that every branch through h of T_b carries the component u_i of $t(h) = t(g)$. In particular, the branches through g are branches through h.

The following examples of (the lower stages of) testing towers may be instructive:

3	4	P $p(x_1)$	
2	3	Q $p(x_0)$	2
	2	P $\forall x\ p(x)$	1
1	1	Q $\exists x\ p(x)$	0
0	0	P $\exists x\ p(x) \rightarrow \forall x\ p(x)$	0

Here x_0, x_1 are eigenvariables; the first column indicates where a new stage begins, the second column numbers the lines (heights), and the rear column indicates which line is to be dissolved next (the value of the function ε). The tower stops with T_3 which is not closed. It can be seen that no other tableau for the line 0 closes, because 1 and 2 will always be dissolved with eigenvariables which must be distinct; instead of this explicit argument, I may also use a general one contained in Lemma 12b below.

		...	
	10	P $\exists x\ p(x)$	8
5	9	P $p(x_1)$	R 6
	8	Q $\forall x\ p(x)$	6
4	7	Q $p(x_1)$	R 4
	6	P $\exists x\ p(x)$	4
3	5	P $p(x_0)$	R 2
	4	Q $\forall x\ p(x)$	2
2	3	Q $p(x_0)$	R 1
	2	P $\exists x\ p(x)$	1
1	1	Q $\forall x\ p(x)$	0
0	0	P $\forall x\ p(x) \rightarrow \exists x\ p(x)$	0

Here R n in the rear column indicates that line n now must be repeated. T_3 is closed already at line 5, but the tower continues through all stages.

...

6	11	P $p(x_0, x_0)$	
	10	Q $\exists x \ \forall y \ p(x,y)$	5
5	9	Q $\forall y \ p(x_1, y)$	R 4
	8	Q $\forall y \ p(x_0, y)$	4
4	7	Q $p(x_0, x_0)$	R 3
	6	P $\forall y \ \exists x \ p(x,y)$	3
3	5	P $\exists x \ p(x, x_0)$	R 2
	4	Q $\exists x \ \forall y \ p(x,y)$	2
2	3	Q $\forall y \ p(x_0, y)$	R 1
	2	P $\forall y \ \exists x \ p(x,y)$	1
1	1	Q $\exists x \ \forall y \ p(x,y)$	0
0	0	P $\exists x \ \forall y \ p(x,y) \rightarrow \forall y \ \exists x \ p(x,y)$	0

Here T_6 is closed already at line 11, but the tower continues through all stages.

...

16	P $p(x_1, x_4)$	EV x_4	$12 \rightarrow 6 \rightarrow 2$
14	Q $\exists x \ p(x, x_2)$	s9 11	
13	Q $p(x_3, x_1)$	EV x_3	$9 \rightarrow 4 \rightarrow 1$
11	P $\forall y \ p(x_1, y)$	s6 8	
10	P $p(x_0, x_2)$	EV x_2	$6 \rightarrow 2$
8	Q $\exists x \ p(x, x_1)$	s4 5	
7	Q $p(x_0, x_1)$	EV x_1	$4 \rightarrow 1$
5	P $\forall y \ p(x_0, y)$	s2 3	
3	Q $\exists x \ p(x, x_0)$	s1 2	
2	P $\exists x \ \forall y \ p(x,y)$	1	
1	Q $\forall y \ \exists x \ p(x,y)$	0	
0	P $\forall y \ \exists x \ p(x,y) \rightarrow \exists x \ \forall y \ p(x,y)$	0	

Here the indications of stages have been omitted, as have been the lines containing repetitions. In the phrases "sn m" and "EVy m" the m indicates that line m is dissolved next, the n indicates that the following repetition of line n is omitted, and EVy indicates that y is an eigenvariable. Beginning with line 10, the lines are repeated with period 6:

$10+6n+4$	Q $\exists x \ p(x, x_{n+2})$	$s6(n+1)+3$	$6(n+1)+5$
$10+6n+3$	Q $p(x_{2+2n+1}, x_{n+1})$	EV x_{2+2n+1}	$6(n+1)+3 \rightarrow \ldots 4 \rightarrow 1$
$10+6n+1$	P $\forall y \ p(x_{n+1}, y)$	$s6(n+1)$	$6(n+1)+2$
$10+6n$	P $p(x_n, x_{2+2n})$	EV x_{2+2n}	$6(n+1) \rightarrow \ldots \rightarrow 6 \rightarrow 2$

The tableau cannot close, because given Q $p(x_{2+2n+1}, x_{n+1})$ then in any later P $p(x_m, x_{2+2m})$ with $m>n$ the eigenvariable x_{2+2m} has to be new.

As testing towers in general are infinite, it will be convenient to extend the notion of a tableau to that of an *itableau* by permitting the underlying 2-ary tree T to be infinite, all other definitions remaining unchanged. Assuming a fixed enumeration of formulas, I make the tree of an itableaux into an oriented tree by counting that one of the (at most two) upper neighbours of a node to be earlier for which the formula it carries has the smaller number.

If an itableau T is closed then it contains a finite closed tableau with the same root. Because remove from every branch B of T all nodes (not in the root piece and) above those at which B closes for the first time; let T^f be the remaining tree. Thus every branch of T^f is finite, and it follows (albeit by an indirect argument) from König's Lemma that T^f is finite.

In particular, given some testing tower, the union $E = \bigcup < E_n \mid n \epsilon \omega >$ of the sets E_n underlying T_n can be made into an ordered set $< E, \lesssim >$ by defining $f \lesssim e$ if there is a T_n containing e (and hence also f) such that $f \lesssim e$ in T_n. Then $< E, \lesssim >$ is a tree T with T_0 as its root piece, and on T I define the functions t und ε as the extensions of the functions t_n and ε_n. Thus T gives rise to an itableau which I may call the *test itableau* of the starting sequence s (or the starting set of signed formulas). Such itableau then also *tests* all nodes g carrying a **UR** formula $t(g) = u$, in that every branch of T through g carries *all* components u_i of u .

I next extend to quantifier logic the semantical connections, established for the propositional case in Chapter **5**.6. Let A be a (2-valued) structure for the language under consideration; recall that a valuation φ is a map from the set of variables of L into A, and that a map φ' is an x-variant of φ, $\varphi' =_x \varphi$, if φ' coincides with φ except at the argument x. A valuation φ extends to the evaluation map $e_A(\varphi \mid -)$ defined on all formulas, and evaluation commutes with substitution (in the sense of Theorem **1**.13.1 (ii)): $e_A(h_\varphi \circ \zeta(x,t) \mid v) = e_A(\varphi \mid sub(x,t \mid v))$. Recall that a structure A is called canonical if its underlying algebra is the term algebra T(L) of the language; in that case, every x-variant φ' of φ can be written as $h_\varphi \circ \zeta(x, \varphi'(x))$. The evaluation maps $e_A(\varphi \mid -)$ now are extended to signed formulas by $e_A(\varphi \mid Pv) = e_A(\varphi \mid v)$ and $e_A(\varphi \mid Qv) = -e_A(\varphi \mid v)$; again the valuation φ into A is said to *falsify* the signed formula u if $e_A(\varphi \mid u) = 0$. The *basic observation* on signed propositional formulas from Chapter **5**.6 I first supplement by :

> If u is **UR** then φ falsifies u if, and in case of a canonical structure only if, φ falsifies all of its components u_i.

Because for $u = Q \forall x v$ the evaluation $e_A(\varphi \mid u) = -e_A(\varphi \mid \forall x v)$ is

$$-\bigcap < e_A(\varphi' \mid v) \mid \varphi' =_x \varphi > \ = \ \bigcup < -e_A(\varphi' \mid v) \mid \varphi' =_x \varphi > .$$

This is 0 if, and only if, $-e_A(\varphi' \mid v) = 0$ for every x-variant φ' of φ. But $h_\varphi \circ \zeta(x,t_i) =_x \varphi$ whence $e_A(\varphi \mid u) = 0$ implies $e_A(\varphi \mid u_i) = e_A(\varphi \mid Q\, sub(x,t_i \mid v)) = e_A(h_\varphi \circ \zeta(x,t_i) \mid v) = 0$. And if A is canonical then every x-variant φ' of φ is of this form. The case of $u = P \exists x v$ is analogous. – A second supplement reads :

Let u be ramified quantified and let φ be a valuation into a structure A. If φ falsifies some component u_y then it also falsifies u. Conversely, if φ falsifies a set C of signed formulas including u, and if y is not free in C, then I find a valuation ψ into A which falsifies C and the the component u_y of u .

For $u = P\,\forall x v$ the evaluation $e_A(\varphi \mid u) = e_A(\varphi \mid \forall x v) = \bigcap < e_A(\varphi' \mid v) \mid \varphi' =_x \varphi >$ is 0 if, and only if, $e_A(\varphi' \mid v) = 0$ for some x-variant φ' of φ. In particular, $h_\varphi \circ \zeta(x,y)$ is an x-variant of φ, and if φ falsifies $u_y = P\,\mathrm{sub}(x,y \mid v)$ then $e_A(\varphi \mid u_y) = e_A(\varphi \mid \mathrm{sub}(x,y \mid v)) = e_A(h_\varphi \circ \zeta(x,y) \mid v)$ shows that φ also falsifies u . Conversely, if φ falsifies u, hence some x-variant φ' falsifies v, then I define ψ by $\psi =_y \varphi$ und $\psi(y) = \varphi'(x)$. As y is not free in C , together with φ also ψ falsifies C . Further, $\chi = h_\psi \circ \zeta(x,y)$ coincides on fr(v) with φ' : the variable x first goes to y and then to $h_\psi(y) = \varphi'(x)$, while all other variables z in fr(v) are in fr($\forall$x v), hence distinct from y , and so they go to $h_\psi(z) = \psi(z) = \varphi(z) = \varphi'(z)$. Consequently, $e_A(\psi \mid u_y) = e_A(\psi \mid P\,\mathrm{sub}(x,y \mid v))$ becomes $e_A(\psi \mid \mathrm{sub}(x,y \mid v)) = e_A(h_\psi \circ \zeta(x,y) \mid v) = e_A(\varphi' \mid v) = 0$. – The case of $u = Q\,\exists x v$ is analogous. Now I extend Lemma 6.7 to

LEMMA 12 (a) If the root piece of an itableau T can be falsified then T is not closed.

(b) If a test itableau is not closed then I can define a canonical structure A such that already the identical valuation into A falsifies its root piece.

The proof of (a) is again indirect: I shall show that, if the root piece of an itableau T can be falsified by a valuation into a structure A, then an entire branch B of T can be falsified – and clearly such branch then cannot be a closed one. So assume that e is a node of T, at least the maximum of the root piece, and that there is a valuation φ into some structure A, falsifying all t(f) for f in ($\leftarrow$, e] ; I define a node r(e) as follows. If e has one upper neighbour e', then r(e) = e' ; as the tableau proceeds from e to e' with help of t(ε(e)) by (t1) or by t(7), in the first case the first supplement shows that φ falsifies ($\leftarrow$, r(e)] ; in the second case the eigenvariable y is not free in the set C of formulas t(f) for f in ($\leftarrow$, e], and so the second supplement shows how to find a valuation falsifying ($\leftarrow$, r(e)] . If e has two upper neighbours then φ, falsifying t(φ(e)), falsifies at least one of the formulas carried by them, and the first one of them I define as r(e). Thus I obtain B recursively as the smallest subset of T containing the root piece and closed under r .

As for (b), let B be a non-closed branch of a test itableau T. In order to obtain a canonical structure A on the set of terms, I shall define, for every m_i-ary predicate symbol p_i of my language, an m_i-ary relation $r^A{}_i$ between terms : for every sequence λ of terms of length m_i there shall hold $r^A{}_i(\lambda) = 1$ if, and only if, $Q\,p_i(\lambda)$ appears on B, and $r^A{}_i(\lambda) = 0$ otherwise. Hence always either $r^A{}_i(\lambda) = 1$ or $r^A{}_i(\lambda) = 0$, and as B does not contain oppositely signed atomic formulas, if $P\,p_i(\lambda)$ appears on B then $r^A{}_i(\lambda) = 0$. The structure A consisting of terms, the identity $\imath$ on the set of variables is a valuation into A for which, by definition of evaluation, $e_A(\imath\,|\,p_i(\lambda)) = r^A{}_i(\lambda)$; hence if $P\,p_i(\lambda)$ appears on B then $e_A(\imath\,|\,P\,p_i(\lambda)) = e_A(\imath\,|\,p_i(\lambda)) = r^A{}_i(\lambda) = 0$, and if $Q\,p_i(\lambda)$ appears on B then $e_A(\imath\,|\,Q\,p_i(\lambda)) = -e_A(\imath\,|\,p_i(\lambda)) = -r^A{}_i(\lambda) = -1 = 0$. So $\imath$ falsifies the signed atomic formulas on B , and I now show by induction on signed formulas u that $\imath$ either falsifies u or u is not on B . If u is propositionally composite then I argue as in Lemma **6**.7. If u is UR and is on B then, T testing its nodes, all components u_i are on B , hence falsified by $\imath$, and so $\imath$ falsifies u by the first supplement. If u is ramified quantified and is on B then u_y is on B, and so, together with u_y, $\imath$ also falsifies u by the second supplement. It so follows that $\imath$ falsifies all of B and, in particular, the root piece.

I thus have shown, in analogy to Theorem **6**.4, that the algorithm producing a test itableau for a sequent s is complete:

THEOREM 8 If there is any closed itableau for a sequent s at all, then the test itableau for s must be closed, and by König's lemma it then gives rise to a closed tableau for s .

But since the test itableau in general will be not finite, the algorithm does not decide in a finite number of steps whether s *does* have a closed tableau. Also, the proof of the theorem relies on an indirect arguments: first, if the test itableau is not closed then the root piece can be falsified, and then any itableau for s will have a non-closed branch.

Again, the concepts here provide a classical semantical interpretation of the calculus MK (with sub). If T is a tableau for a sequent $M \Longrightarrow N$ then it follows from the definitions that a valuation φ into a structure A falsifies the root piece of T if, and only if, it satisfies the (unsigned) formulas in M and does not satisfy the (unsigned) formula $\mathbb{W}N$. Thus φ falsifies the root piece if, and only if, it provides a counterexample to the semantical situation $\mathbb{W}N \in cs(M)$ (as defined in Chapter **1.11**), and Lemma 12 can be restated as

LEMMA 12a (a) If there is a counterexample for $\mathbb{W}N \in cs(M)$ then every itableau for s is not closed.

(b) If a test itableau for s is not closed then there is a counterexample for $\mathbb{W}N \in cs(M)$.

But it is equivalent that no counterexample for $\mathbb{W}N \in cs(M)$ exists and that there holds $\mathbb{W}N \in cs(M)$. Hence also

LEMMA 12b (a) If T is a closed itableau for s then $\mathbb{W} N \in cs(M)$.

(b) If $\mathbb{W} N \in cs(M)$ then any test itableau for s is closed.

Again, the deduction of Lemma 7b from Lemma 7a is heavily indirect. - I now summarize these insights in :

COROLLARY 7 The MK-derivations (with sub) of M-sequents s correspond, by the extended Lemma 1.4, to the closed tableaux for s.

If an M-sequent s : $M \Longrightarrow N$ has an MK-derivation then $\mathbb{W} N \in cs(M)$ (by Lemma 12b (a)).

If $\mathbb{W} N \in cs(M)$ then I find a test itableau for s : $M \Longrightarrow N$. It is closed by Lemma12b (b) and so gives rise to a closed tableau for s which corresponds to an MK-derivation of s.

Thus s : $M \Longrightarrow N$ has an MK-derivation if, and only if, $\mathbb{W} N \in cs(M)$.

Making use of the semantic study of the consequence cs in Book 1, as well as of the translation between sequential derivations and MP-deductions, the conclusion of this Corollary was also observed in Chapter 10.1 .

Finally, I shall use the technique of infinite tableaux in order to prove a *finiteness theorem* for quantifier logic in the form: if a countably infinite set M of formulas is not satisfiable, then a finite subset M' of M can be found which is not satisfiable.

I first extend the notion of a tableau is such a manner that also countably infinite sets M of unsigned formulas can be treated. A *jtableau* for a set M shall be a (in general infinite) 2-ary tree T (without a designated root piece) together with two functions t und ε such that t and ε are defined as before, but

$\varepsilon(e)$ either is a new object $\uparrow$ or is a node such that $\varepsilon(e) \leq e$, the formula $t(\varepsilon(e))$ is composite, and there holds:

(t0) If $\varepsilon(e) = \uparrow$ then e has one upper neighbour e' and t(e') is $Q\,m$ for some m in M

and if $\varepsilon(e)$ is not $\uparrow$ then (t1), (t2), (t7) hold as stated for classical tableaux. Next, I define the *simple extension by* m in M of a finite jtableau T with an employment function δ : it shall arise as T'' , δ'' from the formerly defined direct extension T', δ' by putting a new node e' on top of every maximal node e of T' and then defining $t''(e') = Q\,m$, $\varepsilon''(e) = \uparrow$, $\delta''(e') = 0$. Then (d1) must be changed by replacing b+2 with b+3 , and the other (di) remain in effect.

Assume now that M is a set of unsigned formulas and that β is a bijection from ω onto M. Let T_0 consist of one node r only, with P $\beta(0)$ its value under t , and $\delta^0(r) = 0$ if $\beta(0)$ is composite, $\delta^0(r) = \infty$ otherwise. Let T^{n+1} with δ^{n+1} be the simple extension of T^n with δ^n by $\beta(n+1)$. Then (c1) must be changed by replacing r+2n with 3n, and the other (di) remain in effect. Define the union of the T_n as the *test jtableau* T *for* M (and β). Lemma 11 remains in effect, and so T tests all of its nodes. Of Lemma 12 only part (b) will be needed here:

If the test jtableau for M is not closed then M can be satisfied.

The proof remains unchanged: a non-closed branch B of T determines a canonical structure A such that the identitical valuation falsifies every signed formula on B. By definition of the T_n, every formula from M occurs Q-signed on every branch of T, hence on B, and so it is satisfied in A. – It next follows from the above

If M cannot be satisfied then the test jtableau for M is closed.

Making use of König's lemma, I now again find a finite sub-jtableau T^f of T which is closed. T^f being finite, there is only a finite number of nodes e' with lower neighbours e in T^f such that $\varepsilon^f(e) = \uparrow$ and $t^f(e')$ is a formula Q m with m in M ; let M' be the subset of M containing these m. I transform T^f to a (usual) tableau T^g by (1) adding a rootpiece R with new nodes e_m and $t^g(e_m) = t^f(e')$ for all these m , (2) omitting all those e', (3) defining $\varepsilon^g(e) = \varepsilon^f(e')$ if $\varepsilon^f(e)$ was $\uparrow$, (4) defining $\varepsilon^g(h) = e_m$ if $\varepsilon^f(h)$ was an omitted e'. Then T^g is a closed tableau, and so by Lemma 12(a) its root piece cannot be falsified, meaning that M' cannot be satisfied.

It is a direct and efficient manner in which the method of tableaux establishes the connection between classical sequential calculi and the semantical consequence operation cs. Its limitation is that the language considered must have only countably many terms. Looking at a closed tableau from the aspect of the sequential calculus, it may well be called a *tableau-proof*;looking at it from the semantical side, it is not a proof of verifiability but a counterexample to falsifiability.

References

R.M. Smullyan: A unifying principle in quantification theory. Proc.Nat.Acad.Sci.U.S.A. **49** (1963) 828–832

R.M. Smullyan: First Order Logic. Berlin 1968

Index of concepts and names

Index of symbolic notations

(inversion operators)

(inversion rules)

(modus ponens calculi)